Results from the

Seventh Mathematics Assessment

of the

National Assessment of Educational Progress

edited by

Edward A. Silver
University of Pittsburgh
Pittsburgh, Pennsylvania

Patricia Ann Kenney
University of Pittsburgh
Pittsburgh, Pennsylvania

National Council of Teachers of Mathematics
Reston, Virginia

Copyright © 2000 by
THE NATIONAL COUNCIL OF TEACHERS OF MATHEMATICS, INC.
1906 Association Drive, Reston, VA 20191-9988
www.nctm.org
All rights reserved

Library of Congress Cataloging-in-Publication Data:

Results from the seventh mathematics assessment of the National Assessment of Educational Progress / edited by Edward A. Silver, Patricia Ann Kenney.
   p. cm.
   Includes bibliographical references.
   ISBN 0-87353-488-3
   1. Mathematics—Study and teaching—United States—Evaluation. 2. Mathematical ability—Testing. 3. National Assessment of Educational Progress (Project) I. Silver, Edward A., 1948- II. Kenney, Patricia Ann. III. National Council of Teachers of Mathematics. IV. National Assessment of Educational Progress (Project)

QA13 .R47 2000
510.7'1073—dc21

00-056250

This material is based on work supported by the National Science Foundation under Grant No. RED–9453189. The Government has certain rights in this material. Any opinions, findings, and conclusions or recommendations expressed in this material are those of the author(s) and do not necessarily reflect the views of the National Science Foundation.

The publications of the National Council of Teachers of Mathematics present a variety of viewpoints. The views expressed or implied in this publication, unless otherwise noted, should not be interpreted as official positions of the Council.

Printed in the United States of America

Contents

Preface v

1. The Seventh NAEP Mathematics Assessment: An Overview 1
   Patricia Ann Kenney
   University of Pittsburgh, Pittsburgh, Pennsylvania

2. The State of NAEP Mathematics Findings: 1996 23
   John A. Dossey
   Illinois State University, Normal, Illinois

3. NAEP Findings Regarding Race/Ethnicity: Students' Performance,
   School Experiences, and Attitudes and Beliefs 45
   Marilyn E. Strutchens
   University of Maryland, College Park, Maryland

   Edward A. Silver
   University of Pittsburgh, Pittsburgh, Pennsylvania

4. NAEP Findings Regarding Gender: Achievement, Affect, and
   Instructional Experiences 73
   Ellen Ansell
   University of Pittsburgh, Pittsburgh, Pennsylvania

   Helen M. Doerr
   Syracuse University, Syracuse, New York

5. NAEP Findings on the Preparation and Practices of Mathematics
   Teachers 107
   Douglas A. Grouws
   University of Iowa, Iowa City, Iowa

   Margaret Schwan Smith
   University of Pittsburgh, Pittsburgh, Pennsylvania

6. Whole Number Properties and Operations 141
   Vicky L. Kouba
   University at Albany, State University of New York, Albany, New York

   Diana Wearne
   University of Delaware, Newark, Delaware

7. Rational Numbers 163
   Diana Wearne
   University of Delaware, Newark, Delaware
   Vicky L. Kouba
   University at Albany, State University of New York, Albany, New York

8. Geometry and Measurement 193
   W. Gary Martin
   National Council of Teachers of Mathematics, Reston, Virginia
   Marilyn E. Strutchens
   University of Maryland, College Park, Maryland

9. Data and Chance 235
   Judith S. Zawojewski
   Purdue University, West Layfayette, Indiana
   J. Michael Shaughnessy
   Portland State University, Portland, Oregon

10. Algebra and Functions 269
    Glendon W. Blume
    Pennsylvania State University, University Park, Pennsylvania
    David S. Heckman
    Monmouth Academy, Monmouth, Maine

11. Students' Performance on Extended Constructed-Response Tasks ... 307
    Edward A. Silver
    University of Pittsburgh, Pittsburgh, Pennsylvania
    Cengiz Alacaci
    Florida International University, Miami, Florida
    Despina A. Stylianou
    University of Massachusetts, Dartmouth, Dartmouth, Massachusetts

12. Students' Performance on Thematically Related NAEP Tasks 343
    Patricia Ann Kenney
    University of Pittsburgh, Pittsburgh, Pennsylvania
    Mary M. Lindquist
    Columbus State University, Columbus, Georgia

13. The Performance of Students Taking Advanced
    Mathematics Courses 377
    Jeremy Kilpatrick
    University of Georgia, Athens, Georgia
    Judith Lynn Gieger
    University of Georgia, Athens, Georgia

List of NAEP-Related Publications: Fifth, Sixth, and Seventh
    Mathematics Assessments 411

PREFACE

The National Council of Teachers of Mathematics (NCTM) has a history of preparing and disseminating interpretive reports based on results from mathematics assessments conducted by the National Assessment of Educational Progress (NAEP). NCTM continued its dedication to producing interpretive reports of NAEP results by appointing a steering committee to plan and to secure funds from the National Science Foundation (NSF) for producing the interpretive reports. Members of this steering committee were John A. Dossey, Douglas A. Grouws, Patricia Ann Kenney, Vicky L. Kouba, Mary Montgomery Lindquist, Jack Price, and Edward A. Silver. As a result of the steering committee's efforts, a grant was received from the NSF, with supplemental funds provided by the National Center for Education Statistics (NCES), to support the work of a team of writers of interpretive reports for the fifth, sixth, and seventh NAEP assessments.

The NCTM-NAEP Interpretive Reports Project was directed by Edward A. Silver and Patricia Ann Kenney. The other members of the interpretive team for the 1996 assessment were Ellen Ansell, Glendon W. Blume, Helen M. Doerr, John A. Dossey, Douglas A. Grouws, David S. Heckman, Jeremy Kilpatrick, Vicky L. Kouba, Mary M. Lindquist, W. Gary Martin, J. Michael Shaughnessy, Margaret Schwan Smith, Marilyn E. Strutchens, Diana Wearne, and Judith S. Zawojewski. These members of the interpretive team are the principal authors of the chapters in this monograph, with Cengiz Alacaci, Judith Lynn Gieger, and Despina A. Stylianou as coauthors for two of the chapters.

The interpretive reports in this volume are based on NAEP findings regarding (*a*) the cognitive performance of students at grades 4, 8, and 12 on multiple-choice, short constructed-response, and extended constructed-response items; (*b*) students' responses to a variety of background questions dealing with their attitudes and beliefs concerning mathematics and their participation in various forms of classroom activity; and (*c*) teachers' responses to various background questions dealing with the nature of their mathematics instruction. The results are summarized for the different grade levels, for different content areas, and for subgroups of students by gender and race/ethnicity.

We wish to thank all who helped in the preparation of this volume. In particular, we are grateful to the National Science Foundation and to Larry Suter, our project officer, for providing the funds for the NCTM-NAEP Interpretive Reports Project. We extend our thanks to the National Center for Education Statistics, and especially to Steven Gorman, Arnold Goldstein, and Sharif Shakrani (who has since joined the staff of the National Assessment Governing Board), and to the Center for Assessment at the Educational Testing Service (ETS), especially Jeff Haberstroh and Steve Lazer, for facilitating access to NAEP materials and data and for providing answers to our many questions. For their prompt and detailed answers to technical questions about NAEP data, we are indebted to David Freund and Edward Kulick at ETS. Additional NAEP materials, such as scoring guides and sample responses for constructed-response questions, were obtained with the assistance of the NAEP project staff at National Computing Systems, especially Bradley Thayer and Patrick Bourgeacq.

Over the past three years, members of the project's advisory board— Jane Armstrong, Joan Countryman, John Dossey, Mary Lindquist, Robert Linn, and Jack Price — have offered guidance and support to us and to the writing team. The time devoted by John and Mary to reviewing drafts of chapters is greatly appreciated. We also recognize the careful editorial work done by the NCTM staff, especially Jean Carpenter, Charles Clements, and Raymond Haas (consultant).

Our work on the book at the University of Pittsburgh's Learning Research and Development Center would have been very difficult had it not been for the support of Cengiz Alacaci and Despina A. Stylianou, who faithfully served as our assistants while they were doctoral students; Elizabeth Ann George, who reviewed drafts of chapters during her postdoctoral fellowship year; and Kathy A. Day, who provided a consistently excellent level of clerical support. Finally, the second editor (Patricia Ann Kenney) would like to thank the first editor (Edward A. Silver) for his conscientious support through the ten years they have worked together at LRDC and for facilitating many of the opportunities that were afforded to her during that time.

The Seventh NAEP Mathematics Assessment: An Overview

Patricia Ann Kenney

THE National Assessment of Educational Progress (NAEP) is a congressionally mandated survey of the educational achievement of American students and of changes in that achievement over time. Since 1969, NAEP has been assessing what American students know and can do in a variety of curriculum areas, including mathematics. To provide a context for the achievement results, NAEP also collects demographic, curricular, and instructional background information from students, teachers, and school administrators. The NAEP assessment has evolved to include three components, as shown in figure 1.1: trend NAEP (also called long-term trend NAEP), national NAEP, and state NAEP. The trend assessment, consisting of a set of items that does not change from assessment to assessment, gauges student achievement by age group (9-, 13-, and 17-year-old students). The "main NAEP" components of national NAEP and state NAEP are essentially the same in that they use the same core set of items and questionnaires and focus on grade levels (grades 4, 8, and 12 in national NAEP; grades 4 and 8 in state NAEP) instead of age levels. The primary differences between national and state NAEP are that they use different samples of students, and the national assessment often includes special studies, such as portfolios in writing and thematically related sets of items in mathematics. To date, there have been seven NAEP national mathematics assessments. These took place in the school years ending in 1973, 1978, 1982, 1986, 1990, 1992, and 1996. State-level mathematics assessments were given at grade 8 in 1990 and grades 4 and 8 in 1992 and 1996. This book examines the results for the seventh mathematics assessment, conducted in 1996.[1]

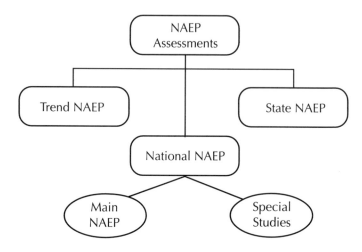

Fig. 1.1. Components of the NAEP assessment. *Source:* Adapted from National Research Council (1999, p. 16).

The purpose of this chapter is to introduce the chapters that follow by summarizing relevant information about the seventh mathematics assessment, with a focus on national NAEP. Aspects of NAEP discussed are the 1996 framework used to guide the development of the grade-level tests; the cognitive items that comprised the grade-level tests; the questionnaires administered to students in the NAEP samples, to teachers whose students were selected to participate in NAEP, and to school administrators whose schools were selected to participate in NAEP; student samples and administration procedures; reporting of NAEP results; and an overview of the special studies included in the 1996 NAEP mathematics assessment. Additional information on the seventh assessment can be found in a number of sources. Some of these include the NAEP mathematics framework document (National Assessment Governing Board [NAGB] 1994), a general overview of NAEP (Ballator 1996), reports of NAEP results from the National Center for Education Statistics (Hawkins, Stancavage, and Dossey 1998; Mitchell, Hawkins, Jakwerth, Stancavage, and Dossey 1999; Mitchell, Hawkins,

---

[1]. Previous interpretive reports of NAEP results published by NCTM have referred to the assessments according to their ordinal designations (for example, the *seventh* mathematics assessment). However, NAEP reports published by the National Center for Education Statistics (NCES) more commonly refer to the assessment by the year of administration (for example, the *1996* mathematics assessment). Because both forms may have been used interchangeably in this book, readers should keep in mind that the seventh assessment is synonymous with 1996.

Stancavage, and Dossey 1999; Reese et al. 1997), and chapter 2 by Dossey in this volume.

THE 1996 NAEP MATHEMATICS FRAMEWORK

Because of NAEP's conscious effort to be sensitive to changes in curriculum and educational objectives, some changes are made in the assessment each time a curriculum subject is measured. Like its predecessors, the seventh mathematics assessment was organized around a set of mathematical processes and content-strand objectives, referred to as a *framework*. This, in turn, guided the development and selection of the test questions, called *items* or *cognitive items* by NAEP. The NAEP mathematics frameworks have evolved over time to reflect the best thinking of the mathematics education community with respect to mathematics content and mathematical processes. At the time that the framework for the seventh assessment was developed, the best thinking was articulated in *Curriculum and Evaluation Standards for School Mathematics* (National Council of Teachers of Mathematics [NCTM] 1989), a document that reflects a vision of what it means to be mathematically literate in today's technological society and that focuses on problem solving, communication of mathematical ideas, critical reasoning, and connections to ideas and procedures both within mathematics and across other content areas.

The mathematics framework for the 1996 NAEP stands as the first update of the framework since the official release of the NCTM *Curriculum and Evaluation Standards* and since the achievement levels Basic, Proficient, and Advanced were first used in official reports of NAEP results. In addition to including ideas from the NCTM *Curriculum and Evaluation Standards,* the framework was developed to maintain ties to the framework used for the fifth (1990) and sixth (1992) assessments, thus enabling a continuation of the six-year, short-term trend assessment of student performance. Among the refinements included in the 1996 framework that distinguish it from the previous frameworks are (Reese et al. 1997, p. 76)—

- a move away from the rigid content-by-process matrix that governed the development of earlier mathematics assessments;
- the creation of a more balanced range of cognitive outcomes by including the NCTM *Standards'* process goals of communication and connections as a complement to the NAEP process categories of conceptual understanding, procedural knowledge, and problem solving, thereby creating a more balanced range of cognitive outcomes;
- a continuation of the move toward including more constructed-response items;

- the creation of families of related items that probed an understanding of mathematics vertically within a content strand or horizontally across strands;
- a revision of some content strands (in particular, Number Sense, Properties, and Operations and Geometry and Spatial Sense) to reflect the NCTM *Standards'* emphasis on topics in these areas.

Figure 1.2 illustrates the evolution of the NAEP mathematics framework from the fifth and sixth assessments to the seventh assessment. Rather than the fifteen-cell matrix (fig. 1.2a) that allowed for the development of the 1990 and 1992 NAEP items according to unique content and process categories, the 1996 framework (fig. 1.2b) is more fluid, showing the relationship between the five content areas and three mathematical abilities within the

MATHEMATICAL ABILITIES	CONTENT AREAS				
	Numbers & Operations	Measurement	Geometry	Data Analysis, Statistics, & Probability	Algebra & Functions
Conceptual Understanding					
Procedural Knowledge					
Problem Solving					

(a)

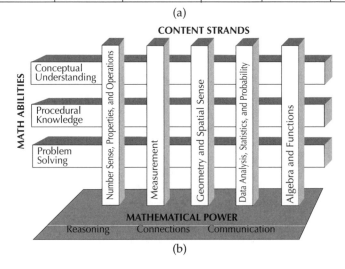

(b)

Fig. 1.2. Frameworks for the Fifth/Sixth and Seventh NAEP mathematics assessments. *Source:* Figure 1.2(a): National Assessment of Educational Progress (1988); Figure 1.2(b): National Assessment Governing Board (1994).

broader context of "mathematical power." According to the framework document, "mathematical power is conceived as consisting of mathematical abilities (conceptual understanding, procedural knowledge, and problem solving) within a broader context of reasoning and with connections across mathematical content and thinking, [and] communication is viewed as both a unifying thread and a way for students to provide meaningful responses to tasks" (NAGB 1994, p. 12).

Brief descriptions of the content areas and ability categories appear next. Additional information can be found in the framework document itself (NAGB 1994) and in the 1996 NAEP mathematics report card (Reese et al. 1997, pp. 75–78).

Content Strands

The five content strands assessed by NAEP were selected to represent topics commonly studied by the majority of students at a given grade level and to reflect the content standards in the NCTM *Standards*. The percent distribution of items by content area suggested in the mathematics framework document is shown in table 1.1. This distribution continued the move toward a more even balance appropriate to the role each strand plays in the curriculum at various grade levels and away from earlier NAEP models in which items assessing number facts and operations made up more than 50 percent of the item pool.

Table 1.1
Percent Distribution of Items by Grade and Content Strand

	Grade 4	Grade 8	Grade 12
Number Sense, Properties, and Operations*	40	25	20
Measurement	20	15	15
Geometry	15	20	20**
Data Analysis, Statistics, and Probability	10	15	20
Algebra and Functions	15	25	25

* At least half of the items in the Number Sense, Properties, and Operations strand at each grade level should involve some aspect of estimation or mental mathematics.
** At grade 12, 25 percent of the items in the Geometry and Spatial Sense strand should involve topics in coordinate geometry.
Source: National Assessment Governing Board (1994).

Next, each of the content strands as defined by NAEP is briefly described. Additional information about each content strand is provided in the chapter that discusses and interprets results from that strand (for example, chapter 9 contains information about the strand Data Analysis, Statistics, and Probability).

Number Sense, Properties, and Operations

This strand focuses on students' understanding of numbers (whole numbers, fractions, decimals, integers, real numbers, and complex numbers), operations, and estimation and their application to mathematical and real-world situations. At grade 4, the focus of this strand is on number sense, involving whole numbers, simple fractions, and decimals through connecting various models and their numerical representations, and on the four basic operations. At grade 8, number sense and operations is extended to include integers and rational numbers. At grade 12, this strand includes real and complex numbers and allows students to demonstrate competency through the precalculus level.

Because of the number of topics covered in the Number Sense, Properties, and Operations strand, two chapters in this book are devoted to reporting NAEP results from this strand. Chapter 5 by Kouba and Wearne discusses whole number properties and operations; Chapter 6 by Wearne and Kouba discusses rational numbers.

Measurement

This strand focuses on an understanding of measurement as a process and on the use of numbers and measures to describe both mathematical and real-world objects. At grade 4, the focus is on such concepts as time, money, temperature, length, perimeter, area, capacity, weight/mass, and angle measure. These concepts are also included for grades 8 and 12, but the focus shifts to more complex problems that involve volume or surface area, and the use of complex measurement formulas.

Geometry and Spatial Sense

This strand is designed to extend beyond the identification of geometric shapes to include transformations and combinations of shapes and beyond the more traditional compass-and-straightedge constructions to include informal constructions and demonstrations along with their justifications. The focus at grade 4 is on modeling properties of shapes under simple combinations and transformations and on drawing figures from verbal descriptions. At grade 8, the focus expands to include properties of angles and polygons and the application of reasoning skills to geometric situations. Topics assessed at grade 12 include transformational geometry and proportional thinking in geometric settings.

Because of the interconnectedness of many concepts in measurement and geometry (for example, area and perimeter of geometric figures such as quadrilaterals and circles), it was decided that NAEP interpretive results for the Measurement and Geometry and Spatial Sense items should be reported in a single chapter. Thus, in this volume results from these two strands are presented in chapter 8 by Martin and Strutchens.

Data Analysis, Statistics, and Probability

This strand emphasizes methods for gathering and representing data, and developing and evaluating arguments based on data analysis. At grade 4, the focus is on understanding number and quantity through problem solving involving data and graphical representations of data, dealing informally with measures of central tendency, and using the basic concepts of chance without formal computation of probability. At grade 8, topics include sampling, the ability to make predictions based on experiments or data, and the use of formal terminology and formulas related to data analysis, statistics, and probability. At grade 12, the emphasis is on application of probability concepts and statistical techniques to model situations and solve problems. In chapter 9, Zawojewski and Shaughnessy report NAEP results from this content strand.

Algebra and Functions

Of the five content categories, this strand is the broadest in scope, covering simple patterns in grade 4 to basic algebra concepts in grade 8 to more sophisticated analyses of functions at grade 12. Topics assessed in grade 4 include recognizing patterns and informally generalizing about them, translating among mathematical representations, and using simple equations. At grade 8, the emphasis is on algebraic notation, the meaning of *variable,* and the solution of simple equations and inequalities algebraically and graphically. By grade 12 the strand has expanded to include algebraic and functional notation and terminology and their applications. Results from the Algebra and Functions strand are reported in chapter 10 by Blume and Heckman.

Mathematical Abilities

The three mathematical abilities, Conceptual Understanding, Procedural Knowledge, and Problem Solving, from the fifth and sixth assessments were retained for use in the seventh assessment. Unlike the recommended distribution of items by content strand in which the percentages differ across grade levels, the recommended distribution of items by ability category was identical for grades 4, 8, and 12: "the overall mixture of assessment items for each grade level [should] include at least one-third of the items measuring each of the abilities" (NAGB 1994, p. 39). In NAEP, the mathematical abilities are not to be considered as separate and distinct factors that uniquely characterize an item, but instead they are presented as descriptions of the way in which students approach and solve problems. Each category defined in the 1996 NAEP framework document (NAGB 1994) is described briefly next.

Conceptual Understanding

In NAEP, conceptual understanding can be demonstrated in a variety of ways. Some of these ways include recognizing, labeling, and generating examples and counterexamples; using and interrelating models, diagrams, and varied representations of concepts; identifying and applying principles; knowing and applying facts and definitions; comparing, contrasting, and integrating related concepts and principles; recognizing, interpreting, and applying symbols and terms used to represent concepts; and interpreting assumptions and relations.

Procedural Knowledge

In NAEP, procedural knowledge includes the various numerical algorithms in mathematics that have been created as tools to meet specific needs. Reading and producing graphs and tables, executing geometric constructions, and performing noncomputational skills such as rounding and ordering are also considered procedures. Students demonstrate procedural knowledge by providing evidence of their ability to select and apply appropriate procedures correctly, by verifying and justifying the correctness of a procedure, and by extending or modifying procedures in problem settings.

Problem Solving

Problem solving includes the ability to use accumulated knowledge of mathematics in new situations. Students demonstrate problem-solving skills by recognizing and formulating problems; determining the sufficiency and consistency of data; using strategies, data, models, and relevant mathematics; using reasoning (spatial, inductive, deductive, statistical, proportional); and judging the reasonableness and correctness of solutions.

COGNITIVE ITEMS

One of the primary functions of an assessment framework is to provide the template for the development of the cognitive items. Using the framework just described, a group of mathematics educators worked with the staff of Educational Testing Service (ETS), the NAEP contractor, to develop the items for the seventh assessment. The process used to develop the items was similar to that described by Carpenter (1989) for the fourth NAEP assessment in 1986. Items on the seventh assessment appeared in either multiple-choice or constructed-response format, and students, while working on certain sets of items, were provided with and permitted to use tools such as rulers, protractors, manipulatives in the form of geometric shapes (for example, squares and isosceles right triangles), and calculators.

Item Formats

As in the fifth and sixth assessments, the seventh assessment used three formats for the cognitive items: multiple-choice, short constructed-response, and extended constructed-response. However, the seventh assessment differed from previous assessments in that there were more constructed-response items and fewer multiple-choice items on each grade-level test. In both 1990 and 1992, about 70 percent of the items on each grade-level test were multiple-choice items and about 30 percent were short or extended constructed-response items. In 1996, however, the distribution changed to 55 percent multiple-choice items and 45 percent constructed-response items.

Again, as in prior assessments, the multiple-choice items administered to fourth-grade students had four choices, and those administered to students in grades 8 and 12 had five choices. Short constructed-response items were open-ended questions that asked students to formulate their own numerical answers or to write a brief explanation of, or justification for, a mathematical situation. For the 1996 assessment, there were two scoring schemes for short constructed-response items. Students' answers to some short constructed-response items, in particular those that required only a numerical answer (for example, see chapter 7, table 7.2), were scored by trained raters according to a right/wrong scheme. Other short constructed-response items, in particular those that were new to the 1996 assessment and that required a written explanation (for example, see chapter 7, tables 7.12, 7.13, and 7.14), were scored according to a partial credit model using three categories: completely correct, partially correct, and incorrect.

Extended constructed-response (ECR) questions, which became a regular feature on the sixth assessment, also appeared on the seventh assessment. These tasks not only required students to construct their own responses, but also provided students with an opportunity to express their mathematical ideas in writing and to demonstrate their depth of understanding. A notable change from the sixth to the seventh assessment was the increase in the number of extended constructed-response questions on main NAEP—seven in grade 4 and eight each in grades 8 and 12 as compared to five and, respectively, six for the previous assessment. Responses to these extended tasks were scored by trained raters using a partial-credit, focused holistic scoring scheme based on the categories incorrect, minimal, partial, satisfactory, and extended. Chapter 11 by Silver, Alacaci, and Stylianou is devoted exclusively to the extended constructed-response tasks and performance on them, and contains a number of examples of ECR tasks (for example, see chapter 11, figure 11.1).

Detailed information about student performance on selected short and extended constructed-response items that were released to the public was included in an official NAEP report, *Student Work and Teacher Practices in Mathematics* (Mitchell, Hawkins, Jakwerth, Stancavage, and Dossey 1999). The content-based chapters (chapters 6–10) and the ECR chapter (chapter 11) in the present volume have used some information from this report. Because we had the opportunity to gather convenience samples of responses to selected short and extended constructed-response questions and to analyze the responses qualitatively, the content-based chapters and the ECR chapter contain new information about performance on the released questions that is not available from the NAEP report.

Despite NAEP's continuing attempt to improve the items on the grade-level tests, the items from the seventh assessment have some of the same limitations as those on prior assessments. In particular, because of the forced-choice format of multiple-choice items and the guessing factor inherent in them, it is difficult to identify with certainty the reasons students selected any particular choice. Authors of the content-based chapters, while acknowledging the guessing factor, have chosen to speculate on misconceptions and error patterns based on performance on multiple-choice items and have tried to link their inferences to relevant literature in the field of mathematics education.

For the extended constructed-response questions, which are positioned as the last item within an intact set of items (called an "item block" by NAEP), a large number of omissions (that is, students leaving the item blank) severely limits the information available from NAEP data and from the convenience sample of students' responses. Because of our attempts to gather and then analyze students' responses to selected constructed-response questions, the authors were able to expand the limited information available from official NAEP sources about student performance on those questions and to make stronger inferences about students' strengths and weaknesses in particular areas such as understanding measures of central tendency (see chapter 9, table 9.2, Mean or Median task), comparing graphs (see chapter 9, table 9.9, Metro Rail Graph task), reasoning about geometric figures (see chapter 11, table 11.3, Compare Geometric Figures task), and extending a pattern (see chapter 11, table 11.6, Extend Pattern of Tiles task). Again, where possible, references to mathematics education research were included to lend support to inferences made by the authors.

Mathematical Tools

While working on some items, students were permitted to use tools such as rulers, protractors, manipulatives in the form of geometric shapes, or calculators. However, the use of these tools was restricted to particular blocks of items; students did not have access to these tools at all times during the assessment. The use of rulers and protractors was permitted on previous NAEP assessments. Items involving a ruler or protractor required students to make measurements (for example, determine the degree measure of a given angle) or to provide a fairly accurate representation of a geometric shape (for example, draw a rectangle of given proportions). The questions involving the use of manipulatives in the form of geometric shapes, new to the sixth assessment, also appeared on the seventh assessment, and asked students to assemble the shapes to create new shapes having particular properties or area. Representations of the geometric shapes can be found in a number of chapters in this volume (for example, chapter 8, table 8.4, item 2). Other kinds of tools provided as part of the seventh assessment included spinners (for a probability activity), number tiles, and three-dimensional models.

Calculators have been included in the NAEP mathematics assessment since the second assessment, administered in 1978. For the fourth assessment, students' ability to use calculators was assessed through the use of "calculator blocks;" that is, on certain blocks of items, students were given a simple four-function calculator, instructed in its use, and then permitted to use the calculator while working on computation and routine application items. Many of the same items in the calculator blocks also appeared in other blocks on which students were not permitted to use calculators. This structure permitted a comparison of performance with and without calculators. (See Kouba, Carpenter, and Swafford 1989 for a discussion of those results.)

As in the fifth and sixth assessments, the seventh assessment was also structured so that students had calculators for particular blocks of items. While working on the items in a calculator block, students in grade 4 were supplied with a Texas Instruments TI-108 calculator (a simple four-function calculator) and students in grades 8 and 12 received a Texas Instruments TI-30 Challenger (a scientific calculator) for their use. Only while working on items in a calculator block were students permitted to use a calculator. In an attempt to ensure that students knew how to use the particular calculators, they were provided with a short introduction to the use of the calculators before beginning the assessment exercises. Further instructions on calculator use were printed on the page preceding the items in a calculator block, and a reference card accompanied the calculators given to eighth- and twelfth-grade students.

Items in the calculator blocks were classified in three ways—calculator-inactive items, calculator-active items, and calculator-neutral items—according to definitions provided in the framework document. The definitions appear in the framework document used to develop the fifth and sixth assessments.

Calculator-inactive items are those whose solution neither requires nor suggests the use of a calculator; in fact, a calculator would be virtually useless as an aid to solving the problem. Calculator-neutral items are those in which the solution to the question does not require the use of a calculator. Given the option, however, some students might choose to use a calculator to perform numerical operations. In contrast, items classified as "calculator active" require calculator use; a student would likely find it almost impossible to solve the question without the aid of a calculator. (NAEP 1988, p. 33)

In the seventh assessment, each calculator block contained all three kinds of items, but calculator-block items did not appear in any other item block within a grade level. This structure stands in contrast to that used in the fourth assessment which ensured that some items were administered twice: both in a designated calculator block and in a block for which students were not permitted to use calculators. Thus, for the seventh assessment, it was not possible to compare item-by-item performance for students who had access to calculators and students who had no such access.

Although NAEP's commitment to allowing students to use various mathematical tools while working on some items is laudable, there are some important concerns associated with the way those tools have been incorporated into the assessment, the most obvious being that the tools were permitted to be used only with particular blocks of items and were not available for others. For the calculators in particular, although there is some political value in permitting students to use calculators on a national assessment of mathematics achievement, and although providing each student with the same calculator addresses equity concerns that have arisen during debates on the role of calculators in assessment (for example, Wilson and Kilpatrick 1989), the restricted use of calculators on NAEP represents an artificial situation inconsistent with the view espoused in the NCTM *Curriculum and Evaluation Standards* (1989, p. 8) that "calculators should be available for all students at all times." Moreover, the continued restriction on calculator use in the main NAEP assessment (that is, the assessment administered to the national and state samples) stands in contrast to a recommendation in the 1996 NAEP framework document: "NAEP should investigate unrestricted use of calculators on all but trend blocks [that is, blocks of items that appeared on the 1990 assessment, the 1992 assessment, or both] or specifically excluded items at each grade level [for example, items that assess estimation or mental mathematics]" (NAGB 1994, p. 17).

However, there is evidence of an effort to move away from restrictions on calculator type and calculator use in NAEP. In 1996, as a part of a special study in advanced mathematics (called the Advanced Study by NAEP) in grades 8 and 12 that used a special sample of students different from that used in main NAEP, students were told that they could bring their own calculators, including graphing calculators; students who did not bring a calculator were provided with the same scientific calculator used in main NAEP (Mitchell, Hawkins, Stancavage, and Dossey 1999). Students were permitted unrestricted use of a calculator on two of the three item blocks that constitute the Advanced Study. Unfortunately, at the time this book was published NAEP had not released any information on the effect that this unrestricted use of calculators may have had on students' performance. (For information on the performance of students taking advanced mathematics courses, see chapter 13 by Kilpatrick and Gieger.)

QUESTIONNAIRES

In addition to cognitive items, the seventh assessment included questionnaires that were administered to students, teachers, and school administrators. Each questionnaire was developed according to relevant categories in five broad educational areas: instructional content, instructional practices and experiences, teacher characteristics, school conditions and context, and conditions beyond school (i.e., home support, out-of-school activities, and attitudes). Results from these three questionnaires were used in several chapters in the volume. In particular, chapter 3 on race/ethnicity, chapter 4 on gender, and chapter 5 on the characteristics of mathematics teachers make extensive use of the questionnaire data. The next sections contain a brief description of each questionnaire; additional information on the questionnaires can be found in Reese et al. (1997, pp. 80–81).

Student Questionnaire

Each student in the grade-level NAEP samples completed a three-part questionnaire. Students were given five minutes to complete each part, with fourth-grade students having more time on the first part because the questions were read aloud to them. The first two parts were administered before the cognitive item blocks. The first part of the student questionnaire included general background questions about race or ethnicity, mother's and father's level of education, reading materials in the home, homework, attendance, and academic expectations. The second part consisted of mathematics background questions involving instructional experiences, courses taken, the use of specialized resources such as calculators, and views on the utility and value of mathematics. The third part of the student questionnaire was administered after students completed the cognitive items, and contained questions about students' motivation to do well on the assessment, their perceptions of the difficulty of the assessment, and their familiarity with the format of the NAEP items (for example, the extended constructed-response questions).

Teacher Questionnaire

The teacher questionnaire was administered to all mathematics teachers of fourth- and eighth-grade students participating in the assessment. Since the twelfth-grade sample in main NAEP purposely included students who were not currently enrolled in a mathematics class, there were no NAEP teacher questionnaires given to twelfth-grade teachers. However, because of the special nature of the Advanced Study, a questionnaire was developed and administered to twelfth-grade teachers whose students were in that special study's sample.

The teacher questionnaire took approximately 20 minutes to complete and consisted of three parts. The first part pertained to the teacher's general background and experience in teaching; the second part focused on the teacher's background in teaching mathematics; and the third part contained questions on information about mathematics instruction. Chapter 5 by Grouws and Smith presents a detailed analysis of results from the grade 4 and grade 8 teacher questionnaire.

School Characteristics and Policies Questionnaire

This questionnaire was given to the principal or other school administrator of each school that participated in NAEP and took about 20 minutes to complete. The questionnaire collected information about backgrounds and characteristics of school principals, school policies, programs, facilities, and demographic characteristics of the students and teachers.

SAMPLES OF STUDENTS AND ADMINISTRATION PROCEDURES

Because ETS was retained as the NAEP contractor for the seventh assessment, the sampling and administration procedures used for this assessment were similar to those developed for the previous three assessments and described by Carpenter (1989). The next sections describe sampling and administration procedures unique to the 1996 main NAEP assessment. (Sampling procedures for the special studies such as the Advanced Study were slightly different and are summarized in another section of this chapter.) More detailed information about sampling and administration procedures can be found in several NAEP reports (for example, Ballator 1996; Reese et al. 1997).

Students taking the 1996 assessment were selected according to a complex procedure that identified a representative national sample for each of grades 4, 8, and 12. All together, 6627 fourth-grade students, 7146 eighth-grade students, and 6904 twelfth-grade students participated in the national assessment, which was administered during the months of February–April 1996 (Reese et al. 1997).

In the NAEP design, no student takes all items at a particular grade level. Instead, the items were divided into blocks of items, that is, sets of items that always remain together. In 1996, each grade-level test had thirteen blocks of items, with the number of items in each block varying from about ten to twenty-five items. The thirteen blocks of items were assembled into twenty-six booklets; each booklet contained the first two parts of the student questionnaire, three blocks of items, and the motivational questionnaire, as well as appropriate instructions. Each student who participated in the assessment completed the questionnaires and items in only one booklet. Students were allowed 15 minutes for each item block, and worked at their own pace through each block.

Of the thirteen blocks, three were carried forward from the fifth (1990) assessment, and five were carried forward from the sixth (1992) assessment, to allow for gauging students' achievement over time. This short-term trend information is reported in some of the content chapters, especially in cases where there were significant increases in performance on clusters of items assessing the same topic (for example, see chapter 10 by Blume and Heckman for short-term trends in performance on pattern items).

REPORTING AND INTERPRETING NAEP RESULTS

NAEP results are reported in two ways that are interconnected: the NAEP scale and the mathematics achievement levels. The NAEP scale, ranging from 0 to 500, reports average overall performance across the three grade levels and is a composite of the five content strands. For example, in 1996 the average mathematics composite scale scores for grades 4, 8, and 12 were 224, 272, and 304, respectively. Scale scores are also used to report NAEP results by various categories such as gender, race/ethnicity, region of the country, type of school, and Title I participation; and for mathematics by individual content strand. Using the NAEP scale facilitates a comparison of results across assessment years as a gauge of students' progress in mathematics achievement. For example, the 1996 average NAEP scale scores for students were significantly higher than those from both 1992 and 1990. However, reporting NAEP results through the use of scale scores has serious limitations in that such reporting provides little information about the kinds of mathematics assessed.

The second reporting method, the NAEP mathematics achievement levels, were authorized in NAEP legislation and adopted by the National Assessment Governing Board. The three achievement levels are called Basic,

Proficient, and Advanced and describe how students should perform on the assessment. Specifically, the generic (non–content-specific) policy definitions for each level are as follows (Reese et al. 1997, p. 8):

- Basic: denotes partial mastery of prerequisite knowledge and skills that are fundamental for proficient work at each grade level
- Proficient: represents solid academic performance for each grade assessed (Students reaching this level have demonstrated competency over challenging subject matter, including subject-matter knowledge, application of such knowledge to real-world situations, and analytical skills appropriate to the subject matter.)
- Advanced: signifies superior performance

As part of the process of setting achievement levels for a content area such as mathematics, panels of teachers, education specialists, and members of the general public produce subject-specific descriptions for each achievement level. The achievement level descriptions for grade-8 mathematics appear in chapter 2 by Dossey, and the other descriptions can be found in Reese et al. (1997, pp. 43, 45). A connection exists between the achievement levels and the NAEP scale in that the scale scores serve as "cutpoints" between the achievement levels (for example, at grade 8 a scale score of 299 defines the cutpoint between Basic and Proficient performance in mathematics) and between Basic performance and that which is considered to be below basic. To better illustrate the connection between the achievement levels and scale scores, NAEP introduced the use of "item maps" as part of reporting results for the seventh mathematics assessment. An item map, an example of which appears in chapter 2 (figure 2.1), is a visual representation that combines scale scores, item descriptions, and achievement level cutpoints in a way that highlights the kinds of questions a student can likely solve at a given level. For example, within the range of scale scores that defines Basic performance (262 to 299) are items that involve mathematics topics such as identifying the solution for a linear inequality, partitioning the area of a rectangle, interpreting the remainder in a division problem, and understanding a sampling technique. (For a complete explanation of how to interpret item maps, see Reese et al. 1997, p. 9.)

Since their introduction in 1992 as the primary reporting method for NAEP results, the achievement levels have gained wide acceptance, but the method of setting the achievement levels (that is, the processes involved in determining the cutpoints for each level on the NAEP scale) continues to elicit a great deal of controversy. This controversy began in the early 1990s, when a number of reports criticized the NAEP achievement levels (for example, U.S. General Accounting Office 1993; National Academy of Education 1993; Silver and Kenney 1993). The controversy continues to the end of the decade with the release of a report by the Committee on the Evaluation of National and State Assessments of Education Progress, which concluded that the process for setting NAEP achievement levels is "fundamentally flawed ... and should be replaced" (National Research Council 1999, p. 7). However, evidence refuting this conclusion appears in NAGB's statement on the use of achievement levels (NAGB 1999) and in a report issued by panel of psychometricians and experts on setting standards (Hambleton et al. 1999). Indications are that the achievement levels will remain the primary reporting mechanism for NAEP results and that their use will continue to spark spirited discussion.

Previous interpretive reports of NAEP mathematics results have found the concept of a scale score or an achievement level description to have little meaning, especially in the case of interpreting performance in content areas. Instead, writers of those reports preferred to examine results on the basis of percent-correct values for clusters of related items and percent-responding values for the choices or score categories on individual items, and then to interpret those results in light of current curriculum and classroom practices. Beginning with the fourth assessment, some chapters in the interpretive reports augmented item-level percent-correct values with other reporting methods such as scale scores and scale anchor descriptions, especially when it was deemed important to report overall performance and performance by subgroups (for example, race/ethnicity and gender).

As for prior interpretive reports of NAEP mathematics results, authors of interpretive reports from the seventh assessment have used NAEP results in a variety of ways, focusing on reporting methods that were deemed the most appropriate within each chapter. For example, the chapters concerned with overall performance (chapter 2 by Dossey), by selected subgroups (chapters 3 by Strutchens and Silver and chapter 4 by Ansell and Doerr) and by grade levels according to teacher characteristics (chapter 5 by Grouws and Smith) make use of scale scores, achievement level results, and the percentage of students or teachers choosing particular responses to questionnaire items. As was the case for the content-based chapters in the previous interpretive reports of NAEP results, the content-based chapters in this volume (chapters 6–10) and the chapter on performance on the extended constructed-response questions (chapter 11 by Silver, Alacaci, and Stylianou) are based primarily on item-level data such as percent-correct values and the percentage of students responding at a given score level.

SPECIAL STUDIES IN THE 1996 NAEP

As illustrated in figure 1.1, in addition to the main assessment at the national level, NAEP periodically conducts special studies in areas that are of interest to educators and policymakers. For example, past special studies have included results from a group assessment in U.S. history (Goodman et al. 1998), the effect of choice of stories in a reading assessment (Campbell and Donahue 1997), and the use of portfolios in writing (Gentile 1992). In mathematics, the introduction of extended constructed-response questions was a special study in 1990, as was a special study in estimation (Mullis et al. 1991).

The seventh mathematics assessment included three special studies: the Estimation Study (a continuation of the study begun in 1990), the Study of Mathematics-in-Context (also called the Theme Study), and the Study of Students Taking Advanced Courses in Mathematics (also called the Advanced Study). Additional information on the 1996 special studies in mathematics can be found in an NCES publication, *Estimation Skills, Mathematics-in-Context, and Advanced Skills in Mathematics* (Mitchell, Hawkins, Stancavage, and Dossey 1999). In the present volume, chapter 12 by Kenney and Lindquist and chapter 13 by Kilpatrick and Gieger present detailed information about the Theme Study and the Advanced Study, respectively. A brief summary of the Estimation Study, which is not discussed elsewhere in this volume, appears next.

The 1996 Estimation Study was administered to a sample of students in grades 4, 8, and 12 and used two blocks of items. One block consisted entirely of multiple-choice items from the 1990 and 1992 assessments and provided short-term trend information on students' progress in estimation; the other block consisted of both multiple-choice and short constructed-response items that were written especially for the special study. Unlike the blocks in the main assessment or the other two special studies, the Estimation items were administered using a paced audio tape that reads the questions to the students and then tells them when to move on to the next question. Reading the questions to students facilitates access to questions for students with limited reading skills, and limiting the time students spend on any one question increases the likelihood that students engage in estimation instead of computation.

Results from the Estimation Study are not described in detail in this volume primarily because none of the items from the two blocks have been released to the public. As mentioned previously, information about results from this special study appears in Mitchell, Hawkins, Stancavage, and Dossey (1999, pp. 9–25). The overall findings from the study based on short-term trend information were as follows (ibid., p. 2):

- Students' performance in Estimation at grades 4 and 12 was stronger in 1996 than in 1990. In particular, fourth-grade students' average estimation scale score in 1996 (206) was significantly higher than that from 1990 (200). For students in grade 12, the 1996 average scale score of 297 was significantly higher than that from 1990 (292).

- Students' performance in Estimation at grade 8 appears to be level across the three years: 270 in 1996, 271 in 1992, and 269 in 1990.

CONCLUSION

NAEP stands as one of the most complex large-scale assessments ever designed and administered on a regular basis. This chapter has attempted to provide information about the NAEP mathematics assessment so that the reader can better understand NAEP and the interpretations offered by the writers of the remaining chapters in this volume. To assist interested readers in learning more about NAEP and about the various reports produced by government sources or through the efforts of NCTM, a bibliography of reports, journal articles, and book chapters based on the fifth, sixth, and seventh NAEP mathematics assessments is included at the end of this volume.

The chapters that follow continue the rich tradition of support by NCTM for projects that involved reporting NAEP results and interpreting them in light of mathematics education research and reform. The NCTM-NAEP interpretive reports tradition began with the first NAEP mathematics assessment (Carpenter et al. 1978) and continued to include the second assessment (Carpenter et al. 1981), the fourth assessment (Lindquist 1989), and the fifth and sixth assessments (Kenney and Silver 1997). NCTM's efforts in the area of interpretation of NAEP results have been recognized by the National Research Council's Committee on the Evaluation of National and State Assessments of Educational Progress as a successful first step in making NAEP results more accessible to a variety of audiences, including teachers, researchers, parents, policymakers, and the public at large (NRC 1999). The report states, "The NCTM [NAEP interpretive reports for mathematics] provide an example of the educationally useful and policy-relevant information that can be gleaned from students' responses in the current assessments, and they point toward the even more useful information that could be provided if the assessments were developed with these analyses in mind" (p. 156), and goes so far as to recommend that NAEP should include a research and development agenda that allows for "producing and presenting more in-depth interpretive information in NAEP reports to make overall results more understandable, minimize improper or incorrect inferences, and support the needs of users who seek information that assists them in determining what to do in response to NAEP results" (pp. 160–61).

It is hoped that continuing the tradition of the NCTM interpretive reports to include results from the seventh NAEP assessment will have an impact on the mathematics education community. In particular, for classroom teachers and teacher educators, the interpretive look at the results from the seventh NAEP contained in this volume will provide valuable information about students' strengths in mathematics and about areas that need additional attention.

REFERENCES

Ballator, Nada. *The NAEP Guide: A Description of the Content and Methods of the 1994 and 1996 Assessments.* Washington, D.C.: National Center for Education Statistics, 1996.

Campbell, Jay R., and Patricia L. Donahue. *Students Selecting Stories: The Effects of Choice in Reading Assessment.* Washington, D.C.: National Center for Education Statistics, 1997.

Carpenter, Thomas P. "Introduction." In *Results from the Fourth Mathematics Assessment of the National Assessment of Educational Progress,* edited by Mary Montgomery Lindquist, pp. 1–9. Reston, Va.: National Council of Teachers of Mathematics, 1989.

Carpenter, Thomas P., Terrence G. Coburn, Robert E. Reys, and James W. Wilson. *Results from the First Mathematics Assessment of the National Assessment of Educational Progress.* Reston, Va.: National Council of Teachers of Mathematics, 1978.

Carpenter, Thomas P., Mary Kay Corbitt, Henry S. Kepner, Jr., Mary Montgomery Lindquist, and Robert E. Reys. *Results from the Second Mathematics Assessment of the National Assessment of Educational Progress.* Reston, Va.: National Council of Teachers of Mathematics, 1981.

Gentile, Claudia. *Exploring New Methods for Collecting Students' School-Based Writing: NAEP's 1990 Portfolio Study.* Washington, D.C.: National Center for Education Statistics, 1992.

Goodman, Madeline, Stephen Lazer, John Mazzeo, Nancy Mead, and Amy Pearlmutter. *1994 NAEP U.S. History Group Assessment.* Washington, D.C.: National Center for Education Statistics, 1998.

Hambleton, Ronald K., Robert L. Brennan, William Brown, Barbara Dodd, Robert A. Forsythe, William A. Mehrens, Jeff Nellhaus, Mark Reckase, Douglas Rindone, Wim J. van der Linden, and Rebecca Zwick. *A Response to "Setting Reasonable and Useful Performance Standards" in the National Academy of Science's Grading the Nation's Report Card.* Iowa City, Iowa: American College Testing, May 1999.

Hawkins, Evelyn F., Frances B. Stancavage, and John A. Dossey. *School Policies and Practices Affecting Instruction in Mathematics: Findings from the National Assessment of Educational Progress.* Washington, D.C.: National Center for Education Statistics, 1998.

Kenney, Patricia Ann, and Edward A. Silver, eds. *Results from the Sixth Mathematics Assessment of the National Assessment of Educational Progress.* Reston, Va.: National Council of Teachers of Mathematics, 1997.

Kouba, Vicky L., Thomas P. Carpenter, and Jane O. Swafford. "Calculators." In *Results from the Fourth Mathematics Assessment of the National Assessment of Educational Progress,* edited by Mary Montgomery Lindquist, pp. 94–105. Reston, Va.: National Council of Teachers of Mathematics, 1989.

Lindquist, Mary Montgomery ed. *Results from the Fourth Mathematics Assessment of the National Assessment of Educational Progress.* Reston, Va.: National Council of Teachers of Mathematics, 1989.

Mitchell, Julia H., Evelyn F. Hawkins, Pamela M. Jakwerth, Frances B. Stancavage, and John A. Dossey. *Student Work and Teacher Practices in Mathematics.* Washington, D.C.: National Center for Education Statistics, 1999.

Mitchell, Julia H., Evelyn F. Hawkins, Frances B. Stancavage, and John A. Dossey. *Estimation Skills, Mathematics-in-Context, and Advanced Skills in Mathematics: Results from Three Studies of the National Assessment of Educational Progress 1996 Mathematics Assessment.* Washington, D.C.: National Center for Education Statistics, 1999.

Mullis, Ina V. S., John A. Dossey, Eugene H. Owen, and Gary W. Phillips. *The STATE of Mathematics Achievement: NAEP's 1990 Assessment of the Nation and the Trial Assessment of the States.* Washington, D.C.: National Center for Education Statistics, June 1991.

National Academy of Education. *Setting Performance Standards for Student Achievement.* Stanford, Calif.: National Academy of Education, 1993.

National Assessment Governing Board. *Mathematics Framework for the 1996 National Assessment of Educational Progress.* Washington, D.C.: National Assessment Governing Board, 1994.

———. *Students Performance Standards on the National Assessment of Educational Progress: Affirmation and Improvements, A Report to Congress on the Long Range Plans for Setting Achievement Levels on NAEP.* Washington, D.C.: National Assessment Governing Board, June 1999. Available at www.nagb.org.

National Assessment of Educational Progress. *Mathematics Objectives: 1990 Assessment.* Princeton, N.J.: Educational Testing Service, National Assessment of Educational Progress, 1988.

National Council of Teachers of Mathematics. *Curriculum and Evaluation Standards for School Mathematics.* Reston, Va.: National Council of Teachers of Mathematics, 1989.

National Research Council. *Grading the Nation's Report Card: Evaluating NAEP and Transforming the Assessment of Educational Progress,* edited by James W. Pellegrino, Lee R. Jones, and Karen J. Mitchell. Washington, D.C.: National Academy Press, 1999.

Reese, Clyde M., Karen E. Miller, John Mazzeo, and John A. Dossey. *NAEP 1996 Mathematics Report Card for the Nation and the States.* Washington, D.C.: National Center for Education Statistics, 1997.

Silver, Edward A., and Patricia Ann Kenney. "Expert Panel Review of the 1992 NAEP Mathematics Achievement Levels." In *Setting Performance Standards for Student Achievement: Background Studies,* pp. 215–81. Stanford, Calif.: National Academy of Education, 1993.

U.S. General Accounting Office. *Educational Achievement Standards: NAGB's Approach Yields Misleading Interpretations.* Report No. GAO/PEMD-93-12. Washington, D.C.: General Accounting Office, 1993.

Wilson, James W., and Jeremy Kilpatrick. "Theoretical Issues in the Development of Calculator-Based Mathematics Tests." In *The Use of Calculators in the Standardized Testing of Mathematics,* edited by John Kenelly, pp. 7–15. New York: College Entrance Examination Board, 1989.

The State of NAEP Mathematics Findings: 1996

John A. Dossey

THE National Assessment of Educational Progress (NAEP) assessments are one of the major policy sources of information on change and progress in school mathematics in the United States and its jurisdictions. This program of information gathering has operated at two different levels since its first national assessment of school mathematics in 1973. Most mathematics educators and policy researchers are aware of the major national and, since 1990, state assessments carried out by the Department of Education's National Center for Education Statistics (NCES). This program provides information on a variety of students' achievement, student and teacher demographic, and school structure variables regarding public and private school mathematics education in grades 4, 8, and 12. The achievement tests and questionnaires are constantly updated with new test items and questions to maintain a contemporary view of mathematics education programs in the nation's schools.

There is a second major program of research associated with NAEP. This is the NAEP trend assessment in mathematics. This program, also referred to as "long-term trend NAEP," measures current students' performance on a set of instruments that was designed for and used in the first NAEP assessment in 1973. These instruments allow the measurement of students' knowledge and skills relative to a 1973 baseline in a setting that allows students to use only paper and pencil in answering the test's items. The trend assessment, therefore, gives a measure of potential slippage with respect to traditional goals for school mathematics in an era of reform.

The NAEP assessments are designed to measure student demographics, mathematics achievement, and beliefs relative to school mathematics, and to monitor trends in these areas across time. By the random sampling of students, NAEP also acquires a matching group—these students' teachers. Through another set of questionnaires, NAEP develops a picture of the mathematics teachers instructing U.S. students, the educational backgrounds of these teachers, and the instructional methods they employ in their classrooms. There is also a set of school questionnaires used for collecting demographic information about the students' schools. While NAEP assessments attempt to evolve and keep abreast of current goals for student achievement and instructional trends, they measure what *does* take place in the sample students' schools rather than what *might* take place in their schools. This means that the NAEP assessment program attempts to target current reforms as much as is feasible but not to be in the vanguard of reform. Consequently, some consider the NAEP tests to be too conservative in their composition and approach.

Besides the well known aspects of the main NAEP assessment administered to all students in the random sample, the National Assessment Governing Board (NAGB) also conducts special studies in estimation and on selected focus topics as part of the basic assessment program with smaller groups of randomly sampled students. Across the years, NAEP has collected information on students' estimation through the use of an audio-paced assessment, one that directs the students' items at a pace that discourages computation and more algorithmic forms of approximation (see Mitchell et al. 1999 for information about the estimation study results). In 1996, in addition to the estimation studies, NAEP conducted special studies to find out what the most curricularly advanced students could do at grades 8 and 12, and that contained some special blocks, called "theme blocks," that presented NAEP items in a different format. These blocks of items were developed so that the questions dealt with an aspect of a single context. For example, one theme block at grade 12 dealt with the topic of whether it was better to buy or lease a car. The outcomes related to the theme blocks and advanced studies are reported in Mitchell et al. (1999) and are also covered in chapter 12 by Kenney and Lindquist and chapter 13 by Kilpatrick and Gieger in this volume.

The special assessments, such as those in estimation and studies in advanced mathematics topics, were efforts to respond to questions and requests from the mathematics education community, from policy makers, or from those studying specific aspects of the way the NAEP assessments themselves operate. Maintaining the stability required for measuring trends while still introducing innovations has led to complexities in NAEP. Thus, NAEP has provided for links to the future and links to the past by conducting separate mathematics assessments for different purposes (Ballator 1996; Calderone, King, and Horkay 1997; Mullis 1991).

THE NAEP ASSESSMENTS

The national NAEP assessment has been given in 1973, 1978, 1982, 1986, 1990, 1992, and, most recently, in 1996. Since 1990, the national NAEP assessment program has included an option allowing states to elect to replicate the national NAEP assessments at the state level. If a state so elects, a special state-based random sample is drawn from students in grade 8 (1990, 1992, and 1996) or grade 4 (1992 and 1996). Both of these programs, national NAEP and the NAEP state assessment, use the same basic core of assessment items and questionnaires. The state assessments do not include the estimation and special study portions.

These assessments are developed following the current NAEP framework (National Assessment Governing Board [NAGB] 1994). The framework for the 1996 NAEP assessment was developed in the early 1990s and reflects, to a considerable degree, the content recommendations of the National Council of Teachers of Mathematics' *Curriculum and Evaluation Standards for School Mathematics* (1989). However, the items on the test do not change rapidly from one assessment to another. New items are phased in as other older items are released to the public to help the overall test to reflect more contemporary goals or curricular content. This gradual change in the assessments is monitored both by NAGB and by a standing committee of mathematics educators and classroom teachers to ensure that the tests both reflect the current framework and, at the same time, maintain the required portion of the previous assessment to allow for statistical linking to measure short-term trends in the NAEP data over time. At present, the short-term trend in national NAEP involves the 1990, 1992, and 1996 mathematics assessments.

In 1990 several changes were made in the format of the national NAEP assessment. These changes were, in part, due to the release of the NCTM *Curriculum and Evaluation Standards.* More constructed-response items were added, and the percentage of items requiring calculator use was increased. In addition to these changes, the overall balance of items was shifted from a disproportionate number of items dealing with the content area of number and operations to a more balanced assessment across the five content areas surveyed: number and operations; measurement; geometry; data analysis, probability, and statistics; and algebra and functions. As a result of these changes in 1990, new short-term trend lines were started for national and state NAEP. To ensure that the student participants in NAEP are representative of students across the nation, or within a state, students are randomly selected based on a stratified, three-stage sampling plan. First, counties across the nation are classified by region and community type and then some are randomly selected. Second, schools—both public and private—are classified, and some are then randomly selected from within the selected regions and communities. Finally, some students are randomly selected from within the selected schools.

In mathematics, as well as in other disciplines such as science, reading, and writing, NAEP's long-term trend assessment program regularly assesses nationally representative samples of students using the methods of past assessments. This program of using the original NAEP assessments to measure trends based on items developed twenty-five years ago provides a measure of possible effects of changes in mathematics education. The trend assessments sample students at ages 9, 13, and 17 rather than at grade levels, as this was the sampling process for the NAEP program in mathematics from 1973 to 1990. The NAEP long-term trend assessment items also reflect the curricular emphasis on basic skills in the 1970s. Further, students are not allowed to use calculators on the NAEP long-term trend assessment in mathematics.

In this chapter, the focus is on the global results emanating from the NAEP national and state assessments with a specific focus on the 1996 national findings. At the same time, the findings of the 1994 and 1996 NAEP long-term trend assessments are reviewed with a focus on what they tell us about changes in basic skills and other aspects of school mathematics based on an assessment developed in 1973. In some sections below, the long-term trend results are reported along with the national results in order to present a more complete picture of NAEP results over time.

The NAEP National Assessment

The 1996 national NAEP assessment was the first assessment conducted using the revised framework (NAGB 1994). This framework made small, but significant changes in both the areas of content assessed and the content included within them from the previous assessment framework (NAEP 1988). These changes brought the content areas and their specific concepts and skills closer to the NCTM *Curriculum and Evaluation Standards.* The name of the content area Number and Operations was changed to Number Sense, Properties, and Operations and its scope was changed to include more topics reflective of number sense and estimation and the concepts of operations. The content area Geometry also had a name change to Geometry and Spatial Sense. This change brought a greater emphasis on spatial topics and those dealing with the effects of performing geometric transformations on shapes.

A second difference between the framework for the 1996 assessment and the 1990 framework was in the relative numbers of items assigned to the five content areas. Table 2.1 shows the percentages of items included in each area for the 1990 and 1992 assessments and for the 1996 assessment. The change columns show the difference in these percentages for each of the content areas. The changes in grades 4 and 8 reflect the increasing emphasis given to the role of algebra and functions at those grade levels, while the changes at grade 12 reflect the increased emphasis on data and chance in the secondary school years. These changes allow for shifts in the NAEP assessments to better measure and provide a basis for reporting on students' achievement in mathematics.

Table 2.1
Percent of Items by Content Area and Percent Change from 1990–92 to 1996

	Grade 4			Grade 8			Grade 12		
	90–92	96	Percent Change	90–92	96	Percent Change	90–92	96	Percent Change
Number Sense, Properties, and Operations	45	40*	–5	30	25*	–5	25	20*	–5
Measurement	20	20	0	15	15	0	15	15	0
Geometry and Spatial Sense	15	15	0	20	20	0	20	20	0
Data Analysis, Statistics, and Probability	10	10	0	15	15	0	15	20	+5
Algebra and Functions	10	15	+5	20	25	+5	25	25	0

* Represents the lower end of the ranges 40–70 percent, 25–60 percent, and 20–50 percent, respectively.
Sources: National Assessment of Educational Progress (1988); National Assessment Governing Board (1994)

Reporting Performance on National NAEP

Students' responses to the items on the national NAEP mathematics assessment are reported in two different fashions. One involves placing both students' performance and items' location onto a 0–500 scale. Such diagrams are called "item maps," like the one for grade 8 shown in figure 2.1, and illustrate the students' performance on items across broad ranges of the NAEP mathematics scale. The short description of individual items given reflects what the items were assessing and their position on the scale indicates their position of relative difficulty for students in the grade level specified (Reese et al. 1997).

Although the NAEP scale provides a picture of students' progress across grades, and between various demographic subgroups, it does not describe what students should know and be able to do in mathematics, nor does it evaluate the students' performance against a nationally held standard of performance. To accomplish this, NAGB established three NAEP achievement levels to describe students' performance in mathe-

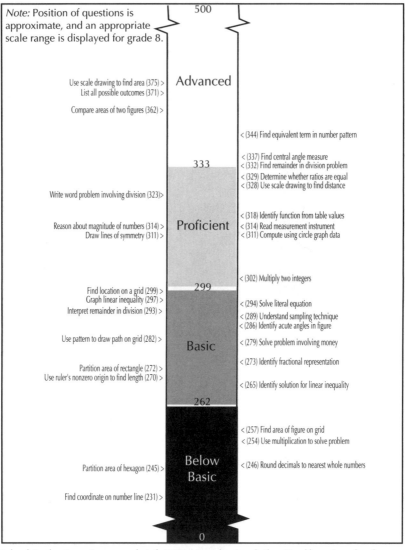

Each grade 8 mathematics question was mapped onto the NAEP 0 to 500 mathematics scale. The position of the question on the scale represents the scale score attained by students who had a 65 percent probability of successfully answering the question. (The probability was 74 percent for 4-option multiple-choice question and 72 percent for 5-option multiple-choice question.) Only selected questions are presented. The eighth-grade mathematics achievement levels are references on the map.

Fig. 2.1. Item map of selected questions on the NAEP scale for grade 8. *Source:* Reese et al. (1997, p. 11)

matics at the grade levels assessed. These NAEP achievement levels are Basic, Proficient, and Advanced. Basic performance denotes partial mastery of the knowledge, content, and skills for desired work at each grade level, but not a level of work that is deemed satisfactory. Proficient performance represents solid academic performance at each grade level. Advanced performance signifies a superior level of work at each of the grades. The performance descriptions of the achievement levels for grade 8 mathematics are in figure 2.2. The NAEP mathematics scale score ranges corresponding to each of these levels are depicted by the different shaded regions in the center of the scale in the item map in figure 2.1.

Overall Performance by Achievement Levels and Scale Scores

The data in table 2.2 reflect the percent of grade 4, 8, and 12 students performing at each of the achievement levels. Note that the percentages of the levels Basic, Proficient, and Advanced are quoted in an inclusive form as the percent "at or above" the level.

Table 2.2
National Percents of Students according to Mathematics Achievement Levels by Grade and by Assessment Year

	1990	1992	1996
Grade 4			
Percent below Basic	50	41*	36*†
Percent at or above			
Basic	50	59*	64*†
Proficient	13	18*	21*†
Advanced	1	2	2
Grade 8			
Percent below Basic	48	42*	38*†
Percent at or above			
Basic	52	58*	62*†
Proficient	15	21*	24*
Advanced	2	3	4*
Grade 12			
Percent below Basic	42	36*	31*†
Percent at or above			
Basic	58	64*	69*†
Proficient	12	15	16*
Advanced	1	2	2

* Indicates significant difference from 1990.
† Indicates significant difference from 1992.
Source: Reese et al. (1997)

BASIC

Eighth-grade students performing at the basic level should exhibit evidence of conceptual and procedural understanding in the five NAEP content strands [Number Sense, Properties, and Operations; Measurement; Geometry and Spatial Sense; Data Analysis, Statistics, and Probability; Algebra and Functions]. This level of performance signifies an understanding of arithmetic operations—including estimation—on whole numbers, decimals, fractions, and percents.

Eighth-graders performing at the basic level should complete problems correctly with the help of structural prompts such as diagrams, charts, and graphs. They should be able to solve problems in all NAEP content areas through the appropriate selection and use of strategies and technological tools—including calculators, computers, and geometric shapes. Students at this level also should be able to use fundamental algebraic and informal geometric concepts in problem solving.

As they approach the proficient level, students at the basic level should be able to determine which of the available data are necessary and sufficient for correct solutions and use them in problem solving. However, these eighth-graders show limited skill in communicating mathematically.

PROFICIENT

Eighth-grade students performing at the proficient level should apply mathematical concepts and procedures consistently to complex problems in the five NAEP content strands.

Eighth-graders performing at the proficient level should be able to conjecture, defend their ideas, and give supporting examples. They should understand the connections between fractions, percents, decimals, and other mathematical topics such as algebra and functions. Students at this level are expected to have a thorough understanding of basic level arithmetic operations—an understanding sufficient for problem solving in practical situations.

Quantity and spatial relationships in problem solving and reasoning should be familiar to them, and they should be able to convey underlying reasoning skills beyond the level of arithmetic. They should be able to compare and contrast mathematical ideas and generate their own examples. These students should make inferences from data and graphs; apply properties of informal geometry; and accurately use the tools of technology. Students at this level should understand the process of gathering and organizing data and be able to calculate, evaluate, and communicate results within the domain of statistics and probability.

ADVANCED

Eighth-grade students performing at the advanced level should be able to reach beyond the recognition, identification, and application of mathematical rules in order to generalize and synthesize concepts and principles in the five NAEP content areas.

Eighth-graders performing at the advanced level should be able to probe examples and counterexamples in order to shape generalizations from which they can develop models. Eighth graders performing at the advanced level should use number sense and geometric awareness to consider the reasonableness of an answer. They are expected to use abstract thinking to create unique problem-solving techniques and explain the reasoning processes underlying their conclusions.

Fig. 2.2. NAEP mathematics achievement level descriptions for grade 8. *Source:* Reese et al. (1997)

These results are discouraging in that approximately one-third of the nation's youth at each grade level is performing at the Below-Basic level. However, a comparison of the data from 1996 with that for previous years reflects increasing students' performance in school mathematics. Nevertheless, the percentage of students reaching the higher levels of achievement falls far short of what one would hope. Most of the increases in students' performance over the time period since 1990 have resulted in gains at the Basic and Proficient levels. Little progress has been made in making equivalent gains at the Advanced level.

As shown in figure 2.3, the performance of the grade level groups on the national NAEP scale in 1996 show significant increases from both 1990 and 1992. Using the fact that there are about 80 scale points from the mean performance for grade 4 to that for grade 12, one can think of 10 scale points as nearly equivalent to a grade placement range. Using this informal metric, the gains observed at each grade level between 1990 and 1996 are approximately one grade level.

Performance by Content Area

Because individual items assess one or more of the content areas, it is possible to examine students' performance in each of the five content areas, as shown in table 2.3. These results also paint interesting pictures of positive change in all content areas across the three NAEP assessments since 1990. In three NAEP content strands—Geometry and Spatial Sense; Data Analysis, Statistics, and Probability; and Algebra and Functions—a pattern

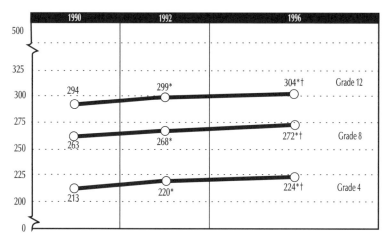

Figure 2.3. Average national NAEP mathematics scale scores, 1996. *Source:* Reese et al. (1997)

of significant increases appears. This pattern, plus some other significant increases in the content strands Number Sense, Properties and Operations, and Measurement over the same six years reflects, perhaps, the broadening of students' opportunity to learn in all five content areas.

Table 2.3
Average NAEP Scale Scores by Content Area, 1990–1996

	1990	1992	1996
Number Sense, Properties, and Operations			
Grade 4	210	217*	221*†
Grade 8	267	272*	274*
Grade 12	293	299*	301*
Measurement			
Grade 4	218	224*	226*
Grade 8	259	267*	270*
Grade 12	292	298*	302*†
Geometry and Spatial Sense			
Grade 4	213	222*	225*†
Grade 8	260	264*	269*†
Grade 12	295	301*	307*†
Data Analysis, Statistics, and Probability			
Grade 4	—	220	225†
Grade 8	263	269*	272*
Grade 12	294	298*	304*†
Algebra and Functions			
Grade 4	214	219*	227*†
Grade 8	261	268*	273*†
Grade 12	296	301*	305*†

* Indicates significant difference from 1990.
† Indicates significant difference from 1992.

Note: No average scale score is available from 1990 for the content strand Data Analysis, Statistics, and Probability because that assessment contained too few items to create a scale.

Source: Mitchell et al. (1999)

Performance by Gender and Racial/Ethnic Group

Analyses of performances by gender and by racial/ethnic groups reflect significant differences in the performance of students. Some studies have suggested that these differences are the result of opportunities afforded the students in school and in their homes and communities; other studies point to the atmosphere of encouragement about education and its roles in students' lives (Eakin and Backler 1993; Mullis, Jenkins, and Johnson 1994; Oakes 1990; Willingham and Cole 1997). Results from national NAEP in 1996 (reported by grade levels [4, 8 and 12]) and from the long-term trend assessment (reported by age groups [9-year-olds, 13-year-olds, 17-year-olds]) provide information on performance by gender and race/ethnicity for different samples of students.

In 1996, the national NAEP results showed that male students outscored female students at all three grade levels. However, this difference was statistically significant only at grade 4 (Reese et al. 1997). According to the long-term trend results, the linear trend for the difference of male and female scores across the twenty-three-year period shows a significant widening of the performance gap at the 9- and 13-year-old levels and a narrowing of the performance gap between genders at age 17. Further, in 1996 there was a statistically significant increase in scale scores for both male and female students in the 9- and 13-year-old groups, but there was no significant change for either males or females in the 17-year-old groups (Campbell, Voelkl, and Donahue 1997).

As in previous assessments, fourth- and twelfth-grade White and Asian/Pacific Islander students and White students in grade 8 scored higher than their Black and Hispanic counterparts in the 1996 national assessment. (The results for eighth-grade Asian/Pacific Islander students were not reported, as there were some questions about the integrity of the data.) Hispanic eighth-grade students outperformed their Black peers. Although scores have improved for many of the racial/ethnic groups since 1990, the differences in performance for White students and their Black and Hispanic counterparts have remained fairly constant. Narrowing these gaps remains a central challenge for mathematics education in the United States in the coming century.

An analysis of long-term trend data indicates that differences between White and Black students' performance have varied over the twenty-three-year period, but were significantly smaller in 1996 than in 1973 at all three age levels. The linear trend over the same time and across all age levels also indicated a significant decrease in differential performance. The same pattern was true for the difference between White and Hispanic students at ages 9 and 13, with the exception of 17-year-olds. There was no significant change for this age group in the overall difference in scale scores or in linear trend of difference for the period from 1973 to 1996. There were positive increases in performance for students of all racial/ethnic groups in each age group with the exception of 17-year-old White students, for whom there was no significant change over the time period from 1973 to 1996 (Campbell, Voelkl, and Donahue 1997).

Performance by Item Type

As the NAEP national assessments changed over the period from 1990 to 1996, more emphasis was given to measuring students' ability to respond to a wider range of item types, especially items that required students to construct their own answers. Table 2.4 presents information on the distribution of different item types across the assessments for the three grade levels (Reese et al. 1997). Estimates developed during the field testing of items for the 1990, 1992, and 1996 assessments suggest that the ratio of students' testing time spent on multiple-choice items to time spent on constructed-response items across the 1990, 1992, and 1996 assessments were 70 to 30, 65 to 35, and 40 to 60, respectively (Chancey O. Jones, conversation with author, 1995). The last ratio reflects the effect of doubling the number of extended constructed-response items for the 1996 assessment.

Table 2.4
Distribution of Items by Item Type, Grade Level, and Assessment Year

	Grade 4			Grade 8			Grade 12		
	1990	1992	1996	1990	1992	1996	1990	1992	1996
Multiple-Choice	102	99	81	149	118	102	156	115	99
Short Constructed-Response	41	59	64	42	65	69	47	64	74
Extended Constructed-Response	—	5	13	—	6	12	—	6	11
Total	143	163	158	191	189	183	203	185	184

Note: No extended constructed-response items were included in the 1990 assessment.
Source: Reese et al. (1997)

Table 2.5 presents information on the performance of students in the 1992 and 1996 national samples on three types of items (multiple-choice, short constructed-response, and extended constructed-response) by grade levels. Examination of the changes in the percents correct for each grade level and for each type of item over time discloses some small increases in performance for multiple-choice items. However, there were decreases in level of performance for students' work on the short and extended constructed-response items. Additional information about students' performance on the extended constructed-response tasks can be found in chapter 11 (Silver, Alacaci, and Stylianou).

The NAEP State-Level Assessment

In 1990, Congress allowed states to draw representative samples and replicate the national assessment at the state level. In that year, state-level

Table 2.5
Correct Response Rates for Three NAEP Item Formats by Grade Level, 1992 and 1996

	Grade 4		Grade 8		Grade 12	
	1992	1996	1992	1996	1992	1996
Multiple-Choice	50	54	56	55	56	60
Short Constructed-Response	42	38	53	49	40	34
Extended Constructed-Response *	16	17	8	9	9	12

*Data for the extended constructed-response questions are for the average percents of satisfactory or extended responses.

Note: Short constructed-response items included in the 1992 assessment were scored dichotomously. New short constructed-response items included in the 1996 assessment were scored on three levels (complete, partial, incorrect) to allow for partial credit.

Source: [1992 data] Dossey, Mullis, and Jones (1993); [1996 data] special computations based on information obtained from the NAEP data almanacs for the cognitive items

NAEP was limited to grade 8, and forty states and jurisdictions participated. The jurisdictions included Guam and the Virgin Islands, the District of Columbia, and the educational programs for children of members of the armed forces stationed both overseas and stateside. In 1992, the state-level NAEP program was extended to grades 4 and 8 and forty-four states and jurisdictions participated. In 1996, this program was made a permanent feature of the NAEP portfolio of activities, along with national and long-term trend programs. In 1996, forty-seven states and jurisdictions participated at grade 4 and forty-four participated at grade 8.

The data from the NAEP state assessments include only students from public schools, as these assessments have been limited to public schools only. The results from the state assessment at grade 4, shown in table 2.6, showed significant scale score differences from 1992 to 1996 for fifteen of the thirty-nine states participating. Using the informal 10 scale point metric for a one grade-placement level, the data suggest that the state scores for Texas and North Carolina indicate an increase of a grade level or more between 1992 and 1996. Several other states show gains of 5 or more NAEP scale points.

As shown in table 2.7, the picture is much the same for the NAEP state assessment at grade 8 with twenty-seven of thirty-two jurisdictions showing significant increases from 1990 to 1996 and thirteen of thirty-seven showing significant increases over the period from 1992 to 1996. Using the informal 10-point metric, the results suggest that from 1990 to 1996 Connecticut, Michigan,

Table 2.6
Average NAEP Mathematics Scale Scores for Selected States Participating in the 1992 and 1996 State Assessments: Grade 4 Public Schools

	1996 Average Scale Score	Change from 1992 Average Scale Score
Maine	232	1
Minnesota	232	4†
Connecticut	232	5†
Wisconsin	231	3
North Dakota	231	2
Indiana	229	8†
Iowa#	229	−1
Massachusetts	229	2
Texas	229	11†
Nebraska	228	2
New Jersey#	227	0
Utah	227	2
Michigan#	226	6†
Pennsylvania#	226	2
Colorado	226	5†
Missouri	225	3
North Carolina	224	11†
West Virginia	223	8†
Wyoming	223	−2
Virginia	223	2
New York#	223	4†
Maryland	221	3
Rhode Island	220	5†
Kentucky	220	5†
Tennessee	219	8†
Arizona#	218	2
Arkansas	216	6†
Florida	216	2
Georgia	215	0
Delaware	215	−3†
Hawaii	215	1
New Mexico	214	1
South Carolina#	213	1
Alabama	212	3
California	209	1
Louisiana	209	5†
Mississippi	208	7†
Guam	188	−4†
District of Columbia	187	−5†

\# Indicates jurisdiction did not completely satisfy the guidelines for school participation rates in 1996. (See Appendix A of Reese et al. 1997 for an explanation of these guidelines.)

† Indicates change since 1992 in average scale scores is significant at a 5-percent level of significance using a multiple comparison procedure based on the 39 jurisdictions.

Source: adapted from Reese et al. (1997)

Texas, North Carolina, and Hawaii had gains equaling or exceeding one grade-placement level of improvement. No state or jurisdiction showed a loss over the six-year comparison period for performance at grade 8.

There is nothing in the NAEP results for grades 4 and 8 that suggests that the recent reform movements in mathematics education have lessened students' performance. Instead, the scores have shown real gains. It is, nevertheless, difficult to assign a cause to these gains. The data on teaching contained in chapter 5 (by Grouws and Smith) suggest that many of the tenets of reform suggested by the NCTM *Curriculum and Evaluation Standards* (NCTM 1989) have yet to be broadly adopted as part of the school mathematics program. Although there has been a change in the focus on problem solving in school mathematics since the appearance of the *Agenda for Action* (NCTM 1980), the quadrivium of problem solving, reasoning, communication, and connections set forth in *Curriculum and Evaluation Standards* has yet to be widely implemented in school classrooms.

The NAEP Long-Term Trend Assessment

Data from the NAEP long-term assessment provide baseline information for interpreting changes in some areas of the mathematics curriculum. Although national NAEP assessments and their frameworks are designed to move with the curriculum and school programs, the NAEP long-term trend assessment test in mathematics has remained unchanged since 1973. This assessment, making use of item blocks written for the 1973 NAEP national assessment, allows mathematics educators and policy analysts to look at the performance of today's students using assessment items designed nearly 25 years ago. The NAEP trend assessment *does not* allow students to use calculators. It is far more focused on students knowing definitions, basic facts, and algorithms than it is on students knowing and being able to apply problem solving strategies, show their reasoning, or communicate about mathematics. As such, the NAEP trend assessment provides valuable information on whether students' fundamentals (for example, paper-and-pencil computation skills, direct application of measurement formulas in geometric settings, and the use of mathematics in daily-living skills involving time and money) have changed as the emphases in the curriculum have changed. These changes include focusing on students' proficiency in constructing their own responses to context-based problems and in knowing when and how to use technology in solving problems. They also reflect the increased expectations of students' knowledge of data analysis, statistics, and probability, and the increased importance of algebra across the curriculum.

Unlike the national NAEP in which data have been collected by grade level since 1990, the NAEP trend program collects its data by age groups of students (9-, 13-, and 17-year-olds) in order to match the sampling program

Table 2.7
Average NAEP Mathematics Scale Scores for Selected States Participating in the 1990, 1992, and 1996 State Assessments: Grade 8 Public Schools

	1996 Average Scale Score	Change from 1992 Average Scale Score	Change from 1990 Average Scale Score
North Dakota	284	1	3
Maine	284	5†	—
Minnesota	284	2	9*
Iowa#	284	1	6*
Montana#	283	—	3
Wisconsin#	283	5	8*
Nebraska	283	5†	7*
Connecticut	280	6†	10*
Massachusetts	278	5	—
Michigan#	277	10†	12*
Utah	277	2	—
Oregon	276	—	5*
Colorado	276	3	8*
Indiana	276	5†	8*
Wyoming	275	0	3*
Missouri	273	2	—
New York#	270	4	9*
Texas	270	6†	12*
Virginia	270	2	5*
Maryland#	270	5	9*
Rhode Island	269	3†	9*
Arizona	268	3	8*
North Carolina	268	9†	17*
Delaware	267	4†	6*
Kentucky	267	4†	9*
West Virginia	265	6†	9*
Florida	264	4	8*
Tennessee	263	4	—
California	263	2	6*
Georgia	262	3	4
Hawaii	262	5†	11*
New Mexico	262	2	6*
Arkansas#	262	5†	5*
South Carolina#	261	0	—
Alabama	257	4	4
Louisiana	252	2	6*
Mississippi	250	4	—
Guam	239	4	7*
District of Columbia	233	-2	1

\# Indicates jurisdiction did not completely satisfy the guidelines for school participation rates in 1996. (See Appendix A of Reese et al. 1997 for an explanation of these guidelines.)
† Indicates change in average scale scores from 1992 is significant at a 5-percent level of significance using a multiple comparison procedure based on the 37 jurisdictions.
* Indicates change in average scale scores from 1990 is significant at a 5-percent level of significance using a multiple comparison procedure based on the 32 jurisdictions.
— Indicates jurisdiction did not participate in 1990 or 1992.
Source: adapted from Reese et al. (1997)

that was used in the original 1973 NAEP mathematics assessment. However, these age groups match well with the current grade level groups.

NAEP Long-Trend Results in Mathematics Proficiency: 1973 to 1996

The line graphs in figure 2.4 depict the overall trend in students' performance on the NAEP scale for each of the three age groups sampled from 1973 to 1996. The dotted lines are added to assist in examining the profiles of performance over time. Statistical analyses of the data indicate that the 9- and 13-year-old groups have made steady progress over the twenty-three-year span and that the 1996 level of performance was statistically greater than that in 1973. The data for the 17-year-olds reflect a decline in students' performance for students in this age group between 1973 and 1982. Since that time there has been an overall pattern of increased performance. However, there is no significant difference between the 1996 score and 1973 score for the 17-year-olds. Comparisons of the scores for each group over the time period from 1990 to 1996 do not show significant increases for any of the three age groups. Analyses conducted to ascertain the significance of the linear component of trend from 1973 to 1996 showed significant improvement for all three groups since the NAEP trend assessment was initiated (Campbell, Voelkl, and Donahue 1997).

A further analysis of the NAEP long-term trend data involved an examination of the percentage of students reaching or exceeding each of the benchmark scale scores of 150, 200, 250, 300, and 350 on the NAEP trend scale. These five levels are accompanied by descriptions that outline the concepts, procedures, and processes associated with performance at each level (see Campbell, Voelkl, and Donahue 1997, p. 56). The results from the 1978 and 1996 assessments are shown in table 2.8.

Table 2.8
Trends in Percent of Students at or above the NAEP Levels of Mathematics Performance, 1978 and 1996

	Age 9		Age 13		Age 17	
	1978	1996	1978	1996	1978	1998
Level 350: Multistep Problem Solving and Algebra	0	0	1	1	7	7
Level 300: Moderately Complex Procedures and Reasoning	1	2^	18	21	52	60^
Level 250: Numerical Operations and Beginning Problem Solving	20	30^	65	79^	92	97^
Level 200: Beginning Skills and Understandings	70	82^	95	99^	100	100
Level 150: Simple Arithmetic Facts	97	99^	100	100	100	100

^ Indicates that the percent in 1996 is significantly different from the percent in 1978.
Source: Campbell et al. (1997)

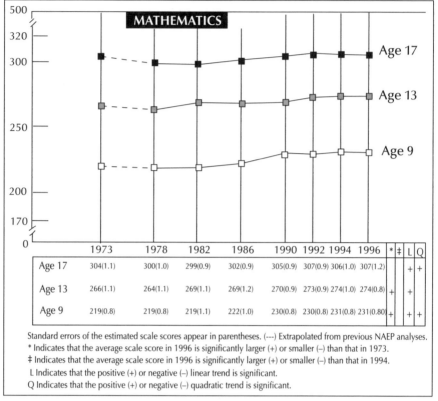

Fig. 2.4. Trend profiles for the NAEP long-term trend assessment, 1973 to 1996. *Source:* Campbell et al. (1997)

Analysis of the data in the table shows that there has been a significant increase in the percentage of students reaching various benchmark levels from 150 through 300 across the period from 1978 through 1996. These analyses indicate that much of the increase in students' long-term trend results come from growth in mathematical topics such as basic number facts and operations and in reading and interpreting graphs, tables, and charts. However, few students at any age achieved the 350 benchmark, a level that indicates substantial ability in elementary algebra and geometry and in multistep problem solving.

To examine these increases closer, an analysis was made of the performance of students in the bottom quartile, middle two quartiles, and upper quartile of each of the age groups over the same periods of time. The resulting growth patterns for each of the three quartile groups paralleled the increases shown in table 2.8. This indicates that the increases were not just an artifact of the performance of the most able students but rather an increase indicative of change in the students in each of the three quartile-defined groups (Campbell, Voelkl, and Donahue 1997).

Mathematics Course Taking by 17-year-olds

Data from the student questionnaires on the NAEP long-term trend assessment provide a picture of students' course taking at the secondary school level. The data in table 2.9 reflect 17-year-olds' responses to a series of questions asking what was the highest mathematics course they had completed or in which they were currently enrolled. The data suggest that there has been an improvement both in students' opportunity to learn and in the retention of students over this six-year period of time. Since 1990, 7 percent fewer 17-year-olds report stopping their study of mathematics with a course in prealgebra and 3 percent fewer 17-year-olds report stopping their study of mathematics with a course in first-year algebra. This has led to a 4 percent increase in students completing coursework through advanced algebra and a 5 percent increase in students completing a precalculus or calculus course. An alternative explanation might be that more students have the opportunity to start their study of algebra in middle school now, and this has resulted in more students reaching the upper levels of the secondary curriculum prior to age 17.

Table 2.9
Percent of Student Reporting Highest Level of Mathematics Taken at Age 17 for Selected Long-Term Trend Assessments

	1978	1990	1994	1996
Prealgebra or General Mathematics	20	15	9	8
Algebra 1	17	15	15	12
Geometry	16	15	15	16
Algebra 2	37	44	47	50
Precalculus or Calculus	6	8	13	13

Sources: Campbell et al. (1997); Campbell et al. (1996); Mullis et al. (1991)

Comparison of the 1996 data with the data from 1978 shows even greater increases in the opportunities afforded by, or the holding power of, the secondary school mathematics curriculum. Part of this increase may be due to increased state and school graduation requirements. Other portions of this increase may be due to an increased national awareness of the importance of mathematics in the future lives and careers of today's secondary school students.

CONCLUSION

The data from the 1996 NAEP national, state-level, and long-term trend assessments all suggest that change is taking place in U.S. school mathematics at grades 4, 8, and 12. What is actually happening is harder to determine from the various forms of NAEP data. It is clear that performance has increased on the NAEP national and state-level assessments. At the same time, results from the long-trend assessment suggest that students are maintaining the traditional paper-and-pencil skills in assessment settings where technology is not present.

Examining the results of teachers' and students' reports of the curriculum taught and the instructional methods employed in the nations' classrooms, one sees small, but significant changes beginning to appear as teachers make tentative steps toward the kinds of classrooms discussed in the NCTM *Standards* documents (NCTM 1989; 1991; 1995). Fullan's (1991) views of change in education note that extended periods of time are required for real change to take hold. Schools, teachers, parents, and students have to move through periods of initiation, implementation, and institutionalization before reform recommendations are forged into a new program, a new curriculum, and a new approach to school mathematics. Many who study educational change consider that the initiation phase alone takes ten years.

Whatever the source for the increased students' mathematical performance, it is gratifying to see students' knowledge and ability to use that knowledge moving in the right directions on the NAEP national and state assessments and then continued increases on the NAEP long-term trend assessments. One can only hope that the next round of NAEP assessments in 2000 will show real movement in the classroom activities of both teachers and students combined with increased students' performance.

REFERENCES

Ballator, Nada. *The NAEP Guide: A Description of the Content and Methods of the 1994 and 1996 Assessments.* Washington, D.C.: National Center for Education Statistics, 1996.

Calderone, John, Laura Mitchell King, and Nancy Horkay, eds. *The NAEP Guide: A Description of the Content and Methods of the 1997 and 1998 Assessments.* Washington, D.C.: National Center for Education Statistics, 1997.

Campbell, Jay R., Clyde M. Reese, Christine O'Sullivan, and John A. Dossey. *NAEP 1994 Trends in Academic Progress.* Washington, D.C.: National Center for Education Statistics, 1996.

Campbell, Jay R., Kristen E. Voelkl, and Patricia L. Donahue. *NAEP 1996 Trends in Academic Progress.* Washington, D.C.: National Center for Education Statistics, 1997.

Dossey, John. A., Ina V. S. Mullis, and Chancey O. Jones. *Can Students Do Mathematical Problem Solving? Results from Constructed-Response Questions in NAEP's 1996 Mathematics Assessment.* Washington, D.C.: National Center for Education Statistics, 1993.

Eakin, Sybil, and Alan Backler, eds. *Every Child Can Succeed: Readings for School Improvement.* Bloomington, Ind.: Agency for Instructional Technology, 1993.

Fullan, Michael. *The New Meaning of Educational Change.* 2nd ed. New York: Teachers College Press, 1991.

Mitchell, Julia H., Evelyn F. Hawkins, Frances B. Stancavage, and John A. Dossey. *Estimation Skills, Mathematics-in-Context, and Advanced Skills in Mathematics: Results from Three Studies of the National Assessment of Educational Progress 1996 Mathematics Assessment.* Washington, D.C.: National Center for Education Statistics, 1999.

Mitchell, Julia H., Evelyn F. Hawkins, Pamela M. Jakwerth, Frances B. Stancavage, and John A. Dossey. *Student Work and Teacher Practices in Mathematics.* Washington, D.C.: National Center for Education Statistics, 1999.

Mullis, Ina V. S. *The NAEP Guide: A Description of the Content and Methods of the 1990 and 1992 Assessments.* Washington, D.C.: National Center for Education Statistics, 1991.

Mullis, Ina V. S., John A. Dossey, Mary A. Foertsch, Lee R. Jones, and Claudia A. Gentile. *Trends in Academic Progress.* Washington, D.C.: National Center for Education Statistics, 1991.

Mullis, Ina V. S., Frank Jenkins, and Eugene G. Johnson. *Effective Schools in Mathematics.* Washington, D.C.: National Center for Education Statistics, 1994.

National Assessment of Educational Progress. Mathematics Objectives: 1990 Assessment. Princeton, N.J.: Educational Testing Service, National Assessment of Educational Progress, 1988.

National Assessment Governing Board. *Mathematics Framework for the 1996 National Assessment of Educational Progress.* Washington, D.C.: National Assessment Governing Board, 1994.

National Council of Teachers of Mathematics. *An Agenda for Action: Recommendations for School Mathematics of the 1980s.* Reston, Va.: National Council of Teachers of Mathematics, 1980.

———. *Assessment Standards for School Mathematics.* Reston, Va.: National Council of Teachers of Mathematics, 1995.

———. *Curriculum and Evaluation Standards for School Mathematics.* Reston, Va: National Council of Teachers of Mathematics, 1989.

———. *Professional Standards for Teaching Mathematics.* Reston, Va.: National Council of Teachers of Mathematics, 1991.

Oakes, Jeannie. *Multiplying Inequalities: The Effects of Race, Social Class, and Tracking on Opportunities to Learn Mathematics and Science.* Santa Monica, Calif.: The RAND Corporation, 1990.

Reese, Clyde M., Karen E. Miller, John Mazzeo, and John A. Dossey. *NAEP 1996 Mathematics Report Card for the Nation and the States.* Washington, D.C.: National Center for Education Statistics, 1997.

Willingham, Warren W., and Nancy S. Cole. *Gender and Fair Assessment.* Mahwah, N.J.: Lawrence Erlbaum Associates, 1997.

3

NAEP Findings Regarding Race/Ethnicity: Students' Performance, School Experiences, and Attitudes and Beliefs

Marilyn E. Strutchens and Edward A. Silver

NAEP provides information about what students know and can do in mathematics, and also about a variety of factors that might influence students' performance. These data are interesting and important when we consider the performance of students in the nation as a whole and also when we consider the performance of various racial/ethnic subgroups within the national student sample. An examination of NAEP data can help us understand both the extent to which there are differences in students' performance across race/ethnicity categories and the extent to which various school-related and personal factors may contribute to group performance differences. Previous analyses of NAEP data have shown that Black and Hispanic students lag significantly behind White students in mathematics

---

At the direction of the U.S. Office of Education, NAEP uses five major race/ethnicity demographic categories: White, Black, Hispanic, Asian/Pacific Islander, and American Indian. For some NAEP analyses the numbers of Asian/Pacific Islander and American Indian students is too small to allow sufficient confidence in the generalizability of the findings. Therefore, in this chapter we omit these groups from our discussion and focus only on the first three categories. Also, although we recognize that there is a diversity of opinion about the appropriate terms to be used when referring to members of the diverse U.S. population racial and ethnic subgroups, here we use the designations Black, White, and Hispanic because they are used by NAEP. In this and in other chapters in the book, NAEP's style of capitalization has been followed for terms designating ethnic and racial groups.

achievement (Johnson 1989; Silver, Strutchens, and Zawojewski 1997). In analyzing these performance disparities, researchers have identified many factors, including students' family background and socioeconomic status, school policies and resources, and classroom instructional practices, that interact in complex ways to influence student performance (Anyon 1995; Oakes 1990; Secada 1992; Tate 1997). In this chapter we present our interpretations of the data collected by NAEP for three major race/ethnicity groups (as described below) regarding not only students' mathematics performance but also factors that might influence students' success in school mathematics.

STUDENTS' PERFORMANCE IN 1996 AND TRENDS OVER TIME

NAEP has been a primary source of information about the relative performance of race/ethnicity groups for nearly three decades. NAEP's reporting of this type of data has been helpful to mathematics education researchers who are interested in equity issues related to race/ethnicity groups. At the direction of the U.S. Office of Education, NAEP uses five major race/ethnicity demographic categories: White, Black, Hispanic, Asian/Pacific Islander, and American Indian. For some NAEP analyses the numbers of Asian/Pacific Islander and American Indian students are too small to allow sufficient confidence in the generalizability of the findings. In this chapter, therefore, we omit these two groups from our discussion and focus only on the first three categories. Although we recognize that there is diversity of opinion about the appropriate terms to be used when referring to members of the diverse U.S. population's racial and ethnic subgroups, here we use the designations White, Black, and Hispanic because they are used by NAEP. In this section, we examine the mathematics performance of White, Black, and Hispanic students on the 1996 NAEP and performance trends over time for these groups.

Mathematical Proficiency in 1996

As explained in chapter 2 by Dossey, students' performance on the 1996 NAEP Mathematics Assessment was expressed in several ways, including both an average proficiency scale score (with scores ranging from 0 to 500) and three achievement levels (Basic, Proficient, Advanced). Table 3.1 provides the average proficiency score and the percentage of students classified at or above the Proficient achievement level (that is, at the Proficient or Advanced level) for all students in each race/ethnicity subgroup. At all grade levels the percentage of all students classified as being at or above the

NAEP FINDINGS REGARDING RACE/ETHNICITY

Highlights

- Substantial performance differences exist among race/ethnicity subgroups at each grade level. The average proficiency score of White students was considerably higher than that of Black or Hispanic students at all grade levels, and Hispanic students performed better than Black students at grades 4 and 8. White students obtained significantly higher proficiency scores at all grade levels in 1996 than in 1990, and Black and Hispanic students performed significantly better in 1996 at grades 4 and 12 but not at grade 8.

- Black and Hispanic students made substantially greater performance gains than White students during the period 1973–1996 on items that assess traditional mathematical concepts and skills, narrowing the gap considerably, but White students performed substantially better in 1996 than did Black and Hispanic students on tasks that called for extended responses and complex problem solving.

- Black and Hispanic students are developing course taking patterns similar to White students, with the largest differences found for courses at the upper and lower ends of the potential course sequence. White students are far more likely than either Black or Hispanic students to report studying the higher-end courses (algebra 2 or precalculus), and Black or Hispanic students are far more likely than White students to report studying the lower-end courses (general mathematics or business mathematics or consumer mathematics).

- Teachers of White, Black, and Hispanic students varied in their similarities and differences across grade levels. Among the more notable differences and similarities are that teachers of White and Hispanic students are more likely to report easy access to needed instructional resources than teachers of Black students and that there is little or no difference with respect to instructional attention to facts and concepts, skills and procedures, and communication, but Black students were less likely than their Hispanic and White counterparts in grade 8 to have teachers who reported a heavy emphasis on reasoning and nonroutine problem solving.

- Overall, there were not many significant differences among race/ethnicity groups related to students' beliefs and attitudes about learning mathematics and the utility of mathematics. But more Black students than White students agreed that concepts are as important as operations, and there are significant differences in the number of White students at the eighth- and twelfth-grade levels agreeing with the statement that all can do well in mathematics if they try compared to the number of Hispanic and Black students.

Proficient achievement level is quite small—only about one in six students at grade 12, one in five at grade 4, and about one in four at grade 8. But the percentage of Black and Hispanic students so classified is alarmingly low; no more than one of every ten Black or Hispanic students at any grade level is classified as being at or above the Proficient achievement level, and the ratio is often as low as 1 in 25.

Table 3.1
Average Scale Scores and Percent of White, Black, and Hispanic Students Classified at or above Proficient Achievement Level

		Average Scale Score	Percent at or above Proficient
Grade 4	Nation	224	21
	White	232	28
	Black	200	5
	Hispanic	206	8
Grade 8	Nation	272	24
	White	282	31
	Black	243	4
	Hispanic	251	9
Grade 12	Nation	304	16
	White	311	20
	Black	280	4
	Hispanic	287	6

Source: Reese et al. (1997)

Substantial differences exist among race/ethnicity subgroups at each grade level. The average proficiency score of White students was considerably higher than that of Black or Hispanic students at all grade levels, and Hispanic students performed better than Black students at each grade level. There is a similar pattern of variation in the percents of White, Black, and Hispanic students classified at or above the Proficient achievement level at each grade level. The general patterns of performance noted for the overall results in table 3.1 were mirrored to a great extent in performance within each mathematical content area (for example, number properties and operations; geometry). That is, in each content area and at each grade level, White students performed better than Black and Hispanic students, and Hispanic students performed better than Black students (Mitchell et al. 1999). These NAEP findings suggest that a substantial gap exists between the performance of White students and that of Black and Hispanic students and that this performance gap persists across all areas of the mathematics curriculum.

Short-Term Trends in Students' Mathematical Proficiency, 1990–1996

Table 3.2 contains average proficiency scale scores obtained on the 1990, 1992, and 1996 NAEP mathematics assessments at each of the three grade levels reported for various demographic subgroups. Looking at performance increases made by students in each of the three race/ethnicity subgroups across the three assessments, we see that White students obtained significantly higher proficiency scores at all grade levels in 1996 than in 1990 and that Black and Hispanic students performed significantly better in 1996 than in 1990 at grades 4 and 12 but not at grade 8.

Table 3.2
Average Scale Scores of White, Black, and Hispanic Students, 1990–1996

		White	Black	Hispanic
Grade 4	1990	220	189	198
	1992	228*	193	202
	1996	232*	200*	206*
Grade 8	1990	270	238	244
	1992	278*	238	247
	1996	282*	243	251
Grade 12	1990	301	268	276
	1992	306*	276*	284*
	1996	311*	280*	287*

* Indicates significant difference from 1990.
Source: Reese et al. (1997)

Long-Term Trends in Students' Mathematical Proficiency, 1973-1996

In addition to looking at short-term trends such as those from 1990 to 1996, NAEP has also examined long-term trends in students' performance by conducting a separate assessment, in which the same (or very similar) sets of items have been administered to age-based samples of students since the 1970s. (See chapter 2 by Dossey for more information about the long-term trend assessment.) Because the set of items on which this analysis is based has been kept secure and has remained relatively stable over time (and constant since 1986), it contains items that can generally be characterized as assessing fairly basic and traditional mathematics concepts and skills (Mullis et al. 1994). Table 3.3 provides the average NAEP mathematics proficiency score for White, Black, and Hispanic students at each of three age levels for all eight time points. (Note that these numbers differ from

Table 3.3
Average Scale Scores of White, Black, and Hispanic Students on the NAEP Long-Term Trend Assessment, 1973–1996

		White	Black	Hispanic
Age 9	1973	225	190	202
	1978	224	192	203
	1982	224	195	204
	1986	227	202	205
	1990	235	208	214
	1992	235	208	212
	1994	237	212	210
	1996	237†	212†	215†
Age 13	1973	274	228	239
	1978	272	230	238
	1982	274	240	252
	1986	274	249	254
	1990	276	249	255
	1992	279	250	259
	1994	281	252	256
	1996	281†	252†	256†
Age 17	1973	310	270	277
	1978	306	268	276
	1982	304	272	277
	1986	308	279	283
	1990	310	288	284
	1992	312	286	292
	1994	312	286	291
	1996	313	286†	292†

† Indicates significant difference from 1973.
Source: Campbell et al. (1997)

those in tables 3.1 and 3.2 because samples of students and sets of items used in the long-term trend assessment are different from that used in NAEP's main assessment.)

On the basis of the data presented in table 3.3, it is evident that Black and Hispanic students performed significantly better in 1996 than in 1973, and gains in proficiency were made by students at all age levels. White 9-year-old and 13-year-old students also performed better, but significant gains have not been made by White 17-year-old students. The gains by all 9-year-old and 13-year-old students suggest that the performance of all students on the traditional mathematical concepts and skills assessed in this portion of the NAEP mathematics assessment has improved significantly between 1973 and 1996. Although critics of education, and especially educational reform, often claim that students are less mathematically skilled now than in the past, these data suggest that there has actually been steady improvement in students' basic knowledge and skill performance over the past quarter century.

It is also noteworthy that Black and Hispanic students made substantially greater performance gains than White students during the period 1973–1996. During this time, Black and Hispanic students have considerably narrowed the performance gap with White students at all age levels. For example, in 1973 the gap between 17-year-old White students and their Black and Hispanic peers was 40 points and 33 points, respectively; in 1996 these gaps had closed to 27 points and 21 points, respectively. Nevertheless, substantial performance gaps remain on these tasks assessing traditional topics. Differences in performance are even more evident when newer types of tasks are considered, such as the extended constructed-response tasks used on NAEP since 1992, as will be seen in the following section.

Students' Performance on NAEP Item Types, 1996

Table 3.4 summarizes the performance of students in 1996 on the three types of tasks (multiple-choice items, short constructed-response, and extended constructed-response) included in the NAEP mathematics assessment at all three grade levels for various demographic subgroups. White students performed considerably better than Black or Hispanic students at each grade level and on each of the three item types. A close examination of the data in table 3.4, however, indicates the possible confounding of these differences with economic condition. At each grade level for which relevant data were available, students who were eligible for the National School Lunch Program or who attended schools that participated in the Title 1 program performed at a lower level than did their more economically advantaged peers. And it is interesting to note that the gaps between these two groups determined by economic condition were generally quite similar to each other at each grade level and were similar in size to the differences between White and non-White students for each item type. Because Black and Hispanic students are overrepresented in low-income categories, the data reported in table 3.4 demonstrate how difficult it is to untangle matters of race/ethnicity and economic condition in these NAEP findings. Although we will continue to report findings in the remainder of this chapter from the perspective of race/ethnicity categories, we urge the reader to keep in mind the role that economic condition may play in providing another interpretive frame for the findings.

In 1996 and 1992, students performed less well on extended constructed-response tasks than on short constructed-response items, which in turn had somewhat lower rates of success than multiple-choice items (Silver,

Table 3.4
Correct Response Rate for Three NAEP Item Formats by Race/Ethnicity and Categories Associated with Economic Conditions

	Multiple Choice			Short Constructed-Response			Extended Constructed-Response		
	Average Percent Correct			Average Percent Correct			Average Satisfactory or Extended		
Grade	4	8	12	4	8	12	4	8	12
Nation	54	55	60	38	49	34	17	9	12
White	58	59	63	43	54	38	21	11	14
Black	42	42	48	23	33	22	6	2	5
Hispanic	45	45	52	27	39	25	9	4	7
Title I									
Participating	41	42	***	22	35	***	5	2	***
Not Participating	57	57	60	42	50	35	20	9	12
National School Lunch Program									
Not Eligible	57	59	61	42	53	36	20	10	13
Eligible	45	46	48	27	38	23	9	4	5
Information Not Available	58	59	62	44	52	37	22	11	12

*** Sample size insufficient to compute average performance.

Strutchens, and Zawojewski 1997). As can be seen in table 3.4, the absolute differences in performance between White and either Black or Hispanic students is smaller for the extended tasks than for the multiple-choice items. Nevertheless, a consideration of the *relative* performance of students in the three groups suggests that the difficulties are more pronounced on the more complex, extended tasks than on the simpler, multiple-choice tasks. Consider for example the relative performance of White and Black students on the three item types. Ignoring the standard errors, which are about the same size for the subgroups, the ratio of Black to White students' performance for multiple-choice, short constructed-response items, and extended constructed-response tasks in grade 8 is 0.71, 0.61, and 0.18, respectively. That is, in 1996 Black eighth-grade students performed about 70 percent as well as White eighth-grade students on multiple-choice items, but they performed only about 20 percent as well on extended constructed-response tasks. It is interesting to note that these performance ratios are almost identical to those reported for the 1992 NAEP (Silver, Strutchens, and Zawojewski 1997). The performance ratios for other grade levels or for White and Hispanic students indicate a similar trend within the 1996 data and between 1992 and 1996. In all cases, the relative performances are much more alike for the multiple-choice and short constructed-response items than for the extended constructed-response tasks.

Although the long-term trend data reported in table 3.3 suggest that the performance gap has been closing over time between White and Black or Hispanic students on items that assess basic level knowledge and skills, the data from the 1996 NAEP assessment reported in table 3.4 suggest the continued existence of performance differences on more complex, extended tasks. As complex, extended tasks become more prevalent in the NAEP assessment and in many state-mandated testing programs, these differences in performance on extended tasks could lead to a widening of the performance gap among race/ethnicity groups in the future.

It is also important to note that a similar pattern of performance differences is evident in the 1996 NAEP data reported for subgroups defined by economic condition. For example, the ratio of Title 1 Participating to Not-Participating students for multiple-choice, short constructed-response items, and extended constructed-response tasks in grade 8 is 0.74, 0.70, and 0.22, respectively. The performance ratios for other grade levels or for students eligible or ineligible for the National School Lunch Program exhibit similar trends across item types in 1996.

In the next section we examine a variety of other data from the NAEP mathematics assessment in order to consider the relationship between these performance results, and a range of factors related to students' school experiences, and attitudes and beliefs. Although causality cannot be argued from the NAEP data, the patterns and relationships are nevertheless suggestive and informative.

FACTORS AFFECTING STUDENTS' PERFORMANCE

Based on the data shown in the "students' performance and trends over time" section of this chapter, there is a clear pattern of differential mathematics performance among White, Black, and Hispanic students. However, interpretations of any observed similarities or differences must be made with caution because of the complex influences that affect mathematics learning, such as school factors that include tracking policies that affect access to mathematics courses, teachers' beliefs about students, assessment or instructional practices and availability of appropriate resources; students' attitudes and beliefs, which include their self-perceptions and expectations regarding their mathematics ability, and their beliefs about mathematics; and family influences that include parental involvement and expectations, socioeconomic status, and cultural customs (Oakes 1990; Silver, Strutchens, and Zawojewski 1997).

In the following sections, we report NAEP findings related to a variety of factors that may influence the mathematics learning of students within and across race/ethnicity groups on the basis of the NAEP data. We begin with school experiences, such as the instructional context (for example, course taking) and the quantity and quality of classroom instruction (for example, time spent, primary emphasis, use of media and practices). Then we report data related to students' attitudes and beliefs about mathematics. Note that although we realize that family influences are important, we will not discuss those issues in this chapter.

School Experiences

Instructional context influences students' mathematics achievement. For example, there is a well-established relationship between mathematics achievement and the amount and nature of mathematics studied in school. Moreover, many researchers believe that the differences in mathematics performance among race/ethnicity groups and those characterized by socioeconomic status can largely be attributed to differences in opportunities to learn in schools, with minorities and groups of low socioeconomic status experiencing lower expectations and weaker standards than their counterparts (Oakes 1990; Robinson 1996; Tate 1995). Thus it is important to examine the type of mathematics students have been exposed to as well as the conditions under which they learned mathematics.

Data regarding the instructional context in which students study mathematics in U.S. schools are collected through questionnaires administered to students, teachers, and school administrators as part of the NAEP assessment. These data illuminate not only the instructional conditions associated with mathematics education in the nation as a whole (see chapter 5 by Grouws and Smith for additional information) but also the particular instructional conditions that undoubtedly influence the observed performance differences by students in the several race/ethnicity categories.

Students' Mathematics Course Enrollment

Previous analyses have shown that differences in the amount and nature of mathematics studied is a contributor to differential achievement by demographic subgroups (Dossey et al. 1994; Naifeh and Shakrani 1996; Welch, Anderson, and Harris 1982). In a review of national trend studies related to students' achievement across race/ethnicity categories, socioeconomic status, and language proficiency, Tate (1997) reported that secondary school students of every racial/ethnic and socioeconomic status group benefited from additional mathematics coursework in high school and that students completing the same number of mathematics courses did not have significant differences in achievement. (See Silver, Strutchens, and

Zawojewski 1997 for an extended discussion of studies related to course enrollment.)

Thus it is important to examine 1996 NAEP data for evidence of differential course taking by students in the race/ethnicity subgroups. Since mathematics is required for students in grades 4 and 8 but optional for students at grade 12, different kinds of information are collected from students at different grade levels. Some data reflect students' attitudes and expectations, and other data reflect students' actual experience.

In 1996, students in all three grade levels were asked to respond to the statement, "If I had a choice, I would not study any more mathematics," whereas in 1992 only students in grades 4 and 8 were asked to respond. White and Hispanic eighth-grade students did not show much change between the two assessments. White and Hispanic fourth-grade students did show change in an "undesirable direction" with more indicating that they agreed with the statement in 1996 than in 1992. On a positive note, the number of Black eighth-grade students agreeing with this statement for the 1996 assessment decreased.

Table 3.5 contains a summary of the data obtained from twelfth-grade students' responses to a question asking how many semesters they had studied mathematics in high school, including the semester in which the survey was taken. However, this data may not present an accurate picture of the changes between 1992 and 1996 in the number of semesters of mathematics taken, due to the implementation of block scheduling in some high schools where students may actually complete an entire course in one semester. On the other hand, the data do present an accurate picture of the differences in the number of semesters of mathematics studied by race/ethnicity groups. The data indicate that Black and Hispanic students are studying fewer semesters of mathematics in high school than White students. Only about one-half of the Black students surveyed in each assessment had studied mathematics for six or more semesters, and many have studied very little mathematics at all. These differential course-taking patterns undoubtedly contribute to differential performance on the NAEP assessment at grade 12, and the relationship would likely be even stronger if the assessment at grade 12 contained more items that assessed topics in advanced high school courses. (See chapter 13 by Kilpatrick and Gieger for more information on advanced mathematics content.)

To understand differences in opportunity to learn important mathematical ideas, one needs to examine not only the number of semesters of mathematics studied but also which courses were taken. Table 3.6 summarizes 1992 and 1996 NAEP data regarding twelfth-grade students' self-reports of having taken a particular course for at least one year. In general, these NAEP data reflect a national trend toward an increase in the number of college preparatory mathematics courses taken by high school students, almost

Table 3.5
Percent of Twelfth-Grade Students Indicating Number of Semesters of Mathematics Taken in High School, 1992 and 1996

	Less than 4 semesters		4 or 5 semesters		6 or more semesters	
	1992	1996	1992	1996	1992	1996
Nation	13	16	19	18	67	66
White	11	14	18	16	70	70
Black	20	24	27	25	51	51
Hispanic	19	16	19	22	60	62

Note: Within each category and year (for example, White in 1992), row percents may not add to 100 because of rounding.

certainly in response to increased graduation requirements in many states (Blank and Gruebel 1993). Moreover, a majority of White, Black, and Hispanic students reported studying three years of college preparatory mathematics, usually algebra 1, geometry, and algebra 2. Even though Black and Hispanic students report taking fewer college preparatory courses than White students the gap is narrower in 1996 than it was in 1992. Since achievement is strongly related to course taking, the continuing differences in course taking undoubtedly contribute to the observed differences in mathematical proficiency among the groups at grade 12.

Table 3.6
Percent of Twelfth-Grade Students Indicating Mathematics Courses Taken for at Least One Year, 1992 and 1996

	Nation		White		Black		Hispanic	
Course	1992	1996	1992	1996	1992	1996	1992	1996
General Mathematics	49	53	45	51	59	53	64	66
Consumer Mathematics	26	20	25	20	32	23	29	20
Prealgebra	56	63	55	62	61	67	57	62
Algebra 1	87	90	87	91	84	90	85	86
Geometry	76	80	78	81	72	81	67	76
Algebra 2	61	70	64	71	49	69	53	61
Trigonometry	20	22	21	23	18	17	13	18
Precalculus	19	24	21	25	13	17	12	16
Calculus	10	12	10	13	6	7	10	9

Several other observations can be made on the basis of the data in table 3.6. There is essentially no difference in the percentages of White, Black, and Hispanic students who reported studying algebra 1. In 1992 White students were more likely than Black students and Hispanic students to report taking a course in geometry, but in 1996 there was virtually no difference in the percents of Black and White students who reported taking a course in geometry. The percentage of Hispanic students reporting taking a course in geometry also increased in 1996.

The largest differences in the percents of White, Black, and Hispanic students' reported course taking is found for courses at the upper and lower ends of the traditional college preparation course sequence, with White students more likely than either Black or Hispanic students to report studying the higher-end courses (trigonometry, precalculus, or calculus), and with Black or Hispanic students more likely than White students to report studying the lower-end courses (general math, business math, or consumer math). However, the differences are smaller in 1996 than they were in 1992, indicating that Blacks and Hispanics are enrolling with greater frequency in higher-level mathematics classes. This may be due to any number of reasons, such as increased graduation requirements, some confusion in reports of course taking generated by a proliferation of first-year algebra courses spread over two years, or the encouragement of students to take higher-level courses to increase their chances of going into mathematics-related fields.

The trajectory for high school course taking is determined to a great extent by when one's first algebra course is taken. In 1992, students in grade 12 were asked when they had first studied algebra, and about 23 percent reported having taken algebra before grade 9. In 1996, 29 percent of students in grade 12 reported taking algebra before the ninth grade. Thus, there was a small increase between the two assessments. Moreover, in 1996 a higher percentage of the White students than Black and Hispanic students reported taking first-year algebra before ninth grade, 30 percent, 27 percent, and 21 percent, respectively (Mitchell et al. 1999). In 1992 and 1996, students in grade 8 were also asked to indicate the type of course in which they were enrolled, and these data are summarized in table 3.7. One-fifth of the nation's students were studying algebra in grade 8 in 1992 and more than one-fourth were enrolled in a prealgebra course. In 1996, 25 percent of the nation's eighth-grade students were enrolled in algebra and 27 percent were enrolled in prealgebra. In 1996, the "algebra enrollment gap" narrowed among Black, White, and Hispanic students, as 56 percent of White eighth-grade students, 45 percent of Black eighth-grade students, and 42 percent of Hispanic eighth-grade students enrolled in either algebra or prealgebra (compared to 52, 36, and 32 percent, respectively, in 1992). Thus, the narrowing of the gap was due to a much larger increase from 1992 to 1996 in algebra or prealgebra course taking for Black and Hispanic students than was the case for White students, for whom the increase was modest.

Table 3.7
Percent of Eighth-Grade Students Indicating Mathematics Course Enrollment, 1992 and 1996

	Algebra		Prealgebra		Eighth-Grade Mathematics		Other Mathematics	
	1992	1996	1992	1996	1992	1996	1992	1996
Nation	20	25	28	27	49	43	3	5
White	22	27	30	29	45	40	3	4
Black	13	20	23	25	60	47	4	8
Hispanic	12	20	20	22	62	52	5	6

Note: Within each category and year (for example, Hispanic in 1992), row percents may not add to 100 because of rounding.

Classroom Instruction

Data available from questionnaires administered to teachers as part of the 1996 NAEP mathematics assessment provide information regarding the teaching practices and professional preparation of teachers at grades 4 and 8, but not at grade 12. Chapter 5 by Grouws and Smith reviews these data comprehensively; we focus here on these data as they pertain to race/ethnicity. In some instances, we comment on differences between the data obtained from responses to the 1996 NAEP questionnaires and those reported previously for the 1992 NAEP. For the interested reader, Silver, Strutchens, and Zawojewski (1997) provide a more extensive report of the relevant 1992 NAEP data than we can provide here.

Instructional support. Intergroup achievement differences may be related to some degree to differences in the kinds and extent of support provided to the teachers of White, Black, and Hispanic students. White and Hispanic students have teachers who are more likely to report easy access to needed instructional resources than teachers of Black students. Teachers of about 70 percent of the White and the Hispanic fourth-grade students reported getting all or most of the resources they needed, but teachers of only about 50 percent of the Black students reported this degree of support. A similar pattern of difference was found for teachers of eighth-grade students, with teachers of 80 percent of the White students and more than 75 percent of the Hispanic students reporting access to all or most of the resources they needed, which was a degree of support reported by less than 70 percent of the teachers of Black students. It should be noted that the responses in 1996 reveal a positive trend for teachers of students in each race/ethnicity category; that is, the percent of students in each subgroup whose teachers reported having access to most or all resources they needed was higher in 1996 than in 1992. Teachers of Hispanic students, in particular, reported far greater access to resources in 1996 than was the case in 1992; at grade 8 the percent of Hispanic students whose teachers reported having access to all or most of the needed resources grew from 58 percent in 1992 to 76 percent in 1996.

A different pattern of access to instructional resources appears to hold for human resources. The assistance of mathematics specialists at the fourth-grade level reportedly is available more often to teachers of a larger percent of Black students (57 percent) than to teachers of either Hispanic (46 percent) or White (40 percent) students. In grade 8, access to the advice and support of a mathematics curriculum specialist is reported by teachers of larger percentages of Black (60 percent) and Hispanic (57 percent) students than by teachers of White students (45 percent).

Instructional time. Instructional time is another factor that might relate to mathematical proficiency. Thus, it might be expected that White students, who attain higher levels of mathematical proficiency in NAEP than Black and Hispanic students, receive more mathematics instruction. However, data obtained from the 1996 NAEP questionnaires do not indicate a straightforward correspondence between the relative performance of White, Black, and Hispanic eighth-grade students and the amount of instructional time reported by their teachers. Students in each race/ethnicity group are equally likely to have teachers who report an alarmingly low amount of mathematics instructional time each week; about 20 percent of the students in each group have teachers who report no more than 2.5 hours of mathematics instructional time each week. At the opposite extreme, it is Black students, and not White students, who are more likely to have teachers who report spending more than 4 hours of instructional time each week (44 percent versus 31 percent), with Hispanic students having teachers who report about the same amount of time as do the teachers of White students (34 percent).

For students and teachers at grade 4, it was also the case that the instructional time data did not reveal a direct correspondence to the achievement data. It is interesting to note, however, that the amount of time spent on mathematics instruction appeared to be higher for all groups in grade 4 than in grade 8. The percent of students in each group whose teachers reported providing 4 or more hours of instructional time per week was about twice as large at grade 4 than at grade 8.

Although there were variations across groups in teachers' reports of time spent on mathematics instruction, teachers of students in each group reported quite similar amounts of time during the school week to prepare for mathematics instruction. For each group of students, most teachers reported spending 3–4 hours of class preparation a week at grade 4 and more than 5 hours at grade 8.

In addition to instructional time and teacher planning and preparation time, another potential influence on students' learning of mathematics is homework. As was the case for instructional time, the data regarding homework do not correspond directly to the trends in students' achievement for White, Black, and Hispanic students. At grade 4, 84 percent of Black students had teachers who reported assigning 15–30 minutes of homework a night, in contrast to 91 percent of the Hispanic or White students. But at grade 8, Black students were as likely as White students to have teachers who assigned 15–30 minutes of homework a night (87 percent for each group), and more likely than Hispanic students (81 percent). The general trend in 1996 for teachers of students in each race/ethnicity group was to report more assigned homework at grades 4 and 8 than had been the case in 1992. These data conform to the overall achievement gains for all groups during the same period at grade 4, but they fail to correspond to the pattern at grade 8.

Instructional emphasis. Differences in performance might also be attributed to corresponding differences in opportunity to learn specific content topics, but the 1996 NAEP data indicate only a few instructional differences of this type. Teachers were asked to report their instructional emphasis (that is, none, a little, some, a lot) on several different content strands in the curriculum: number and operations, measurement, geometry, and probability and statistics. Their responses indicated that instructional emphasis is similar at the eighth-grade level across the three student populations. There is some difference, however, in the topic area of algebra and functions. A higher percent of White eighth-grade students have teachers who report at least some (that is, more than a little) emphasis on this topic (93 percent) than is the case for teachers of Black students (85 percent), who in turn give more emphasis than teachers of Hispanic students (81 percent). At grade 4, the reported emphasis on number and operations, algebra and functions, and probability and statistics is similar, but White and Hispanic fourth-grade students are more likely to have teachers who report at least some emphasis on geometry and measurement than are Black students.

Differences in instructional opportunities might also be reflected in teachers' responses to questions about their emphasis on various types of mathematical ideas (for example, facts, skills, concepts) and cognitive processes (for example, reasoning and problem solving). The general picture that emerges from the 1996 NAEP data is that there is little or no difference among groups with respect to teachers' reported instructional emphasis on facts and concepts, skills and procedures, and communication. There was, however, a difference noted with respect to the processes of reasoning and nonroutine problem solving. For this instructional topic, there was no difference across groups at grade 4, but Hispanic and White students were more likely (49 percent and 55 percent, respectively) than their Black counterparts (40 percent) in grade 8 to have teachers who reported a heavy emphasis on reasoning and nonroutine problem solving.

Instructional media and practices. Variations in the ways that students gain access to mathematical ideas, and the tools they use to explore mathematics and to solve problems, are also potential sources of difference in instructional practice that could affect students' proficiency. In this regard, NAEP data pertaining to classroom use of textbooks, worksheets, a variety of instructional strategies, and technology (that is, calculators and computers) can be informative.

Patterns of textbook use are similar across race/ethnicity groups in grade 4, but in grade 8, teachers of White students are more likely to report using a textbook every day, in contrast to teachers of Black and Hispanic students, who report less regular use of a text. At grade 4, the pattern of use of worksheets was also similar across the three populations, but Black and Hispanic students in grade 8 were more likely to have teachers who used worksheets on a weekly or daily basis (78 percent and 73 percent, respectively) than were White students (62 percent).

At grades 4 and 8, teachers of students in all populations reported similar use of written reports or mathematics projects, pair or small group work, discussions among students about solutions to problems, and students' presentations of ideas. At grade 4, teachers of all populations were also similar in their reported use of rulers, counting blocks, and geometric pieces, but Black and Hispanic students were more likely to have teachers who reported use of such tools and manipulative materials at grade 8 than were White students.

One area in which there were notable differences in instructional practice is the use of calculators on daily work and tests. Several different questions asked teachers about students' access to, and use of, calculators at home and in class. Several interesting findings and some potentially important intergroup similarities and differences can be noted in the pattern of teachers' responses at grade 8.

- When asked if they provided instruction in the use of calculators, teachers of about 80 percent of the eighth-grade students across all three race/ethnicity groups reported doing so.

- When asked if students have access to calculators that are owned by the school, teachers of Black and Hispanic students in grade 8 were more likely to respond affirmatively (88 percent and 84 percent, respectively) than teachers of White students (77 percent) were.

- When asked how frequently students were allowed to use a calculator in class,

    (*a*) White students in grade 8 were far more likely to have teachers who reported allowing calculator use *at least once each week* (80 percent) than Black or Hispanic students (60 percent and 71 percent, respectively) were;

(b) Black and Hispanic students in grade 8 were more likely to have teachers who reported *no use* of a calculator in class (17 percent and 11 percent, respectively) than White students (8 percent) were;

(c) White students in grade 8 were far more likely to have teachers who reported allowing *unrestricted* use of a calculator in class (51 percent) than Black or Hispanic students (35 percent and 38 percent, respectively) were.

- When asked whether or not students were allowed to use calculators on tests, White students in grade 8 were far more likely to have teachers who reported allowing the use of a calculator *on tests* (72 percent) than Black or Hispanic students (51 percent and 56 percent, respectively) were.

At grade 4, it was also the case that about 80 percent of the students in each group had teachers who reported providing instruction in calculator use. Also, there was no difference noted across the populations in the unrestricted use of calculators in class nor in the use of calculators on tests. White students were somewhat more likely to have teachers who reported frequent (that is, at least weekly) use of calculators in instruction (35 percent) than Black and Hispanic students (25 percent and 27 percent, respectively) were, but the difference was not as great as that reported above for grade 8.

Some intergroup differences were noted also for computer use. At grade 4, a larger percent of Black students (11 percent) than of Hispanic or White students (7 percent and 4 percent, respectively) had teachers who reported that their students did not have access to a computer in school. But the situation was reversed in grade 8; Black students were less likely to have teachers who reported a complete lack of access to a computer (18 percent) than Hispanic and White students were (27 percent and 25 percent, respectively). Among eighth-grade students, Black students were more likely to have teachers who reported using computers at least weekly (19 percent) than teachers of Hispanic or White students were (10 percent and 9 percent, respectively). But Black students were about twice as likely to have teachers who reported frequent use of these computers for the purpose of drill and practice (26 percent) than Hispanic or White students were (14 percent and 14 percent, respectively). In contrast, White students in grade 8 were twice as likely to have teachers who had them use computers for simulations and applications than Black students were (14 percent versus 7 percent), and the percent of Hispanic students whose teachers reported such computer use was midway between the reported use by Black and White students.

Instruction and assessment. Data regarding the assessment practices of teachers of White, Black, and Hispanic students in grades 4 and 8 highlights another area in which there are some notable intergroup differences. Black fourth-grade students were more likely than their Hispanic or White counterparts to have teachers who reported giving tests to students at least weekly (45 percent versus 32 percent and 30 percent, respectively). A similar pattern was found for Black eighth-grade students when compared to Hispanic or White students (54 percent versus 44 percent and 44 percent, respectively).

Overall, it appears that the frequency of assessment does not vary much among teachers of the various student populations, but the form of assessment does vary, at least in grade 4. As the data in table 3.8 indicate, Black and Hispanic fourth-grade students are more likely than White students to have teachers who frequently use multiple-choice tests to assess students' progress. In contrast, the intergroup differences at grade 8 in 1996 were not as great.

Table 3.8
Percent of Fourth- and Eighth-Grade Students Whose Teachers Report Frequent or Rare Use of Multiple-Choice Tests, 1992 and 1996

| | Frequent[a] | | | | Rare[b] | | | |
| | Grade 4 | | Grade 8 | | Grade 4 | | Grade 8 | |
	1992	1996	1992	1996	1992	1996	1992	1996
Nation	49	48	34	35	51	52	66	65
White	47	44	32	33	53	56	68	67
Black	58	65	46	40	42	35	54	60
Hispanic	49	53	38	38	51	47	62	62

[a] Once or twice a week or once or twice a month
[b] Once or twice a year or never

Table 3.8 also provides some indication of short-term trends with respect to the use of multiple-choice tests. A comparison of 1996 data with that reported from 1992 indicates that the frequency of use among teachers of Black and Hispanic fourth-grade students actually increased during this period of time. In contrast, the trend between 1992 and 1996 was toward less frequent use of multiple-choice tests by teachers of Black eighth-grade students; whereas usage by teachers of Hispanic and White students remained about the same during this time period. The findings from the 1992 NAEP were consistent with the results obtained in a large-scale study of the use of standardized tests in mathematics (Madaus et al. 1992), in which it was noted that teachers who had at least 60 percent Black or Hispanic student enrollment in their classrooms were far more likely to spend classroom time using multiple-choice testing and other means of testing low-level cognitive objectives than their counterparts who had a majority of White students in their classrooms. The 1996 NAEP data suggest that this pattern of excessive teacher attention to assessing low-level rather than high-level cognitive processes and skills may be less prevalent at grade 8, but the increased use of such testing in grade 4 is disturbing. The frequent use of multiple-choice testing is unlikely to address the poor performance by Black and Hispanic students on extended constructed-response tasks, which was discussed earlier in this chapter.

An alternative to multiple-choice testing that has become increasingly popular in recent years is portfolio assessment. A question about frequency of use of portfolio assessment was asked for the first time in 1996. These data suggest that this alternative form of assessment is being frequently used with non-White fourth-grade students. In fact, Black and Hispanic fourth-grade students were more likely to have teachers who reported using portfolio assessment at least once a month than White students were (56 percent and 52 percent, respectively, versus 40 percent). Although the overall frequency of use of portfolio assessment was lower in grade 8, a similar pattern of difference across the populations was noted. Eighth-grade Black and Hispanic students were more likely to have teachers who reported using portfolio assessment at least once a month when compared with White students (34 percent and 37 percent, respectively, versus 26 percent). Given the other data available from NAEP about teachers' assessment practices, it is not clear how to reconcile this finding regarding greater use of portfolio assessment by teachers of Black and Hispanic students. If one assumes that teachers' responses indicate the use of portfolios as collections of students' journal writings, extended projects, and solutions of nonroutine problems, then the results may signal an important shift in emphasis in the classrooms of Black and Hispanic students, perhaps due to an effort to integrate the improvement of reading and writing with instruction in other school subjects, including mathematics. But it is also possible that teachers interpreted the word *portfolio* simply to mean a collection of traditional mathematics worksheets, in which case the data would signal no such shift in instructional emphasis in these classrooms. Regrettably, given the other data available from NAEP regarding the nature of classroom instruction and assessment, the latter is a more reasonable and likely interpretation of the reported use of portfolios.

Students' Attitudes and Beliefs

In the previous section we examined NAEP data to look at mathematics instruction across race/ethnicity groups, and in this section we look at students' attitudes and beliefs towards mathematics. Many mathematics education researchers believe that instruction affects students' attitudes and beliefs toward mathematics and that the converse is also true, that is students' attitudes influence mathematics instruction (McLeod 1992; Reyes

1980). In fact, McLeod (1992) contended that affective issues play a central role in mathematics learning and instruction. Moreover, the National Council of Teachers of Mathematics (1989) expressed the importance of affective issues by including two goals related to affect in their *Curriculum and Evaluation Standards for School Mathematics:* (1) students should learn to value mathematics; and (2) students should become confident in their ability to do mathematics.

In 1996, as in previous NAEP mathematics assessments, students were given questionnaires to determine their attitudes and beliefs related to mathematics. Nine statements were given to eighth-grade and twelfth-grade students, and eight statements were given to fourth-grade students. Students were asked to indicate the extent to which they agreed with each statement. In grades 8 and 12 students selected one of five choices (Strongly Agree, Agree, Undecided, Disagree, or Strongly Disagree) and, in grade 4, one of three choices (Agree, Undecided, or Disagree).

Most of the attitude or belief questions asked by NAEP in 1996 were also asked in 1992, and so comparisons in students' attitudes and beliefs about mathematics based on two assessments can be made. The 1996 NAEP mathematics assessment also included some new attitude or belief questions that focused on students' beliefs relative to the changes in teaching and learning mathematics advocated by *Curriculum and Evaluation Standards for School Mathematics* (NCTM 1989).

Table 3.9 gives a summary of the percents of students who agree with each of four statements regarding students' attitudes toward mathematics and their perceptions of themselves as learners of mathematics. For students in grades 8 and 12, this reflects a sum of the percent selecting "Agree" and the percent selecting "Strongly Agree" for 1996 NAEP mathematics assessment. In 1996, there was a slightly larger percent of Black students across all three grade levels who agreed that they liked mathematics than White and Hispanic students. On the other hand, there was a smaller percent of Hispanic students who agreed that they were good at mathematics across all three grade levels. As was noted in prior NAEP assessments, older students were less likely than younger students to like mathematics or express confidence in their ability to do mathematics across all three race/ethnicity groups.

On both the 1992 and 1996 NAEP assessments, more than two-thirds of the fourth-grade students agreed that they liked mathematics and a similar proportion agreed that they were good at it, but only about one-half of the twelfth-grade students expressed these views. These findings are important because studies of students at various age or grade levels, ranging from elementary school to college, have consistently shown that students' beliefs about their competence in mathematics are positively related to their achievement in mathematics, with correlations ranging from

Table 3.9
Percent of Students Agreeing with Statements regarding Perceptions of Themselves with Respect to Mathematics

Statement	Grade	Percent Agreeing			
		Nation	White	Black	Hispanic
I like	4	69	67	75	70
mathematics.	8	56	55	61	53
	12	50	48	55	50
I am good at	4	66	66	67	60
mathematics.	8	63	65	65	54
	12	53	54	53	46
I understand most	4	78	81	73	70
of what goes on in	8	77	78	78	71
mathematics class.	12	66	66	71	62
Everyone can do	4	89	88	91	88
well in mathematics	8	73	67	87	86
if they try.	12	50	43	65	67

Note: For students in grades 8 and 12, "Agreeing" represents the sum of the percent selecting "Agree" and the percent selecting "Strongly Agree."

.3 to .4 for both female and male students (Hart 1989). Moreover, confidence is also related to elective enrollment in mathematics courses (McLeod 1992). The data from NAEP contradicts findings related to the correlations between attitudes and achievement. Based on NAEP data, Black students tend to have slightly more positive attitudes than White students regarding mathematics, but White students tend to perform better on the mathematics sections of NAEP.

Another attitude or belief item suggests one plausible explanation for students' decreasing liking for mathematics and their diminished sense of confidence over the years. Older students were less likely than younger students to indicate in 1992 and 1996 that they understood classroom mathematics instruction. This pattern was observed for students in all race/ethnicity categories in both assessments.

Along with decreases in students' confidence over the years, there is also a decrease in the belief that everyone can do well in mathematics if they try. This statement is important because it provides information about whether or not students believe that mathematics is an appropriate area of study. It is interesting to note that there are significant differences in the percent of White students agreeing with this statement at the eighth- and twelfth-grade levels compared to the percent of Hispanic and Black students agreeing with statement. About two-thirds of Black and Hispanic students in grade 12 agreed with this statement versus a little over two-fifths of the

White students at this grade. These attitude differences may be due, at least in part, to different mathematics course-taking experiences by students. As shown by the course enrollment data (see table 3.6), more White twelfth-grade students are enrolled in higher-level classes that are more difficult than the courses that many Black and Hispanic students may be taking. Thus, students' beliefs about whether everyone can do well in mathematics may be based on their course experiences. Also, students' attitudes may reflect differences among racial/ethnic groups in their views regarding ability and achievement.

In 1992 students were asked two questions regarding the nature of learning mathematics and the utility of school mathematics, and in 1996 students were asked those two questions along with two others. One question new to the 1996 questionnaire regarding the importance of mathematical concepts and ideas was not given to students in grade 4. Table 3.10 contains the percents of students who agreed with each of these statements. Again, for students in grades 8 and 12, the data reflect a sum of the percent selecting "Agree" and the percent selecting "Strongly Agree."

Table 3.10
Percent of Students Agreeing with Statements regarding the Nature and Utility of Mathematics

Statement	Grade	Percent Agreeing			
		Nation	White	Black	Hispanic
Learning mathematics is mostly memorizing facts.	4	54	51	63	61
	8	40	36	58	49
	12	35	29	55	44
Describing mathematical concepts and ideas is as important as doing mathematical operations such as addition and multiplication in solving problems.	4	—	—	—	—
	8	61	59	69	61
	12	62	59	71	63
There is only one correct way to solve a problem.	4	17	13	27	24
	8	8	6	12	12
	12	6	6	7	7
Mathematics is useful for solving everyday problems.	4	69	72	62	63
	8	80	81	82	77
	12	70	69	73	68

Note: For students in grades 8 and 12, "Agreeing" represents the sum of the percent selecting "Agree" and the percent selecting "Strongly Agree."

As has been noted in prior NAEP assessments, a large percentage of students appears to view the learning of mathematics as primarily involving memorization, with younger students more likely than older students to hold this view among all three race/ethnicity groups. An instructional emphasis on facts and procedures in the elementary grades appears to dominate the perceptions of a majority of students in grade 4; whereas students in grade 8 and especially grade 12 are more likely to have encountered mathematics that calls for more complex thinking. However, there have been some positive changes from the 1992 mathematics assessment to the 1996 mathematics assessment.

In 1992 nearly two-thirds of Black students, even in grades 8 and 12, viewed the learning of mathematics as primarily involving memorization; however, the 1996 assessment revealed that there was a decrease in the number of Black students who agreed with this statement at grades 8 and 12. Nevertheless, there remains a large difference between the numbers of Black and White students who agree with this statement, 58 percent and 36 percent, respectively, at the eighth-grade level and 55 percent and 29 percent respectively, at the twelfth grade level, perhaps suggesting that Black students at grades 8 and 12 have more frequent experience than White students with instruction emphasizing rote learning rather than problem solving or critical thinking. Also disturbing is the fact that the proportion of Hispanic students agreeing with this view is higher than that of White students in grades 8 and 12, which may be reflective of the cognitive level of the instruction received by the students.

Although the role of memorization is clearly prominent in all students' views of mathematics, especially Black and Hispanic students, responses to another statement from the 1996 assessment suggest that students also view concepts and multiple solution methods as being important in mathematics. About 60 percent of the nation's eighth-grade and twelfth-grade students agreed with the statement that concepts are as important as operations, and less than 10 percent agreed that there is only one way to solve a problem. Black students were more likely than either White or Hispanic students to agree that concepts are as important as operations, but they were also more likely to agree that there is one way to solve every mathematics problem. The belief in a single method to solve every problem was strongest within each race/ethnicity group at grade 4, and White students were less prone to hold this belief than were Black or Hispanic students.

Despite the finding that many students associate mathematics with memorization, students at all grade levels and race/ethnicity categories appear to view mathematics as having utility. Across the grades assessed by NAEP for 1996, 69 percent of the fourth-grade students, 80 percent of the eighth-grade students, and 70 percent of the twelfth-grade students agreed that mathematics is useful for solving everyday problems. The percentages for each race/ethnicity were quite close to the data for the nation. It is interesting that across all race/ethnicity subgroups more eighth-grade students agreed with this statement than did fourth-grade and twelfth-grade students. Here again, the instruction that students receive at different grade levels may be affecting how they perceive mathematics.

In presenting the factors related to students' school experiences and attitudes or beliefs, we could not make any broad generalizations due to the fact that information on individual students is not available. Perhaps even trying to make a profile of a typical high-achieving student or a low-achieving student would do more harm than good because of the complex nature of the interactions of all of the variables related to students' performance. However, as we reflect back on this section, certain questions are inevitable: Why do teachers of White and Hispanic students have more access to needed instructional materials than teachers of Black students? Why are Black eighth-grade students less likely to have teachers who reported a heavy emphasis on reasoning and nonroutine problem solving? If Black students have such positive attitudes toward mathematics, why do they lag significantly behind White students in mathematics achievement? If teachers of Black students report having more access to a mathematics specialist, why are Black students not surpassing White and Hispanic students in mathematics achievement? Why do fewer Hispanic and Black students take college preparatory courses despite raised standards? Perhaps if the answers to these questions were simple, then we could find a solution to the disparities related to mathematics education that exist among the different race/ethnicity groups.

SUMMARY AND CONCLUDING REMARKS

In this chapter, we used 1996 NAEP data (and occasionally also data from 1992) as a lens through which to view both the extent of differences in students' achievement across race/ethnicity categories and the relation of various school-related factors and students' attitudes and beliefs to these group-achievement differences. To our dismay, we found that substantial achievement differences continue to exist among race/ethnicity subgroups at each grade level. The average proficiency of White students was considerably higher than that of Black or Hispanic students at all grade levels, and Hispanic students performed better than Black students at grades 4 and 8.

Even though the long-term trend data from 1973–1996 suggest that the performance gap has been closing over time between White and Black or Hispanic students on items that assess basic-level knowledge and skills, the data from the 1996 NAEP assessment suggest the existence of significant performance differences on more-complex, extended tasks. We found a similar pattern of performance differences in the 1996 NAEP data reported for subgroups defined by economic condition. If the current trend in educational programs toward more demanding standards and expectations continues, and if this trend is reflected in the inclusion of more-complex, extended tasks in the NAEP assessment and in state-mandated testing programs, then these differences noted in 1996 NAEP on performance on extended tasks could signal that we are on the precipice of a performance gap among race/ethnicity groups and subgroups defined by economic condition that is likely to become even wider. Also, we urge readers to consider that the race/ethnicity categories may be serving largely as proxies for socioeconomic categories.

Although it is not possible for us to specify the extent to which various school-related and other factors contribute to group achievement differences, the data certainly suggest that poverty is at least as prominent as race/ethnicity in its relation to students' achievement. Moreover, the data also suggest that for Hispanic and Black students in particular, receiving more instructional time and having more adults in the classroom may actually mean receiving less in terms of quality instruction and mathematical content.

Societal structures as well as school structures affect how students learn and what they are given the opportunity to learn. Thus, it is important for policymakers to think about the larger picture. That is, the gap will only become narrower when expectations are raised for all students and when all students are placed in schools and supported by communities that will encourage them and provide them with means to use their talents and skills once they have developed them.

REFERENCES

Anyon, Jean. "Race, Social Class, and Educational Reform in an Inner-City School." *Teachers College Record* 97 (Fall 1995): 69–94.

Blank, Rolf. K, and Doreen Gruebel. *State Indicators of Science and Mathematics Education 1993.* Washington, D.C.: Council of Chief State School Officers, 1993.

Campbell, Jay R., Kristin E. Voelkl, and Patricia L. Donahue, *NAEP 1996 Trends in Academic Progress.* (Report NCES 97-985). Washington, D.C.: National Center for Education Statistics, 1997.

Dossey, John A., Ina V. S. Mullis, Steven Gorman, and Andrew S. Latham. *How School Mathematics Functions: Perspectives from the NAEP 1990 and 1992 Assessments.* (Report No. 23-FR-02). Washington, D.C.: National Center for Education Statistics, 1994.

Hart, Laurie E. "Classroom Processes, Sex of Student, and Confidence in Learning Mathematics." *Journal for Research in Mathematics Education* 20 (May 1989): 242–60.

Johnson, Martin L. "Minority Differences in Mathematics." In *Results from the Fourth Mathematics Assessment of the National Assessment of Educational Progress,* edited by Mary M. Lindquist, pp. 135–48. Reston, Va.: National Council of Teachers of Mathematics, 1989.

Madaus, George F., Mary M. West, Maryellen C. Harmon, Richard G. Lomax, and Katherine A. Viator. *The Influence of Testing on Teaching Math and Science in Grades 4–12* (Report of Grant No. SPA8954579 funded by the National Science Foundation). Boston: The Center for the Study of Testing, Evaluation and Educational Policy, Boston College, 1992.

McLeod, Douglas, B. "Research on Affect in Mathematics Education: A Reconceptualization." In *Handbook of Research on Mathematics Teaching and Learning,* edited by Douglas A. Grouws, pp. 575–96. New York: Macmillan, 1992.

Mitchell, Julia H., Evelyn F. Hawkins, Pamela M. Jakwerth, Frances B. Stancavage, and John A. Dossey. *Student Work and Teacher Practices in Mathematics.* Washington, D.C.: National Center for Education Statistics, 1999.

Mullis, Ina V. S., John A. Dossey, Jay R. Campbell, Claudia A. Gentile, Christine O'Sullivan, and Andrew S. Latham. *NAEP 1992 Trends in Academic Progress.* (Report No. 23-TR-01). Washington, D.C.: National Center for Education Statistics, 1994.

Naifeh, Mary, and Sharif Shakrani. *Math Matters: The Relationship between High School Mathematics Course-Taking and Proficiency on the NAEP Assessment.* Washington, D.C.: National Center for Education Statistics, 1996.

National Council of Teachers of Mathematics. *Curriculum and Evaluation Standards for School Mathematics.* Reston, Va.: National Council of Teachers of Mathematics, 1989.

Oakes, Jeannie. "Opportunities, Achievement, and Choice: Women and Minority Students in Science and Mathematics." In *Review of Research in Education, Vol. 16,* edited by Courtney B. Cazden, pp. 152–222. Washington, D.C.: American Educational Research Association, 1990.

Reese, Clyde M., Karen E. Miller, John Mazzeo, John A. Dossey. *NAEP 1996 Mathematics Report Card for the Nation and the States: Findings from the National Assessment of Educational Progress.* (Report NCES 97-488). Washington, D.C.: National Center for Education Statistics, 1997.

Reyes, Laurie H. "Attitudes and Mathematics." In *Selected Issues in Mathematics Education,* edited by Mary M. Lindquist, pp. 161–84. Berkelely, Calif.: McCutchan Publishing, 1980.

Robinson, Sharon, P. "With Numeracy for All: Urban Schools and the Reform of Mathematics Education. *Urban Education* 30 (January 1996): 379–94.

Secada, Walter G. "Race, Ethnicity, Social Class, Language, and Achievement in Mathematics." In *Handbook of Research on Mathematics Teaching and Learning,* edited by Douglas A. Grouws, pp. 623–60. New York: Macmillan, 1992.

Silver, Edward A., Marilyn E. Strutchens, and Judith S. Zawojewski. "NAEP Findings Regarding Race/Ethnicity and Gender: Affective Issues, Mathematics Performance, and Instructional Context." In *Results from the Sixth Mathematics Assessment of the National Assessment of Educational Progress,* edited by Patricia Ann

Kenney and Edward A. Silver, pp. 33–59. Reston, Va.: National Council of Teachers of Mathematics, 1997.

Tate, William F. "Mathematics Communication: Creating Opportunities to Learn." *Arithmetic Teacher* 1 (February 1995): 344–49, 369.

———. "Race-Ethnicity, SES, Gender, and Language Proficiency Trends in Mathematics Achievement: An Update." *Journal for Research in Mathematics Education* 28 (December 1997): 652–79.

Welch, Wayne W., Ronald E. Anderson, and Linda J. Harris. "The Effects of Schooling on Mathematics Achievement." *American Educational Research Journal* 19 (Spring 1982): 145–53.

4

NAEP Findings Regarding Gender: Achievement, Affect, and Instructional Experiences

Ellen Ansell and Helen M. Doerr

THE rich history of the relationship between mathematics and gender provides the backdrop for the analyses of the NAEP mathematics results discussed in this chapter and establishes the perspectives from which we have chosen to select, organize, and interpret these results. The seminal work of Sells (1980) clearly illuminated how mathematics serves as a crucial filter for women, because disproportionately few women were entering the highest ranks of professional mathematicians in academia and were significantly underprepared or underrepresented in mathematically-oriented undergraduate programs of study. Recent studies reveal that an increasing percentage of women are enrolled in the scientific and engineering fields for which mathematics is an essential area of study (National Science Foundation 1999). However, while women accounted for 55 percent of all bachelor's degrees awarded in 1995, women earned only 35 percent of the bachelor's degrees in mathematics and computer science. Similarly, women account for only 22 percent of the science and engineering workforce, while comprising 46 percent of the total workforce. The wish to afford women equal access to learning opportunities and to scientific and technical careers drives a careful examination of issues related to mathematics and gender. Secada (1989) and others (Fennema and Leder 1990; American Association of University Women [AAUW] 1991; Rogers and Kaiser 1995) have argued that the principle of equity demands that our educational system provide a fair and just mathematics education for all students, regardless of gender, socioeconomic background, race, religion, or ethnicity.

The past two decades of research in mathematics and gender have examined gender-related differences in mathematical achievement, attitudes toward mathematics, perceptions about mathematics, mathematical course taking, motivation, and learning styles. The research has examined the differential impact on learners of teachers' attitudes and beliefs, the use of technology such as calculators, and varied instructional approaches. Yet more recently, theorists such as Gilligan (1982, 1992) and Belenky et al. (1997) have claimed that the analysis of differences is often grounded in a view that takes a male perspective as the standard and tends to measure female performance in light of that standard. These feminist theorists have argued that listening closely to the expression of the voices of girls and young women suggests different approaches to such fundamental issues as moral and intellectual development. Although it is not our intent to offer a critique or an analysis of the large body of research on mathematics and gender, we do wish to lay out how our framework for the analysis of the NAEP data has been guided by that work.

We provide an analysis and critique of the findings from the NAEP data suggesting where the findings support or contribute to established research results and where they indicate that further research is needed. We have organized our analysis of the data into three broad areas: students' achievement; students' beliefs about, and attitudes toward, mathematics; and students' perceived experiences in the mathematics classroom. Although the focus of most sections is on the three grade levels assessed in NAEP, interpretation of the twelfth-grade results is limited because more than one-third of the students in the grade 12 sample were not currently enrolled in any mathematics course. The implications of this limitation are discussed in each section of our analysis.

STUDENTS' ACHIEVEMENT

As in previous NAEP assessments, results from the 1996 assessment were reported in terms of average scale scores. Figure 4.1 shows the trend in NAEP scale scores for males and females from 1990 to 1996. The 1996 average mathematics scale scores of both males and females in grades 4 and 12 showed a significant increase over those from 1992 and 1990. Eighth-grade females (but not males) also showed a significant increase in average scale scores over both the 1992 and 1990 results. Figure 4.1 also provides information about performance between gender groups within an assessment year. On a positive note, the significant gender difference in grade 12 has been narrowing since 1990 and disappeared in 1996. However, over the same period of time, the gender gap increased for grade 4 and in 1996 for the first time showed significant difference. For grade 8, NAEP results have not shown a significant difference in average scale scores by gender.

Highlights

- In 1996, male and female students in grades 4 and 12 and females in grade 8 (but not males) showed a significant increase in average scale score over the 1992 and 1990 results. The only significant gender difference in overall performance by scale score was found in grade 4, in favor of males. Males and females did equally well on two-thirds of the fourth-grade items and three-fourths of the eighth-grade items.

- In grade 4, of the five NAEP content strands, males had significantly higher average scale scores than females in three strands—Measurement; Geometry and Spatial Sense; and Number Sense, Properties, and Operations. An item-level analysis of percent-correct values revealed some historically common, research-based patterns of difference such as males performing better than females on items that required spatial visualization, the use of measurement tools such as rulers, and working with rational numbers.

- Both fourth- and eighth-grade males performed better than females on a disproportionate number of items that did not require an explanation versus those that did. With respect to item difficulty, for both grades 4 and 8 there was no difference in performance between males and females on the most difficult items. However, at both grade levels, females' performance on the least-difficult items was higher than that for their male counterparts, and males' performance was higher on moderately difficult items.

- The vast majority of students have a positive self-concept with respect to mathematics in grade 4; however, their self-concept declines through their middle school and high school years. This trend is more precipitous for females than males. At all grade levels, significantly more males than females agree that they are good at mathematics.

- Although the usefulness of mathematics is clear to students at all grade levels, there is a decrease with increasing grade in students' intention to continue to study more mathematics. This decrease is more striking for females than for males. In a related result, course enrollment data suggest that females tend to drop out of mathematics at the precalculus and calculus level despite their significantly higher enrollments in the first three years of mathematics study.

- Males and females report different experiences with respect to the kind and extent of work they do in mathematics class, the kind and extent of their communication in mathematics class, and their use of computers and calculators.

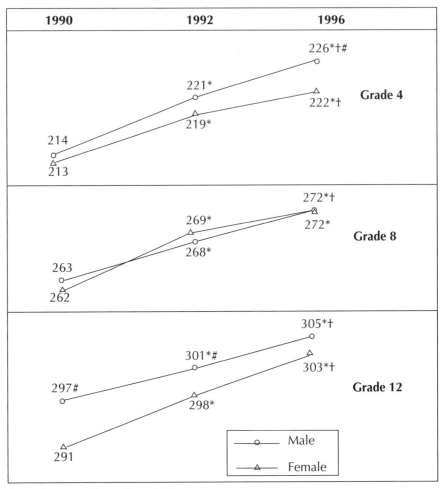

* Significantly different from 1990
† Significantly different from 1992
\# Significantly higher than average score for females within year

Fig. 4.1. Average mathematics scale scores by gender groups, 1990 to 1996

In both the 1990 and 1992 NAEP mathematics assessments, there were significant gender differences in overall performance reported only for twelfth-grade students in favor of males (Reese et al. 1997). These earlier reports corroborated Friedman's (1989) finding that gender difference in achievement favoring males shows itself consistently only at the high school level. In 1996, however, the only significant gender difference in overall performance was reported for grade 4, with males outperforming females (scale scores of 226 and 222, respectively). This result diverges from the literature on gender differences that has reported few, if any, gender differences prior to age ten. Further, past differences at this age level have generally reported higher performance for females than for males (Friedman 1989).

Although the overall average scale score indicated a gender difference at grade 4, the interpretation of that difference is not straightforward. The analysis of standardized test data has shown that, in general, when the average scale score is significantly higher for males, the greatest difference tends to occur above the median of the distribution of scores; conversely, on those tests where the average scale score is higher for females, the greatest difference tends to be found below the median (Han and Hoover 1994). It has also been found that males tend to have greater variability in scores than females (Friedman 1995). These results make comparisons based on the average scale score potentially misleading and suggest that comparing the entire distributions for males and females would be more informative (Martin and Hoover 1987; Becker and Forsyth 1990). Working within the limitations of the data available from the 1996 NAEP mathematics assessment reported as average scale scores (and not in terms of the median of the distribution), this kind of analysis is not possible. However, the NAEP data do allow analysis at the item level, a practice that has recently become more common as researchers move away from comparisons of overall mean differences (Garner and Englehard 1999).

One possible way to understand the gender difference in overall average scale score in grade 4 is to look for patterns of difference at the item level. Table 4.1 shows the number of mathematics items administered to students in grades 4, 8, and 12 and their distribution with respect to gender differences. The results in the table are based on percent-correct performance on multiple-choice and short constructed-response items scored as right or wrong and on satisfactory or extended performance on extended constructed-response tasks and on a significance level of .05. Males and females did equally well on about two-thirds of the fourth-grade items and about three-fourths of both the eighth- and twelfth-grade items. Given that there was an overall gender difference in performance at grade 4 and not at the other two grade levels, it is not surprising that there is a smaller percentage of items on which males and females performed similarly in grade 4 than in grades 8 or 12. What is surprising, however, is the similarity of the distribution of items that elicited differential performance in grades 4 and 12. Fourth-grade males had a significantly higher performance than females on 80 percent of the items; at grade 12, males performed significantly better than females on 76 percent of the items. Given no overall difference in performance by gender in grade 12, one might expect a distribution more similar to that found in grade 8, where females performed significantly better on 56 percent of the items that elicited differential performance. This seeming inconsistency between the distribution of items and average scale scores for grades 4 and

12 may be explained in large part by the complex way in which NAEP data are analyzed and reported. (For a detailed explanation of how percent-correct values and scale scores are determined in NAEP, see Allen et al. 1997.) This complexity also suggests that an analysis of patterns at the item level could be informative even if there is no gender difference in overall average scale score.

Table 4.1
Distribution of Items according to Gender Differences in Performance

	Grade 4 [148 items]	Grade 8 [179 items]	Grade 12 [176 items]
No Significant Difference	99	136	130
Significant Difference	49	43	46
Favoring Males	39	19	35
Favoring Females	10	24	11

Note: The item counts in the table for each grade level may differ from totals that appear in other sources (for example, Reese et al. 1997). The difference is a result of the way in which items were counted for the purposes of the analysis. For example, for items with multiple parts, each part was counted separately.

To investigate gender-related performance patterns at the item level, we consulted past research. According to this research, gender differences in mathematics achievement vary with content area and cognitive level of the content (Frost, Hyde, and Fennema 1994; Leder 1992; Seegers and Boekaerts 1996). More specifically, studies over the past two decades have sought to describe and explain gender differences in mathematics achievement by focusing on such areas as the relation of differences in spatial visualization and mathematics performance (Battista 1990; Friedman 1995; Tartre 1990); the relation of differences in verbal ability and mathematics performance (Tartre and Fennema 1995); differences in performance on computational tasks versus conceptual, problem-solving, and application tasks (Aiken 1986; Han and Hoover 1994); and differences in performance based on item format (Garner and Engelhard 1999; Lane, Wang, and Magone 1996). The framework used to construct the NAEP mathematics assessment includes two areas that research suggests would lend themselves to analyses related to gender differences: the five NAEP content strands and the item type (multiple-choice or constructed-response). Preliminary analyses within these two areas revealed that item difficulty may also warrant investigation as a contributor to gender differences. Thus, in addition to analyses based on content area and item type, we examined item difficulty, defined by the percentage of correct responses to each item.

In the fourth-grade data, we looked for patterns within the forty-nine items on which there were gender differences in performance (about 33 percent of the items administered). Although in grade 8 there was no gender difference in the overall average scale score, there was significant difference in performance by gender on about one-fourth of the administered items. We were interested to see if patterns that emerged in the fourth-grade data existed in this eighth-grade data as well. Therefore, we looked for similar patterns of difference at the item level in the performance of grade 8 males and females on the forty-three items that elicited gender differences. Because of the discrepancy in the number of items on which males and females performed better (thirty-nine to ten, respectively, in grade 4), comparisons were based on percentage distributions of items across categories within gender rather than on frequencies across gender within category.

The results in table 4.1 also suggest that a similar comparison based on the twelfth-grade, item-level results might be of interest. However, differential course taking associated with students in grade 12 (discussed later in this chapter), precludes meaningful interpretation of gender differences that may be found in performance at that grade level. For this reason we do not discuss the twelfth-grade results, based on item-level performance here. Each of the next three sections—content strands, item types, and item difficulty—begins with an analysis of the fourth-grade data followed by a brief comparison with the eighth-grade results.

Content Strands

The NAEP framework includes five content strands: Number Sense, Properties and Operations; Measurement; Geometry and Spatial Sense; Data Analysis, Statistics, and Probability; and Algebra and Functions. This configuration provides an avenue for interpretation of gender differences. For grade 4, three content strands had modest but statistically significant gender differences according to NAEP scale scores: Number Sense, Properties, and Operations (226 for males, 222 for females); Measurement (223 for males, 219 for females); and Algebra and Functions (230 for males, 225 for females). For grade 12, only Measurement (306 for males, 299 for females) and Geometry and Spatial Sense (309 for males, 305 for females) had significant gender differences by scale score. There were no significant scale score gender differences across the content strands for grade 8. Analysis at the content-strand level does not offer specific information to interpret the presence or absence of gender differences. In fact, patterns related to gender differences can appear at the level of individual items within a strand regardless of the presence or absence of a gender difference in average content-strand scale score.

Table 4.2 shows the distribution of the fourth- and eighth-grade items across the five content strands used in the NAEP framework according to the presence or absence of a significant gender difference. Across the content

Table 4.2
Distribution of Items by Content Strand and according to Gender Difference in Performance, Grades 4 and 8

	No Significant Difference	Significant Difference		
		Total	Favoring Males	Favoring Females
Number Sense, Properties, and Operations				
Grade 4 [60 items]	39	21	16	5
Grade 8 [50 items]	34	16	7	9
Measurement				
Grade 4 [28 items]	18	10	10	0
Grade 8 [27 items]	23	4	4	0
Geometry and Spatial Sense				
Grade 4 [25 items]	18	7	4	3
Grade 8 [32 items]	28	4	1	3
Data Analysis, Statistics, and Probability				
Grade 4 [17 items]	14	3	3	0
Grade 8 [33 items]	25	8	2	6
Algebra and Functions				
Grade 4 [18 items]	10	8	6	2
Grade 8 [37 items]	24	13	7	6

Note: The item totals in the table for each grade level may differ from totals that appear in other sources (for example, Reese et. al 1997). The difference is a result of the way in which items were counted for the purposes of the analysis. For example, for items with multiple parts, each part was counted separately.

strands, the percentages of fourth-grade items for which there was no significant difference in performance ranged from 56 percent for Algebra and Functions to 82 percent for Data Analysis, Statistics, and Probability. However, in all five content strands and within those items for which there was a significant difference, fourth-grade males were more likely to perform significantly better than their female counterparts.

The results in table 4.2 also show that three of the five content strands in grade 4 had gender differences on more than one-third of their items. The percentages of items that showed gender differences were, in order from least to greatest, Data Analysis, Statistics, and Probability (18 percent); Geometry and Spatial Sense (28 percent); Number Sense, Properties, and Operations (35 percent); Measurement (36 percent); and Algebra and Functions (44 percent). This comparison reveals that the Algebra and Functions content strand had the greatest percentage of items for which there was differential performance by gender. Males performed significantly better than females on six of the eight items. The content strand with the greatest number of items—Number Sense, Properties, and Operations—had gender differences for 35 percent of its sixty items. Males performed better than females on 76 percent of the twenty-one items that showed significant difference by gender. The results for grade 4 in table 4.2 suggest that although males and females performed similarly on a majority of the items in all five content strands, in the instances where there was a significant gender difference, males were more likely to outperform females.

The results for grade 8 had a slightly higher range than grade 4 with respect to the percentage of items on which the performance of males and females did not differ significantly—between 65 percent (twenty-four of thirty-seven items) for Algebra and Functions and 88 percent (twenty-eight of thirty-two items) for Geometry and Spatial Sense. Correspondingly, in four of the five strands, the percent of grade 8 items for which there was a difference in performance by gender was less than the corresponding percent for the grade 4 results; Data Analysis, Statistics, and Probability was the exception to this. The percents for grade 8 were, from least to greatest: Geometry and Spatial Sense (13 percent); Measurement (15 percent); Data Analysis, Statistics, and Probability (24 percent); Number Sense, Properties, and Operations (32 percent); and Algebra and Functions (35 percent). However, a different picture than that for grade 4 emerged with respect to the relative performance of males and females. As noted previously, in all five strands fourth-grade males performed significantly better on more items than females. The results were more mixed at grade 8, with the only similar performance pattern between the two grades appearing in the Measurement strand. Here, as was the case in grade 4, on all four Measurement items that had a significant performance difference, males outperformed females. The pattern of significantly different performance was about the same for males and females in the Number Sense, Properties, and Operations and Algebra and Functions strands, but females outperformed males on more items in the Geometry and Spatial Sense and Data Analysis, Statistics, and Probability strands.

A comparative analysis within each content strand of the items on which males performed better than females revealed some historically common, research-based patterns of difference in the Measurement, Geometry and Spatial Sense, and Number Sense, Properties, and Operations strands. These patterns are described next. Because there were no readily discernible patterns in the Data Analysis, Statistics, and Probability or Algebra and Functions strands that could be linked to research, they are not discussed further.

Measurement

As stated previously, there was a significant difference in average scale scores for fourth-grade males and females, with males outperforming females. At the item level, we found that fourth-grade males and females performed equally well on eighteen (64 percent) of the Measurement items. This included performance on all items that involved time in either a calculation or estimation situation, the measurement or calculation of area, and the selection of appropriate instruments for measuring length, temperature, and weight. However, males were more successful on all ten items for which there was a gender difference. Of these ten items, five items required students to read or use a given measurement instrument, and three of the remaining items focused on the choice of appropriate units. Males performed better than females when asked to read instruments (for example, a thermometer or a speedometer). Additionally, although males and females were equally able to use rulers to measure lengths, males outperformed females when the use of the ruler required iterative measurements (that is, finding the length of an item that was longer than the given ruler) or indirect calculation (that is, finding the length of an item placed in the middle of the ruler). It has been suggested that differences associated with the choice, reading, and use of measurement devices might be attributed to gender-typed leisure activities in which boys and girls have different experiences with tools and measurement (Leder 1990). Thus, NAEP results suggest that males in the elementary grades are more likely than their female counterparts to have or to benefit from having such experiences with measurement tools.

Similar to the pattern in fourth-grade students, males outperformed females on any Measurement item for which eighth-grade males' and females' performance was significantly different. However, unlike the fourth-grade pattern, there was no discernible distinction between the kind of information elicited from those items on which males performed better than females and those items on which males and females performed equally well.

Geometry and Spatial Sense

The average scale scores for the Geometry and Spatial Sense content strand did not show a significant gender difference at either grades 4 or 8, and there was a relatively small percentage of items for which there was a significant gender difference: 28 percent in grade 4 and 12 percent in grade 8. There were no obvious patterns in the item-level performance of eighth-grade males and females. However, when the seven fourth-grade items were examined for the kind of geometric knowledge they elicited, some typical gender-related patterns were found.

Table 4.3 shows two released NAEP items which are representative of items on which fourth-grade males performed significantly better than females (item 1) and those for which females performed significantly better than males (item 2). Of the four items (including item 1 in the table) on which males performed significantly better than females, all required spatial visualization with or without manipulatives to aid in the process. In contrast, of the three items (including item 2 in the table) on which females performed significantly better than males, all involved the identification and analysis of characteristics of given shapes, but required no spatial manipulation of the figures. Further, none of the four of the Geometry and Spatial Sense items on which males performed significantly better than females required a verbal response, whereas two of the three items (including item 2 in the table) on which females performed better than males required verbal responses. This possible confounding factor of verbal responses is discussed later in the section on item type.

It is important to note that one-half (nine of eighteen) of the grade 4 geometry items without significant performance differences by gender also required the use of spatial visualization. However, if the difficulty of an item is defined as percent correct, then the difficulty of the spatial visualization items with no gender-related performance differences tended to be lower than that for spatial visualization items for which males significantly outperformed the females. That is, the mean percent correct for the set of spatial visualization items with no gender differences was 57 percent, with a range of 26 to 91 percent correct. In contrast, for the four spatial visualization items on which males outperformed females, the mean percent correct was 18 percent, with a range of 8 to 28 percent. In other words, in grade 4, males performed better on the *most difficult* items involving spatial visualization in their solutions.

Spatial visualization skills have long been studied as an area to explain gender differences in mathematics achievement in general (Friedman 1995), and particularly in geometry (Battista 1990). Tartre (1990) and Battista (1990) both reported studies suggesting spatial visualization skills may play a larger role in the mathematics performance of females than of males. The 1996 NAEP data suggest that spatial visualization skills may still be an important factor in grade 4, but perhaps not in grade 8, due to the lack of discernible differences in the grade 8 geometry items requiring spatial visualization.

Number Sense, Properties, and Operations

There was an overall difference in performance by gender for the Number Sense, Properties, and Operations strand, with fourth-grade males having a significantly higher average scale score than females. Still, male and female fourth-grade students performed equally well on thirty-nine of the sixty items (65 percent). These items covered a variety of topics including single-

Table 4.3
Geometry Items with a Significant Gender Difference in Performance, Grade 4

Item	Percent Correct	
	Males	Females
[Two sets of these shapes were available for use as manipulatives.]		
1. Use the two pieces labeled Q to make a 4-sided shape that is not a square. Trace the shape and draw the line to show where the 2 pieces meet.	19*	12
2. Laura was asked to choose 1 of the 3 shapes N, P, and Q that is different from the other 2. Laura chose shape N. Explain how shape N is different from shapes P and Q.	56	63*

* Indicates a statistically significant difference between males and females.

and multistep word problems using varied operations on whole and rational numbers, computation items, and items that involved the application of number theory concepts.

Analysis of the twenty-one items for which there was a significant gender difference in performance revealed a pattern with respect to two areas cited in the literature. Past studies have reported that females tend to perform better than males on computational tasks (Frost, Hyde, and Fennema 1994); however, males tend to perform better than females on items that involve rational number (Seegers and Boekaerts 1996). The NAEP data showed that females performed better than males on five of the twenty-one items on which there was significant difference in performance by gender. All five items involved operations with whole numbers, and three items involved numerical computation and were not word problems. In contrast, in the sixteen items on which males performed significantly better than females, there was a conspicuous absence among them of any strictly computational problems.

There was no significant difference in average scale scores for grade 8 males and females. However, the same pattern with respect to items involving rational number found in the fourth-grade data was apparent in the grade 8 data. That is, there was a relative absence of rational number items within the group of items on which females performed better than males (one of nine) and a preponderance of such items (four of seven) on which males performed better females. This corroborates research that has found a performance difference between males and females in the middle school grades on fraction, ratio, and percent items (Seegers and Boekaerts 1996). This differential distribution of items involving rational number was the only apparent similarity of gender-related results for grades 4 and 8 in the Number Sense, Properties, and Operations strand.

Summary

The analysis within content strands of items according to the presence or absence of gender differences in percent correct performance uncovered some patterns that have been traditionally associated with gender. Fourth-grade performance on NAEP items that required reading and using measurement instruments, that entailed spatial visualization, and that involved straight computation with whole numbers corroborates established patterns of gender difference. These patterns of difference were, for the most part, not found in the results for grade 8.

Item Type

One feature of NAEP items that may be pertinent to gender differences is whether the item requires a verbal explanation. Although research on gender differences in achievement has had inconsistent results with respect to the relationship of verbal skills to mathematics achievement (Tartre and Fennema 1995), females tend to do better than males on language-oriented tasks (Han and Hoover 1994). A recent study of performance assessment items administered to middle school students (Lane, Wang, and Magone 1996) found that completeness of verbal explanations was a key factor in male and female performance differences. In particular, females displayed more complete explanations whereas males were more likely to provide no explanation to support their solution. These studies suggest that there may be a differential performance by gender on those items that do or do not require verbal explanation.

The NAEP assessment included both multiple-choice and constructed-response items. There were two types of constructed-response items: short constructed-response and extended constructed-response. The constructed-response items differed in the complexity of their demands; that is, for the most part short constructed-response questions required a numerical answer, a drawing, or a brief explanation, whereas extended constructed-response questions asked students for extensive work or an elaborate explanation. Table 4.4 shows the average percentage correct for multiple-choice and short and extended constructed-response items for the national NAEP sample and for gender groups. The results show that within grade level the average percentage of correct response for both males and females decreased from multiple-choice to extended constructed-response items, a pattern similar to that for the nation. With respect to gender categories, males and females appear to have done equally well within each type of item and nearly the same as the national averages.

Table 4.4
Correct Response Rate for Three NAEP Item Formats by Gender Categories, Grades 4 and 8

	Multiple Choice			Short Constructed Response			Extended Constructed Response		
	Average Percent Correct			Average Percent Correct			Average Satisfactory or Extended		
	Nation	Male	Female	Nation	Male	Female	Nation	Male	Female
Grade 4	54	55	53	38	39	37	17	17	17
Grade 8	55	55	55	49	49	49	9	9	9

As noted above, not all NAEP constructed-response items required a verbal explanation. To investigate the conjecture concerning possible gender differences in performance based on verbal explanations, the set of NAEP items administered to fourth- and eighth-grade students was sorted into three categories: multiple-choice items, constructed-response items (short or extended) that required a verbal explanation, and constructed-response items that did not have such a requirement but instead asked students for a numerical answer or a drawing. Table 4.5 shows the distribution of items by item format and according to significant gender differences in performance. Fourth-grade males performed better than females on a disproportionate number of items that did not require verbal explanations versus items that had such a requirement; in contrast, females performed better than males on an equal number of items that required explanation and those that did not. The distribution pattern in grade 8 was similar to that in grade 4; eighth-grade females performed better than males on about the same number of items that required and did not require explanations, whereas males were far more likely to perform better than females on items that did not require explanation than on those that did. These results tend to support the differential relation of verbal skills to mathematics achievement by gender.

Item Difficulty

The relation of item difficulty to performance by gender has been established on the basis of a variety of definitions of the term *difficulty*. For example, Seegers and Boekaerts (1996) discussed the relative difficulty of items from simple to more complex, using the terms *complexity* and *difficulty* levels interchangeably. In a comparison of performance by gender on items designed to differ on this dimension, they found that gender differences in

Table 4.5
Distribution of Items by Item Format according to Gender Difference in Performance, Grades 4 and 8

	Grade 4			Grade 8		
	Multiple Choice	Constructed Response		Multiple Choice	Constructed Response	
		Explanation	No Explanation		Explanation	No Explanation
No Significant Difference	58	7	34	82	15	37
Significant Difference	26	3	20	28	7	10
Male	20	1	18	13	2	6
Female	6	2	2	15	5	4

performance increased with item difficulty. In particular, although there was no gender difference in performance on the items of least complexity, males performed significantly better on mid-level items and on those items that were in the most difficult category.

Another common measure of the relative difficulty of an item involves the percentage of respondents who correctly answered it. To look for possible gender differences in the NAEP items for grades 4 and 8, the items were sorted into three levels of difficulty: percent-correct rates from 0 to 33 percent, from 34 to 67 percent, and from 68 to 100 percent. Our assumption was that the percent-correct responses to an item were inversely related to its level of difficulty. That is, these three ranges define categories from most (0 to 33 percent) to least difficult (68 to 100 percent), respectively.

Table 4.6 shows the distribution of items across these difficulty categories according to the presence or absence of a performance difference between genders at the .05 significance level. In grade 4, the distribution of items overall was such that about 50 percent fell in the mid-range of moderately difficult, about 20 percent in the least difficult level, and about 30 percent in the most difficult level. The items on which males and females performed equally well were distributed across difficulty levels in a way that was similar to the distribution of items overall. The items for which there was a significant performance difference, however, were skewed toward the most difficult problems; that is, the majority of the items for which there were significant performance differences were in the most difficult and moderately difficult categories. The distribution across difficulty of those items on which females performed better than males and those on which males performed better than females was also very different. Males and females performed better than one another on the same number of the least difficult items (4 items each); however, these four items accounted for only 10 percent of the items on which males performed better than females (four of thirty-nine items), whereas it was 40 percent of the items on which females performed better than males (four of ten items).

Table 4.6
Distribution of Items by Difficulty according to Gender Difference in Performance, Grades 4 and 8

	Grade 4			Grade 8		
	Most difficult [46 items]	Moderately difficult [72 items]	Least difficult [30 items]	Most difficult [51 items]	Moderately difficult [75 items]	Least difficult [53 items]
No Significant Difference	25	52	22	42	56	36
Significant Difference	21	20	8	9	19	17
Male	17	18	4	4	13	4
Female	4	2	4	5	6	13

In grade 4, more than twice as many of the items for which there were significant gender differences fell into the most difficult category than into the least difficult. The relative distribution with respect to difficulty of such items was the opposite for students in grade 8. The eighth-grade items whose performance by males and females was significantly different were almost twice as likely to be in the least difficult than in the most difficult category. This seems to verify that males tend to perform better than females on more-difficult items. If females tend to perform better than males on the least-difficult items, and since females in grade 8 performed significantly better than males on more items than males performed better than females, one would expect the distribution of items to be skewed in the direction of the less difficult items. On the 1996 NAEP, the distribution for the most difficult items was nearly identical: males outperformed females on four items in the most difficult category, and females outperformed males on five items in that category. However, the distribution of the moderately difficult and least difficult items was different for males and females. Males outperformed females on more moderately difficult items than on the least difficult items; the pattern for females was just the opposite, with their performance higher on more least difficult items than on moderately difficult items.

Summary

The results of the 1996 NAEP mathematics assessment indicated far more similarities than differences with respect to the achievement of fourth- and eighth-grade males and females. However, the differences in content, item type, and difficulty evident in an item-level analysis raise concern about the persistence of gender-related patterns that may have serious consequences. From this analysis we cannot comment on specific consequences of these differences, nor does the NAEP data available allow for analyses that might help us investigate gender-related patterns more extensively. However, further investigation is warranted.

Past research on gender-related patterns in achievement have looked to explain some of these patterns through an analysis of their relationship to students' attitudes and beliefs about themselves as learners of mathematics (Leder 1990). These studies suggest that the interactions of attitudes and achievement are "complex and unpredictable" (McLeod 1992, p. 582). Rather than try to untangle this relationship, in the remainder of the chapter we use the NAEP questionnaire data to look at gender-related patterns in students' attitudes and beliefs about mathematics and in students' instructional experiences in the mathematics classroom. These areas of persistent gender differences are important regardless of any connection they may have to achievement.

STUDENTS' ATTITUDES AND BELIEFS

Although research on affect continues to be treated as peripheral (McLeod 1992), its importance to students' learning and using mathematics is clearly expressed by students and teachers, and its centrality is conveyed in the explicit inclusion of affective goals with respect to mathematics education standards (NCTM 1989, 1991, 1995). The inclusion of affective measures in the questionnaire that students complete as part of the NAEP assessments has been an important source of information about their beliefs and attitudes with respect to mathematics and, in particular, for establishing the trends and persistence of gender differences. We grouped the questions from the 1996 NAEP assessment into three general areas shown by past research to be of consequence: students' self-concept in relation to mathematics, their view of the nature of mathematics, and their perception of the usefulness of mathematics. The first two areas are discussed in this section of the chapter; the third area, concerning perceived usefulness of mathematics, is discussed in a later section. Statistically significant differences at the .05 level are noted for each group of questions. Unfortunately, the complexity of NAEP results based on weighted data often precluded the use of useful statistical tests (for example, chi-square) that could have facilitated the investigation of students' attitudes and beliefs in relation to their experiences in the mathematics classroom.

Self-Concept Related to Mathematics

Students indicated the extent of their agreement with four statements related to their conceptions of themselves as learners of mathematics: (1) "I like mathematics," (2) "I understand most of what goes on in mathematics class," (3) "I am good at mathematics," and (4) "Everyone can do well in mathematics if they try." Students' attitude data for grades 8 and 12 were reported in five categories (Strongly Agree to Strongly Disagree); the data for grade 4 were reported in three categories (Agree to Disagree). In order to facilitate cross-grade comparisons in this section of the chapter, the data for grades 8 and 12 were collapsed into three categories to match the categories from grade 4. Table 4.7 shows the percentages of males and females who responded to each of the four statements.

With regard to the statement concerning attitude toward mathematics ("I like mathematics"), overall results for the national sample revealed that more than two-thirds of fourth-grade students reported that they like mathematics; however, with increasing grade level, the percentage of agreement decreased. Only about 55 percent of eighth-grade students and about 50 percent of twelfth-grade students agreed that they like mathematics. The data in table 4.7 show that although the rate of agreement was the same for males and females in grade 4, significantly more eighth- and twelfth-grade males than females agreed with that statement. Eighth-grade students regardless of gender were equally likely to be undecided as to disagree with the statement. In contrast, more twelfth-grade students disagreed that they like mathematics than were undecided. With respect to disagreement with the statement, there was a significant difference between the percents for females and males, 37 percent and 29 percent. The overall picture is one of students with a positive attitude toward mathematics in the middle of their elementary school experience whose interest declines through their middle and high school years. This change was more precipitous for females than for males.

Fourth- and eighth-grade males and females were similar in their assessment of their own understanding of what goes on in mathematics class. More than three-fourths of the students by gender at both grade levels agreed that they understand most of what goes on in mathematics class and, on average, fewer than 10 percent disagreed with this statement. Gender differences in students' evaluation of their understanding of mathematics class did not appear until grade 12. When compared to the results for grades 4 and 8, a smaller percentage of twelfth-grade students agreed that they understand most of mathematics class (70 percent of males and 63 percent of females) and significantly more males than females shared this sentiment. It must be noted, however, that 36 percent of the twelfth-grade respondents were not currently enrolled in mathematics courses; therefore, caution must be used in interpreting this result.

Table 4.7
Percent of Males and Females Responding to Statements regarding Self-Concept in Mathematics

	Disagree		Undecided		Agree	
	Male	Female	Male	Female	Male	Female
I like mathematics.						
Grade 4	15	14	16	17	69	70
Grade 8	22	24	20	23	58*	53
Grade 12	29	37*	18	16	53*	48
I understand most of what goes on in mathematics class.						
Grade 4	6	7	15	15	79	78
Grade 8	9	10	13	13	78	77
Grade 12	15	21*	15	16	70*	63
I am good at mathematics.						
Grade 4	9	12*	21	27*	70*	61
Grade 8	13	17*	19	25*	68*	59
Grade 12	21	30*	21	23	59*	47
Everyone can do well in mathematics if they try.						
Grade 4	4*	2	9	8	88	90
Grade 8	11	12	14	17*	75*	71
Grade 12	27	32*	20	22	54*	46

* Indicates a statistically significant difference between males and females within grade level and response category.
Note: Row percents by gender within grade level may not add to 100 because of rounding.

Although a gender difference in students' own assessment of their understanding in mathematics class did not appear until grade 12 in the NAEP results, there was a gender difference at all three grade levels when students were asked if they are good at mathematics. At all three grade levels, significantly more males than females agreed that they are good at mathematics, and females were significantly more likely than males to disagree with the statement. In grades 4 and 8, females were also significantly more likely than males to be undecided.

There is a surprising difference in the pattern of students' responses to the statement about being good at mathematics and the statement about understanding what goes on in mathematics class. Significantly fewer students, male and female, at all three grade levels agreed that they are good at mathematics than students who agreed that they understand most of what goes on in mathematics class. In addition, males and females at grades 4 and 8 assessed their understanding of most of their mathematics class equally, but males were significantly more likely than females to see themselves as good at mathematics. Why should fewer students think they are good at mathematics than understand mathematics class, and why should there be no gender difference with respect to students' assessment of their understanding yet a gender difference in their evaluation of how good they are at mathematics? One possible explanation is that an individual's perception of how well she or he understands mathematics is based on a self-evaluation, whereas how good one is at mathematics is based on an assessment of oneself relative to others or on how others (such as teachers and peers) are assessing the learner. This may correspond to the difference between self-efficacy and self-concept of ability. The former, usually used with respect to specific tasks rather than a domain in general, is a judgment of performance competence relative to the goals of the task, whereas the latter is a "social comparison between perceived self-competence and perceived peer competence" (Seegars and Bockaerts 1996, p. 217).

The dramatic pattern of decrease in agreement as grade level increases seen in students' assessment of themselves as being good at and liking mathematics is also apparent in students' response to the statement that "everyone can do well in mathematics if they try." The vast majority (about 90 percent) of fourth-grade males and females agreed with this statement. This decreases to fewer than three-quarters agreeing in grade 8 and agreement by only half the students in grade 12. In both grades 8 and 12 males are significantly more likely than females to agree with the statement, and twelfth-grade females are more likely than males to disagree. This pattern of decline and the concurrent trends in course taking suggest that a fruitful area for further research might be an investigation of the relationship between attitudes and course taking.

Students' confidence is one aspect of their self-concept. If one likes, understands, and is good at mathematics, it is likely that there is a congruent belief in one's competence or confidence in mathematics (McLeod 1992). Confidence may also be evidenced by the degree to which one commits oneself to a response. In particular, when responding to five categories involving extremes such as strong agreement and strong disagreement, one might expect a relationship between one's confidence and one's use of the extreme categories versus the middle categories. There seems to be such a pattern in the way males and females responded to middle and extreme categories in the NAEP student questionnaire data.

The questionnaire included a total of nine items that elicited information about eighth- and twelfth-grade students' attitudes and beliefs. (The five questions not shown in table 4.7 are discussed later in the chapter and can be found in tables 4.9 and 4.10.) These items were used to compare the number of instances in which males or females in grades 8 and 12 had a significantly higher percentage response with respect to the extreme categories (strongly agree or strongly disagree), the more moderate categories (agree or disagree), and the noncommittal category (undecided). The results of this analysis appear in table 4.8. At both grades 8 and 12, males were far more likely than females to respond at the extremes, and females were more likely to respond in the agree or disagree category. At grade 8, any significant difference in the percentage of males and females choosing the undecided category corresponded to more female than male responders. Eighth-grade females were more likely than males to be undecided.

Table 4.8
Distribution of Responses within Response Categories for Which There Were Significant Gender Differences, Grades 8 and 12

	Strongly Agree or Strongly Disagree	Agree or Disagree	Undecided
Grade 8			
Male	10	1	0
Female	0	5	4
Grade 12			
Male	9	3	1
Female	1	7	0

A View of the Nature of Mathematics

The student questionnaires at all three grade levels included two statements that elicited students' views of the nature of mathematics: (1) "There is only one correct way to solve a mathematics problem," and (2) "Learning mathematics is mostly memorizing facts." The eighth- and twelfth-grade questionnaires had an additional statement: "Describing mathematical concepts and ideas is as important as doing mathematical operations such as addition and multiplication in solving problems." Table 4.9 shows the extent to which male and female students agreed with these statements. Although only 63 percent of fourth-grade students disagreed that there is only one correct way to solve a mathematics problem, between 75 and 80 percent of students in grade 8 and about 80 percent in grade 12 disagreed with the statement. This is perhaps an encouraging overall trend in light of the view of the nature of mathematics espoused by the NCTM *Curriculum and Evaluation Standards* (1989), but there are some gender related differences in students' views of the nature of mathematics. Across the grades, significantly more males than females indicated a view that is not in line with the NCTM *Standards*, whereas females in grades 8 and 12 were significantly more likely to concur with the *Standards*-oriented view.

A little more than one-third of the twelfth-grade students, both male and female, agreed that learning mathematics is mostly memorizing facts.

Unfortunately, more than half of fourth-grade students and more than 40 percent of eighth-grade students agreed with this statement. In fact, only one-fifth of students in grade 4, one-third of students in grade 8, and two-fifths of twelfth-grade students disagree with the statement that learning mathematics is mostly memorizing facts. The view of mathematics as mostly memorizing facts is related to gender only at grade 8. Here, more males agree with the statement, and more females are undecided.

With respect to the third statement in table 4.9, a majority of both males and females at grades 8 and 12 agreed that describing concepts is as important as doing operations, although more males than females disagree with this view. Nearly one-third of the students at both grade levels were undecided as to the importance of concepts relative to operations. Females in grade 8 were more likely than males to be undecided.

Although the level of students' agreement or disagreement varied across the three statements related to views of the nature of mathematics, there is a general trend across grade level in students' responses to each of the three statements; according to this trend, with increase in grade level students' views of the nature of mathematics become more closely aligned with the NCTM *Standards.* It seems that fourth-grade students' views of the nature of mathematics are the furthest from the *Standards*-based view, and this view becomes more prominent from grade 8 to grade 12.

The trend of decline in students' attitudes about mathematics across elementary, middle, and high school is dramatic, especially for females. The relationship of students' report of the extent to which they like and understand mathematics and their view of whether all can do well in mathematics if they try warrants further investigation. The shift across grade level in students' views of the nature of mathematics toward the view espoused by the *Standards* as more mathematics is taken is positive; however the still-low percents of agreement with this view, even at the higher grade levels, should make us think about how the curriculum and its implementation at the lower grades contributes to this view.

STUDENTS' PERCEIVED EXPERIENCES IN MATHEMATICS

The reform of mathematics education, such as embodied in the NCTM's curriculum, teaching, and assessment *Standards* (1989, 1991, 1995), envisions classrooms in which students are actively engaged in learning while working collaboratively with partners or in small groups, where writing, discussing, and communicating about mathematics is the norm, where calculators and computers are appropriately used in the learning of mathematics, and where portfolios of students' work are used to assess their learning. Both conventional wisdom and research suggest that such classroom

Table 4.9
Percent of Males and Females Responding to Statements regarding Beliefs about the Nature of Mathematics

	Disagree		Undecided		Agree	
	Male	Female	Male	Female	Male	Female
There is only one correct way to solve a mathematics problem.						
Grade 4	63	63	19	22*	18*	15
Grade 8	76	80*	15	14	10*	6
Grade 12	79	85*	13	11	8*	4
Learning mathematics is mostly memorizing facts.						
Grade 4	21	22	24	26	55	53
Grade 8	30	33	26	30*	45*	37
Grade 12	44	44	22	22	35	34
Describing mathematical concepts and ideas is as important as doing mathematical operations such as addition and multiplication in solving problems.						
Grade 8	9*	5	29	35*	62	60
Grade 12	12*	8	28	29	61	63

* Indicates a statistically significant difference between males and females within grade level and response category.
Note: Row percents by gender within grade level may not add to 100 because of rounding.

experiences lead to greater achievement by students and benefit all learners, regardless of gender, race, ethnicity, or socioeconomic status. However, studies (Sadker and Sadker 1994; AAUW 1991) have shown that even though males and females are in the same classrooms, their experiences can be dissimilar. These studies indicate that in many classrooms, males receive greater attention, are encouraged to become more-autonomous learners, are asked more follow-up questions, and are asked more high-level and conceptually oriented questions than are females. The differential effects of course taking further suggest that students' school experiences of mathematics could be differentiated along gender as well as socioeconomic lines (Oakes 1985; Page 1991). The 1996 NAEP assessment data provide some insights into students' perceptions of their experiences in the mathematics classroom. In addition, some corroborating results are available from similar data reported by the students' teachers from the NAEP teacher questionnaire. (Additional analyses based on data from the teacher questionnaire can be found in chapter 5 by Grouws and Smith.) The NAEP student background questionnaire asked a range of questions about students' classroom experiences at grades 4, 8, and 12. We discuss them in relation to two major areas: students' persistence in mathematics course taking, and students' experiences of mathematics classroom instruction.

Persistence in the Study of Mathematics

The student background questionnaire for grade 12 asked students about the mathematics courses they had taken. At all three grades, students were asked if, given the choice, they would continue to study mathematics and about their perception of the usefulness of mathematics for solving everyday problems. The responses to these latter two items are outlined in table 4.10. Course-taking patterns have long been a gender-related concern, and one past explanation for differential course-taking patterns was the gender difference in the perceived usefulness of mathematics (Meyer and Koehler 1990). Therefore, these items are discussed together as they relate to the persistence of students' mathematics education.

Table 4.10 shows that a large majority of all students at all three grade levels see the usefulness of mathematics for solving everyday problems; this result is consistent with well-established results in the literature (Hyde, et al. 1990). Interestingly, the NAEP results peak at grade 8 to about 80 percent of males and females agreeing with the statement and then return in grade 12 to the same 70 percent level found at grade 4. Although the usefulness of mathematics appears to be clear to students at all levels, there is a decline with increasing grade level in students' intentions to continue to study mathematics, if given the choice. Nearly three-fourths of fourth-grade students and close to two-thirds of eighth-grade students indicate that they disagreed with the statement that "If I had a choice, I would not study any more mathematics." However, fewer than 50 percent of twelfth-grade students, both males and females, disagreed. At grade 4, significantly more females than males indicated disagreement, but at grade 12, significantly more females than males say they would not study more mathematics. These are troubling statistics with respect to already established differences in males' and females' trends in course-taking patterns.

As shown in Figure 4.2, significantly more females than males have studied first-year algebra, geometry, and second-year algebra. Despite this unequal enrollment in the first three years, equal percentages of males and females go on to study precalculus and calculus. This suggests that although more females than males are enrolling in the lower-level courses that are preparatory for more advanced secondary mathematics, males show greater persistence in continuing to study mathematics. The greater persistence of

Table 4.10
Percent of Males and Females Responding to Statements regarding Persistence in Mathematics

	Disagree		Undecided		Agree	
	Male	Female	Male	Female	Male	Female
Mathematics is useful for solving everyday problems.						
Grade 4	14	14	16	18	70	68
Grade 8	8	7	12	12	80	81
Grade 12	14	13	16	17	70	70
If I had a choice, I would not study any more mathematics.						
Grade 4	69	76*	16	15	15*	9
Grade 8	65	66	19	20	17	15
Grade 12	49	46	24*	20	27	34*

* Indicates a statistically significant difference between males and females within grade level and response category.
Note: Row percents by gender within grade level may not add to 100 because of rounding.

males in the study of high school mathematics is also reflected in students' responses when asked the number of semesters of mathematics they have studied: 43 percent of males reported that they studied eight or more semesters of mathematics, whereas only 39 percent of female respondents did so.

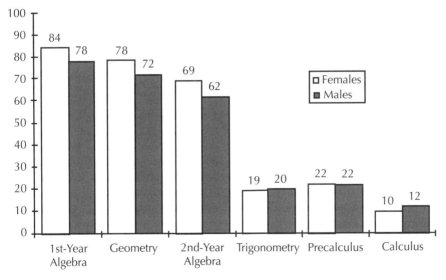

Fig. 4.2. Percent of twelfth-grade students by gender according to mathematics courses studied for one year

Conversely, 20 percent of female respondents reported having studied six semesters of math, but only 16 percent of males indicated that they studied six semesters of math. These figures show that more females than males have studied the equivalent of three years of mathematics, but more males than females go on to study the equivalent of four years of mathematics.

Together, these results suggest that despite significantly higher enrollments in the first three years of mathematics study, females tend to drop out of mathematics at the precalculus and calculus level. Even larger differences in enrollment occur among different racial/ethnic and socioeconomic groups, as detailed in chapter 3 by Strutchens and Silver. The current underrepresentation in mathematics and related technical fields of women of color warrants further research and study of potential interaction effects among gender and ethnic and socioeconomic groups.

Perceptions of Classroom Experience

The student questionnaire also asked for information about students' experience of classroom instruction. We clustered these questions around three topics: (1) the use of instructional materials, (2) classroom communication, and (3) the use of computers and calculators. These topics are also addressed on the questionnaire given to the teachers of students in grades 4 and 8, with some minor wording differences. There was no teacher questionnaire for grade 12 in the main NAEP sample. Given that only 64 percent of the students in the twelfth-grade sample reported being currently enrolled in a mathematics course and that these courses included a range at and below the calculus level, we did not find it useful to analyze with respect to gender the perceived experiences of twelfth-grade students within their mathematics classes. Any differences found could easily be attributed to differences in course taking or differential experiences when taking a course below grade level. Hence, in the sections that follow we examine the classroom experiences of fourth- and eighth-grade students.

The Use of Instructional Materials

Both fourth- and eighth-grade students, regardless of gender, appear to spend a significant portion of classroom time doing textbook problems and worksheets. Fifty-seven percent of the fourth-grade males and females reported that they do mathematics problems from their textbooks every day. This increased to 75 percent in grade 8. At the same time that textbook use increased, the presence of worksheets decreased from 47 percent of the students in grade 4 reporting daily use to 30 percent in grade 8. In both grades, the vast majority of students reported that they take tests either once or twice a month or once or twice a week. As would be expected, there are no gender differences in these classroom practices.

One somewhat surprising difference in perceived classroom experience appears in grade 4: significantly more females than males reported doing 10 or more mathematics problems by themselves every day (55 percent for females versus 50 percent for males). Similarly, significantly more males reported never doing ten or more mathematics problems by themselves (17 percent versus 13 percent). In ordinary classroom practice, one would not expect teachers to assign practice sheets to males and females differentially. However, it could be that females are more compliant in doing drill types of problems for homework or that the females perceive that more of this type of work is assigned to the class as part of their mathematics learning. This difference is also found in grade 8 where significantly more males than females reported never doing ten or more mathematics problems by themselves (27 percent versus 21 percent). Unfortunately, there are no corresponding data on this topic on the teacher questionnaires that would allow us to corroborate the students' perceptions of instructional materials use with their teachers' perceptions.

Classroom Communication

Communication in the mathematics classroom can be measured by questions about students' experiences in writing, reporting, talking, discussing, and working with partners or groups. These areas are addressed on both the student and teacher questionnaires. The results from some items about classroom communication from the student questionnaires for grades 4 and 8 appear in table 4.11. The results show that when asked how often they write a few sentences about how they solved a mathematics problem, there were no significant differences between males and females in any frequency category. However, the number of students (males and females alike) who reported that they never or hardly ever write about solving a mathematics problem increased from grades 4 to 8 (35 percent to 48 percent, respectively). These numbers are quite high given the current emphasis on communication and writing. There were also no gender differences in grade 8 with respect to the frequency with which students reported that they write reports or do mathematics projects; this question did not appear on the questionnaire given to fourth-grade students.

Students also communicate in writing when they take mathematics tests. There is a surprising gender difference among grade 8 students when they were asked: "This year in school, how often have you taken mathematics tests where you were asked to provide detailed solutions to problems you had not worked on before?" In grade 8, significantly more males than females (24 percent and 18 percent, respectively) reported that they were asked to provide such detailed solutions on tests once a week. Significantly more females than males (28 percent and 23 percent, respectively) reported that this occurred only once or twice a year. Since it is unlikely in ordinary

Table 4.11
Percent of Males and Females Responding to Statements regarding the Frequency of Written and Oral Communication in Mathematics Class, Grades 4 and 8

	Never or Hardly Ever		Once or Twice a Month		Once or Twice a Week		Almost Every Day	
	Male	Female	Male	Female	Male	Female	Male	Female
Write a few sentences about how you solved a mathematics problem								
Grade 4	37	33	17	18	27	27	20	22
Grade 8	49	47	18	20	20	20	13	13
Write reports or do mathematics projects								
Grade 4	—	—	—	—	—	—	—	—
Grade 8	60	63	29	28	8	7	3	2
Talk to the class about your mathematics work								
Grade 4	51	48	12	12	17	18	20	22
Grade 8	40	37	13	12	17	17	29	33
Solve mathematics problems with a partner or in small groups								
Grade 4	43	41	24	21	24	28	9	10
Grade 8	30	26	27	28	27	29	16	17
Discuss solutions to mathematics problems with other students								
Grade 4	35	32	18	18	28	30	18	20
Grade 8	22*	17	16*	13	29	29	33	41*

* Indicates a statistically significant difference between males and females.
Note: Row percents by gender within grade level may not add to 100 because of rounding.

classroom practice that teachers would differentially (based on gender) test students, it is difficult to explain the source of this difference. It may be that the male and female students have different perceptions about the kinds of tasks they are asked to perform.

Spoken communication takes place in both whole-class and small-group settings. As reported in table 4.11, there were no significant gender differences in grades 4 or 8 in the reported frequency with which students talk to the class about mathematics work. Also, there were no significant gender differences in how often students solved mathematics problems with partners or in small groups. Taking males and females together, more than half of the fourth-grade students (62 percent) and about half of the eighth-grade students (52 percent) reported that they either never or only once or twice a month talk with their class about mathematics work. Students in grade 4 and grade 8 reported more frequent discussion of mathematics problems with other students than they reported talking with the class about their mathematics work. With respect to discussing solutions to mathematics problems with other students, there were no gender differences in grade 4; however, the eighth-grade females report being more frequently engaged in discussion with other students than their male peers were. Significantly more males (22 percent) than females (17 percent) reported never discussing mathematics problems with other students, whereas significantly more females (41 percent) than males (33 percent) reported discussing math problems every day.

The results from the teacher questionnaire, however, suggest that most students do discuss mathematics problems with their peers, with 40 percent of eighth-grade students having teachers who report that such discussion happens almost every day. Lacking NAEP data to look at gender by ethnicity or socioeconomic status, we cannot examine how these classroom interactions (or lack thereof) might differentially impact males or females within these groups.

The Use of Computers and Calculators

The last component of the perceived classroom experience addressed by the NAEP results is the use of computers and calculators for mathematics instruction. The NAEP questionnaire asked about computer use at school and at home and about the use of calculators for classwork, homework, and on tests. It is clear from the data that calculator use—not computer use—dominates the mathematics classroom. In grade 8, 69 percent of the students have teachers who reported never using a computer, whereas only 2 percent reported everyday use. This is in the same general direction as the results from the student questionnaire, and it could help to explain the gender difference found in the responses to this question. Significantly more males than females (15 percent and 10 percent, respectively) reported using a computer every day for mathematics in grade 8. It is possible that this greater use by the males resulted from work in the open computer labs in the schools or through home computer use. There were no gender differences found in computer use in grade 4.

There were no gender differences in reported frequency of calculator use for classwork or on tests in grades 4 or 8. However, significantly more males (32 percent) than females (28 percent) reported using calculators for their mathematics homework every day in grade 4. For both grades, reported calculator use for classwork, homework, and on tests increased substantially from grades 4 to 8. The vast majority of students (regardless of gender) in grade 8 reported having calculators available to them for their mathematics work. In grade 8, significantly more females (92 percent) than males (87 percent) reported that they do not use graphing calculators for their mathematics work.

Summary

The students' perceptions of their classroom experiences in grades 4 and 8 suggest a pattern of instruction that is more characterized by the use of worksheets, textbooks, and calculators and less by the use of projects, writing, and computer technology. At all grade levels for all students, the calculator is used far more than the computer for mathematics instruction. However, at grade 8, significantly more males than females report using a computer every day for mathematics. In grades 4 and 8, more females than males report doing ten or more math problems by themselves every day, which may be an indication of greater drill or routine work being done by female students. In grade 8, significantly more males than females reported that they are asked to provide detailed solutions once a week and significantly more females reported that they are rarely asked to provide detailed solutions. Although written communication among all eighth-grade students is somewhat limited, with nearly two-thirds of all students reporting that they never write reports, females are engaged in more discussions about mathematics with other students than males are.

CONCLUSION

Results of the 1996 NAEP mathematics assessment give us reason to be positive about the mathematics education of male and female students. The equal overall achievement of males and females in grades 8 and 12 and the increase in performance over the 1990 and 1992 results of all females at all grade levels and males in grades 4 and 12 are positive trends. A more detailed analysis of the fourth- and eighth-grade data showed that there were many items on which males and females had similar performance levels and few items on which one gender group outperformed the other. However, an analysis of items on which males and females performed significantly differently relative to items on which their performance was similar revealed that patterns long associated with gender differences in Measurement, Geometry and Spatial Sense, and Number Sense, Properties, and Operations still persist. There was also persistence in the pattern of performance by gender with respect to item difficulty and the relationship between gender and items that involve a verbal explanation.

Most fourth-grade students, both male and female, reported that they enjoy mathematics and have confidence in themselves as learners of mathematics. Nearly all believed that mathematics is something that all can do well, though fewer believe that they are good at math, and more males than females report such confidence. They recognized the usefulness of mathematics and intend, given the choice, to continue to study mathematics. Yet by grade 12 only about half the students, and significantly fewer females than males, reported that they like or are good at mathematics or that all can do well at mathematics. Although these students still recognize the usefulness of mathematics, far fewer indicate they would choose to continue to study mathematics, a trend more pronounced for females than males. This is confirmed in the course-taking patterns reported by male and female students. Significantly more twelfth-grade females than males have taken three years of preparatory mathematics classes, yet there are equal percentages of males and females at the precalculus and calculus levels. Together these data indicate a tendency of females not to persist in the study of mathematics despite the equal overall performance level of eighth- and twelfth-grade males and females.

The shift in attitude from grade 4 to grade 12 is paralleled by changes in students' perceived experiences of the mathematics classroom. Experiences with respect to oral and written communication and the use of technological tools are, in many respects, not consistent with those recommended by the NCTM *Standards* documents (1989, 1991, 1995) and vary for males and females. The discursive community advocated by the NCTM *Standards* (1989, 1991) with respect to oral and written communication seems to be experienced most by fourth-grade students and declines in grades 8 and 12. The mathematics classroom seems to be a place that is more active and engaging for males than for females.

The NAEP assessment contributes important information about the state of mathematics achievement, attitudes, and experience with regard to male and female students in grades 4 and 8 and, to a more limited degree, in grade 12. However, the data as presently available limit the opportunities to untangle the complexity of the differential achievement, attitudes, and experience of males and females. In addition to the limitations noted throughout the chapter with regard to the extent of twelfth-grade enrollment in mathematics courses, relational analyses that are potentially of great import are not available. For example, in addition to relating demographic variables such as gender, race, and socioeconomic status, research indicates that there may be clusters of items whose relation would help us better understand the achievement of males and females. We hope that future NAEP assessments and analyses of NAEP data will take such research into consideration so that it can be a useful tool in the continuing efforts of mathematics educators to understand and better the mathematics education of all students.

REFERENCES

Aiken, Lewis. "Sex Differences in Mathematical Ability: A Review of the Literature." *Educational Research Quarterly* 10 (1986): 25–35.

Allen, Nancy L., Frank Jenkins, Edward Kulick, and Christine A. Zelenak. *Technical Report of the NAEP 1996 State Assessment Program in Mathematics.* Washington, D.C.: National Center for Education Statistics, 1997.

American Association of University Women. *Shortchanging Girls, Shortchanging America: A Call to Action.* Washington, D.C.: American Association of University Women, 1991.

Battista, Michael. "Spatial Visualization and Gender Differences in High School Geometry." *Journal for Research in Mathematics Education* 21 (January 1990): 47–60.

Becker, Douglas F., and Robert A. Forsyth. "Gender Differences in Academic Achievement in Grades 3 through 12: A Longitudinal Analysis." Paper presented at the annual meeting of the American Educational Research Association, Boston, April 1990. (ERIC document ED323259)

Belenky, Mary F., Blythe M. Clinchy, Nancy R. Goldberger, and Jill M. Tarule. *Women's Ways of Knowing: The Development of Self, Voice, and Mind.* New York: Basic Books, 1997.

Fennema, Elizabeth, and Gilah C. Leder, eds. *Mathematics and Gender.* New York: Teachers College Press, 1990.

Friedman, Lynn. "Mathematics and the Gender Gap: A Meta-analysis of Recent Studies on Sex Differences in Mathematical Tasks." *Review of Educational Research* 59 (Summer 1989): 185–213.

———. "The Space Factor in Mathematics: Gender Differences." *Review of Educational Research* 65 (Spring 1995): 22–50.

Frost, Laurie A., Janet S. Hyde, and Elizabeth Fennema. "Gender, Mathematics Performance, and Mathematics-Related Attitudes and Affect: A Meta-analytic Synthesis." *International Journal of Educational Research* 21 (May 1994): 373–85.

Garner, Mary, and George Englehard. "Gender Differences in Performance on Multiple-Choice and Constructed Response Mathematics Items." *Applied Measurement in Education* 12 (1999): 29–51.

Gilligan, Carol. *In a Different Voice: Psychological Theory and Women's Development.* Cambridge, Mass.: Harvard University Press, 1982.

———. *Meeting at the Crossroads: Women's Psychology and Girls' Development.* Cambridge, Mass.: Harvard University Press, 1992.

Han, Lei, and H. D. Hoover. "Gender Differences in Achievement Test Scores." Paper presented at the annual meeting of the National Council on Measurement in Education, New Orleans, April 1994. (ERIC document ED369816)

Hyde, Janet, Elizabeth Fennema, Marilyn Ryan, Laurie Frost, and Carolyn Hopp. "Gender Comparisons of Mathematics Attitudes and Affect: A Meta-analysis." *Psychology of Women Quarterly* 14 (September 1990): 299–324.

Lane, Suzanne, Ning Wang, and Maria Magone. "Gender-related Differential Item Functioning on a Middle-School Mathematics Performance Assessment." *Educational Measurement: Issues and Practice* 15 (Winter 1996): 21–27, 31.

Leder, Gilah C. "Gender Differences in Mathematics: An Overview." In *Mathematics and Gender,* edited by Elizabeth Fennema and Gilah C. Leder, pp. 10–26. New York: Teachers College Press, 1990.

———. "Mathematics and Gender: Changing Perspectives." In *Handbook of Research on Mathematics Teaching and Learning,* edited by Douglas A. Grouws, pp. 597–622. New York: Macmillan, 1992.

Martin, David J., and H. D. Hoover. "Sex Differences in Educational Achievement: A Longitudinal Study." *Journal of Early Adolescence* 7 (Spring 1987): 65–83.

McLeod, Douglas B. "Research on Affect in Mathematics Education: A Reconceptualization." In *Handbook of Research on Mathematics Teaching and Learning,* edited by Douglas A. Grouws, pp. 575–96. New York: Macmillan, 1992.

Meyer, Margaret R., and Mary Schatz Koehler. "Internal Influences on Gender Differences in Mathematics." In *Mathematics and Gender,* edited by Elizabeth Fennema and Gilah C. Leder, pp. 60–95. New York: Teachers College Press, 1990.

National Council of Teachers of Mathematics. *Assessment Standards for School Mathematics.* Reston, Va.: National Council of Teachers of Mathematics, 1995.

———. *Curriculum and Evaluation Standards for School Mathematics.* Reston, Va.: National Council of Teachers of Mathematics, 1989.

———. *Professional Standards for Teaching Mathematics.* Reston, Va.: National Council of Teachers of Mathematics, 1991.

National Science Foundation. *Women, Minorities, and Persons with Disabilities in Science and Engineering.* NSF 99-338. Arlington, Va.: National Science Foundation, 1999.

Oakes, Jeannie. *Keeping Track: How Schools Structure Inequality.* New Haven, Conn.: Yale University Press, 1985.

Page, Reba. *Lower-Track Classrooms: A Curricular and Cultural Perspective.* New York: Teachers College Press, 1991.

Reese, Clyde, Karen E. Miller, John Mazzeo, and John A. Dossey. *NAEP 1996 Mathematics Report Card for the Nation and the States.* Washington, D.C.: National Center for Education Statistics, 1997.

Rogers, Pat, and Gabriele Kaiser, eds. *Equity in Mathematics Education: Influences of Feminism and Culture.* Washington, D.C.: Falmer Press, 1995.

Sadker, Myra, and David Sadker. *Failing at Fairness: How America's Schools Cheat Girls.* New York: Maxwell Macmillan International, 1994.

Secada, Walter G., ed. *Equity in Education.* New York: Falmer Press, 1989.

Seegers, Gerard, and Monique Boekaerts. "Gender-Related Differences in Self-Referenced Cognitions in Relation to Mathematics." *Journal for Research in Mathematics Education* 27 (March 1996): 215–40.

Sells, Lucy. "The Mathematics Filter and the Education of Women and Minorities." In *Women and the Mathematical Mystique,* edited by Lynn H. Fox, Linda Brody, and Dianne Tobin, pp. 66–75. Baltimore: Johns Hopkins University Press, 1980.

Tartre, Lindsay. "Spatial Skills, Gender, and Mathematics." In *Mathematics and Gender,* edited by Elizabeth Fennema and Gilah C. Leder, pp. 27–59. New York: Teachers College Press, 1990.

Tartre, Lindsay, and Elizabeth Fennema. "Mathematics Achievement and Gender: A Longitudinal Study of Selected Cognitive and Affective Variables Grades 6–12." *Educational Studies in Mathematics* 28 (April 1995): 199–217.

5

NAEP Findings on the Preparation and Practices of Mathematics Teachers

Douglas A. Grouws and Margaret Schwan Smith

ONE of the most important influences on what students learn is the teacher. For it is what a teacher knows and can do that influences how she or he organizes and conducts lessons, and it is the nature of these lessons that ultimately determines *what* mathematics students learn and *how* they learn it. In this chapter we present a portrait of our nation's teachers and identify, to the extent possible, teacher characteristics and instructional practices that appear to be connected to differential levels of students' performance.

This portrait is based on the responses to a two-part questionnaire in which teachers were first asked to provide information related to their background and training and then to their instructional practices. Questionnaires were administered to teachers at grades 4 and 8 who had one or more students participating in the 1996 NAEP. The sample included 964 teachers at grade 4, 87 percent of whom completed the questionnaire, and 1825 teachers at grade 8, 84 percent of whom completed the questionnaire. The teacher questionnaire data are reported in terms of the percent of students who have teachers who fall into a particular category. This wording is necessary, since NAEP data are based on a representative sample of *students*, not teachers.

In discussing the analysis of these data, we make comparisons when possible to the results of the 1992 NAEP questionnaire analysis and to other studies of teachers' backgrounds and practices in order to corroborate the results reported herein, to highlight differences over time, and to better understand issues raised in the current analysis. It is important to note, however, that since changes were made in the 1996 NAEP questionnaire, it will not always be possible to make comparisons. In particular, additional questions were added to the questionnaire, answer choices from which teachers could select a response option were modified, and in some cases, different types of analysis were available from Educational Testing Service (ETS), the NAEP contractor. We will also draw on data collected from the Third International Mathematics and Science Study (TIMSS) (Beaton et al. 1996), the TIMSS video study at grade 8 (Stigler and Hiebert 1997), and the results of the 1993 National Survey of Science and Mathematics Education (NSSME) (Weiss 1995; National Science Foundation [NSF] 1996).

The chapter is organized into two sections, which correspond to the two parts of the questionnaire. In the first section—Who Are Our Nation's Teachers?—we present information on the background and training of teachers. In the second section—What Happens in Mathematics Classrooms?—we present findings regarding the instructional practices used by teachers in their classroom.

WHO ARE OUR NATION'S TEACHERS?

In this section we examine a set of background characteristics that include teaching experience, race/ethnicity and gender, type of certification, and type of degree and majors. In addition, we discuss the mathematical and pedagogical preparation of teachers.

Years Teaching

As shown in table 5.1, about one-fourth of students at grades 4 and 8 have teachers with 5 years or less of experience in teaching mathematics. In addition, about 25 percent of the students at grade 4 and 20 percent of the students at grade 8 have teachers with between 6 and 10 years of mathematics teaching experience. Hence, about 50 percent of fourth-grade students and about 45 percent of eighth-grade students have teachers with 10 years or less of experience in teaching mathematics.

The data in table 5.1 also show that mathematics teaching experience is related to students' performance. Specifically, students at grade 4 who had teachers with 25 years or more of experience and students who had teachers with between 6 and 10 years of experience had significantly higher NAEP average scale scores than students who had teachers with 5 years or less of teaching experience. At grade 8 students' performance is also linked to teaching experience. Students with teachers who have 11 to 24 years of experience in teaching mathematics have significantly higher average scale scores than students whose teachers have 5 years or less of experience in teaching mathematics. Although these data suggest a relationship between

THE PREPARATION AND PRACTICES OF MATHEMATICS TEACHERS

Highlights

- The majority of students at both grades 4 and 8 are taught by white females. In addition, about 50 percent of students at both grade levels have teachers with more than 10 years of teaching experience.

- Teachers spend very little time participating in ongoing professional development activities related to mathematics teaching (or teaching more generally), such as district or school in-service programs and college or university courses.

- Teachers' knowledge of the *Curriculum and Evaluation Standards for School Mathematics* (NCTM 1989) appears to be related to students' performance at grade 8. Students at grade 4 are more than twice as likely as students at grade 8 to have teachers who report having little or no knowledge of the *Standards*.

- Teachers at grade 8 are more likely than teachers at grade 4 to be certified to teach mathematics or to have majored in mathematics or mathematics education. More than four-fifths of students at grade 8 have teachers who are certified to teach mathematics and nearly one-half of the students have teachers who have majored in mathematics. In addition, certification in mathematics and majoring in mathematics are linked to students' performance at grade 8.

- The amount of time allocated to mathematics instruction decreases considerably from grades 4 to 8, with one-fifth of the students at grade 8 receiving 30 minutes or less mathematics instruction a day.

- Instruction continues to be textbook-based although the daily use of textbooks has decreased significantly from 1992. Instructional practices that emphasize oral and written communication of mathematical ideas are not daily features of most classrooms, and there is a continued emphasis on skills and procedures. Calculators are used more frequently at grade 8 than 4, whereas computers are used more frequently at grade 4 than 8. Nearly 70 percent of students at grade 8 never use computers.

- Assessment practices are dominated by the use of problem sets. Innovative assessments are used with less frequency overall and are more likely to be used by teachers at grade 4 than at grade 8.

Table 5.1
Teachers' Reports on Number of Years of Mathematics Teaching Experience

	Grade 4		Grade 8	
	Percent of Students	Average Scale Score	Percent of Students	Average Scale Score
5 Years or Less	26	219	27	269
6 to 10 Years	26	227*	20	272
11 to 24 Years	33	224	37	276*
25 Years or More	15	229*	17	277

* Indicates value is significantly higher than the value for 5 Years or Less.
Note: Column totals for percent of students may not add to 100 because of rounding.

teaching experience and students' performance, as Lindquist (1997) pointed out in her analysis of the 1992 NAEP data, more experienced teachers are often assigned classes that contain the most well prepared students and this may in part account for the observed performance differences.

Race/Ethnicity and Gender

As shown in table 5.2, almost 90 percent of all fourth-grade students and 60 percent of all eighth-grade students, regardless of race or gender, are taught by female teachers. In addition, the majority of all students at both grades 4 and 8, regardless of race, are taught by White teachers. Only about 30 percent of all Black students at each grade level are taught by Black teachers, and about 10 percent of Hispanic fourth-grade students and 5 percent of Hispanic eighth-grade students are taught by Hispanic teachers. Data from NAEP indicate that there is a slight, but not significant, increase from 1992 to 1996 in the percentage of Black students with Black teachers and a slight, but not significant, decrease from 1992 to 1996 in the percentage of Hispanic students with Hispanic teachers.

Type of Certification

The type of certification teachers hold is usually related to the college degree they have earned and the courses they have taken but can also be affected by state policies (Hawkins, Stancavage, and Dossey 1998). For example, the availability of certification in some areas (for example, elementary school mathematics) varies from state to state, as does the type of certification required for teaching mathematics at specific grade levels.

Although teachers may hold certification in more than one area (for example, elementary mathematics and elementary education), the data reported in table 5.3 represents three mutually exclusive categories of certification:

Table 5.2
Teachers' Reports on Their Gender and Race/Ethnicity

	Teachers' Gender		Teachers' Race/Ethnicity		
	Male	Female	White	Black	Hispanic
Grade 4	Percent of Students		Percent of Students		
White	12	88	95	2	1
Black	9	91	65	31	1
Hispanic	12	88	79	8	9
Male	12	88	88	7	2
Female	11	89	88	8	2
Grade 8	Percent of Students		Percent of Students		
White	40	60	96	2	1
Black	36	64	63	34	1
Hispanic	38	62	76	11	5
Male	40	60	89	7	1
Female	39	61	88	8	1

mathematics, which includes certification in elementary, middle/junior high, or secondary mathematics; *education,* which includes certification in elementary or middle/junior high education (but excluding certification in mathematics education); and *other,* which includes certification in fields other than those included in mathematics and education (Hawkins, Stancavage, and Dossey 1998).

Table 5.3
Teachers' Report on the Type of Teaching Certification They Hold

	Mathematics		Education but not Mathematics		Other	
	Percent of Students	Average Scale Score	Percent of Students	Average Scale Score	Percent of Students	Average Scale Score
Grade 4	32	225	67	224	1	223
Grade 8	81	276*	18	265	2	248

*Indicates value is significantly higher than values for Education but not Mathematics and Other.
Source: Hawkins, Stancavage, and Dossey (1998)

As shown in table 5.3, the majority of students at grade 8 (approximately 80 percent) have teachers who are certified in mathematics education at some level, compared to about 30 percent of students at grade 4 who have teachers with mathematics certification. By contrast, the majority of students at grade 4 (nearly 70 percent) have teachers who are certified in education but not mathematics. This difference in certification between teachers at grade 4 and 8 is not surprising; many states require mathematics certification to teach at grade 8, and in most states, elementary certification is sufficient for teaching mathematics at grade 4.

Although there is no apparent relationship between the type of certification and students' performance at grade 4, this is not the case at grade 8. Specifically, students at grade 8 who were taught by teachers who were certified in mathematics had significantly higher NAEP average scale scores (276) than students whose teachers were certified in education (265) or in some other field (248).

Type of Degree and Majors

Degrees provide some information on teachers' education. According to the 1996 NAEP results, about 60 percent of students in grades 4 and 8 have teachers who hold bachelor's degrees, and 40 percent of the students at these grades have teachers with either master's or specialist's degrees. Very few students have teachers with doctorates.

Teachers' majors at the undergraduate and graduate level provide additional information regarding their preparation for teaching mathematics. Although teachers may have dual majors, or different majors at the undergraduate and graduate levels, teachers' majors were sorted into four mutually exclusive categories: *mathematics,* which includes teachers with an undergraduate or graduate major in mathematics; *mathematics education,* which includes teachers with an undergraduate or graduate major in mathematics education but not in mathematics; *education,* which includes teachers with an undergraduate or graduate major in education at any level, but not mathematics education; and *other,* which includes teachers who had majors in areas other than those described by previous categories (Hawkins, Stancavage, and Dossey 1998).

As shown in table 5.4, 83 percent of students at grade 4 were taught by teachers who majored in elementary education, whereas only 32 percent of the students at grade 8 had teachers with this major. At grade 8, 49 percent of the students had teachers who majored in mathematics and 13 percent had teachers who majored in mathematics education. In comparison, only 9 percent of the students at grade 4 had teachers with majors in mathematics and 4 percent had teachers with majors in mathematics education. This suggests that teachers of eighth-grade students had a stronger preparation in mathematics than the teachers of fourth-grade students.

Preparation in mathematics appears to be related to students' performance at grade 8 but not at grade 4. Students at grade 8 who were taught by teachers who majored in mathematics had significantly higher NAEP average achievement scores (278) than students who were taught by teachers with majors in education or some other field, with average scale scores of 269 and 267, respectively. Some other studies have also demonstrated the

Table 5.4
Teachers' Reports on Their Undergraduate or Graduate Major

	Grade 4	Grade 8
Mathematics		
Percent of Students	9	49
Average Scale Score	220	278*+
Mathematics Education but not Mathematics		
Percent of Students	4	13
Average Scale Score	235*	270
Education but not Mathematics or Mathematics Education		
Percent of Students	83	32
Average Scale Score	225	269
Other		
Percent of Students	4	7
Average Scale Score	208	267

* Indicates value is significantly higher than the value for Other.
*+ Indicates value is significantly higher than the values for Education but not Mathematics or Mathematics Education and Other.
Source: Hawkins, Stancavage, and Dossey (1998).

link between teachers' knowledge and students' performance. For example, Armour-Thomas et al. (1989), in a study comparing high-achieving and low-achieving elementary schools with similar student characteristics in New York City, found that differences in teachers' qualifications accounted for more than 90 percent of the variance in students' achievement in mathematics and reading. These data together suggest that mathematics preparation may be particularly important in the upper grades, where the content is more complex than what is encountered in primary grades.

It is also interesting to note that the percent of students at grade 8 with teachers who majored in mathematics or mathematics education (62 percent) is considerably less than the percent of students whose teachers held certification in mathematics (81 percent; see table 5.3). This is most likely due to the fact that in many states certification in mathematics does not require majoring in mathematics. At the national level, 28 percent of public high school mathematics teachers (grades 9–12) do not have at least a minor in mathematics (National Commission on Teaching and America's Future [NCTAF] 1997). Changes have been enacted in some states, such as Maine (NCTAF 1997), that require teacher certification and a disciplinary major in order to meet licensing requirements. It will be interesting to watch the impact of these changes on teacher preparation and ultimately on students' performance in subsequent NAEP assessments.

Mathematical Background

The *Curriculum and Evaluation Standards for School Mathematics* (NCTM 1989) places an increased emphasis in the elementary and middle grades on mathematics topics such as geometry and measurement, probability and statistics, and algebra. The National Council of Teachers of Mathematics (1991), the National Board for Professional Teaching Standards (1997), and many research studies (for example, Simon and Schifter 1991) have argued that teachers need an opportunity to learn the mathematics that they will be expected to teach.

Table 5.5 shows the percent of students in grades 4 and 8 whose teachers studied particular mathematics content during their preservice preparation or as part of their in-service professional development. Approximately 90 percent of students at each grade level had teachers who had some exposure (that is, the complement of those who had little or no exposure) to number systems and measurement either during preservice or in-service training, although a considerably smaller percent of students at grades 4 and 8 had teachers who had actually taken courses in these content areas. It is most likely that these topics were covered as part of a methods or content course specifically intended for elementary or middle school teachers. Weiss (1995) reported that number sense/numeration and measurement were two of three topics that mathematics teachers in grades 1–4 indicated that they felt well qualified to teach. This may be due to the fact that the majority of teachers had exposure to these topics in their preservice and in-service preparation.

Although more than four-fifths of students in grade 4 had teachers who had some exposure to geometry and three-fourths had teachers with some exposure to probability and statistics, almost twice as many eighth-grade students as fourth-grade students had teachers who had taken courses in these content areas. A slightly higher percent of fourth-grade students had teachers who had courses in college algebra, although this percent was still less than that for eighth-grade students. This may reflect the fact that college algebra is a course that is offered at most colleges and universities.

Abstract algebra, linear algebra, and calculus are courses that are often taken as part of a mathematics or mathematics education major. As a result, it is not surprising that the percent of students at grade 4 who had teachers with exposure to these content areas was considerably less than both the percent of students at grade 4 who had teachers who had taken other mathematics courses and the percent of students at grade 8 who had teachers who had exposure to these topics. This is consistent with the fact that there are more eighth-grade students than fourth-grade students who have teachers who majored in mathematics or mathematics education at the undergraduate and graduate levels and are certified to teach mathematics at some level.

Table 5.5
Teachers' Reports on Their Exposure to Mathematics Content Areas

	One or More College Courses	Part of a College Course	In-Service Training	Little/No Exposure
		Percent of Students		
Number Systems and Numeration				
Grade 4	43	33	33	11
Grade 8	50	35	20	7
Measurement in Mathematics				
Grade 4	37	33	39	13
Grade 8	37	38	26	11
Geometry				
Grade 4	34	31	34	16
Grade 8	64	21	22	8
Probability/Statistics				
Grade 4	36	25	22	26
Grade 8	68	20	17	7
College Algebra				
Grade 4	45	12	3	35
Grade 8	74	13	7	9
Calculus				
Grade 4	13	8	1	69
Grade 8	68	10	2	19
Abstract/Linear Algebra				
Grade 4	13	11	3	66
Grade 8	68	10	2	19

Note: Row percents may not add to 100 because the categories are not mutually exclusive. For example, one could select both "Part of a College Course" and "In-Service Training."

The discrepancy between the percent of students at grade 8 with teachers who are certified to teach mathematics (81 percent) and the percent of students at grade 8 with teachers who have completed courses listed in table 5.5 (between 37 and 74 percent) raises questions about the number and content of courses required for mathematics certification.

Pedagogical Background

The *Professional Standards for Teaching Mathematics* (NCTM 1991) identifies "knowing mathematical pedagogy" as one of the essential components for the professional development of teachers of mathematics. This standard states that the preservice and continuing education of mathematics teachers should develop teachers' knowledge of and ability to use and evaluate instructional materials and resources, ways of representing mathematical concepts and procedures, methods for facilitating discourse, and means for assessing students' learning. The NAEP questionnaire provides some information on teachers' pedagogical learning opportunities in initial preparation and in ongoing professional development experiences.

Participation in Mathematics-Related Activities

The first formal opportunity to learn about mathematics pedagogy often occurs in a mathematics methods course during preservice preparation. In the NAEP sample, 84 percent of the students at grade 4 and 44 percent of the students at grade 8 had teachers who had one or more college courses in elementary mathematics methods. Virtually all teachers had some exposure to elementary mathematics methods either in a course, as part of a course, or through in-service training. The low percent of students at grade 8 who had teachers with one or more courses in elementary mathematics methods may reflect the fact that many eighth-grade teachers have certification in middle or secondary mathematics and may have taken methods courses that focus specifically on middle or high school rather than at the elementary school level. This conjecture is supported by the fact that in the 1992 NAEP, 58 percent of the students at grade 8 had teachers who reported having had a course in middle school mathematics methods (Lindquist 1997). Unfortunately, this option concerning coursework in middle school mathematics methods was not included on the 1996 NAEP questionnaire.

Learning opportunities for practicing teachers are also available through ongoing professional experiences. As shown in table 5.6, the majority of students at grades 4 and 8 (more than 80 and 90 percent, respectively) have teachers who are receiving some ongoing support. Students at grade 4, however, were twice as likely as students at grade 8 to have a teacher with no such recent experience. In addition, nearly three-fourths of the students at grade 4 and a little more than half of the students at grade 8 have teachers who have had less than 16 hours of professional development in the last year. Data from the NSSME provide an even more discouraging view of teacher professional development opportunities, with 60 percent of the teachers in grades 5–8 and 70 percent of the teachers in grades 1–4 reporting that they had 15 hours or less of in-service education in the last three years (NSF 1996). Taken together, these data lead one to question whether the amount of time spent in professional development is adequate to support teachers in adopting the types of practices being advocated by reformers, in learning and becoming proficient in using new technologies, and in learning mathematics in ways that enhance their teaching of a range of topics.

Table 5.6
Teachers' Reports on the Amount of Time Spent in Professional Development Workshops Related to Mathematics or Mathematics Education in the Last Year

	None	Less than 6 Hours	6 to 15 Hours	16 to 35 Hours	More than 35 Hours
			Percent of Students		
Grade 4	16	30	27	15	12
Grade 8	6	20	29	20	25

In addition to opportunities presented by in-service training, college courses provide another way for teachers to continue to learn mathematics content and pedagogy. Approximately one-fourth of the students at each grade level had teachers who had taken one or more college courses in mathematics or mathematics education during the last two years. Although few teachers appear to be pursuing this more traditional professional development option, this may be due to the fact that course taking is often associated with obtaining permanent certification in some states or with completing a master's degree program. Given that approximately 50 percent of students have teachers who have been teaching more than 10 years, and 40 percent of students have teachers with master's or specialist's degrees, teachers may have no additional motivation for pursuing additional coursework.

Teachers were also asked about their opportunities to learn about a number of specific topics related to mathematics that have received increased attention in recent years. As shown in table 5.7, at least 70 percent of the students at grades 4 and 8 have teachers who have taken college or in-service courses on the topics of estimation, problem solving, the use of manipulatives, the use of calculators, and understanding students' thinking about mathematics.

Table 5.7
Teachers' Reports on College or In-Service Course on Specific Topics in Mathematics and in Special Areas

	Grade 4	Grade 8
	Percent of Students	
Estimation	76	75
Problem solving in mathematics	90	93
Use of manipulatives (e.g., counting blocks or geometric shapes) in mathematics instruction	90	88
Use of calculators in mathematics instruction	71†	78
Understanding students' thinking about mathematics	70	71
Gender issues in the teaching of mathematics	45†	47
Teaching students from different cultural backgrounds	49	51

† Indicates significant difference from 1992.

As shown in table 5.7, 71 percent of fourth-grade students had teachers who have taken a college or in-service course on the use of calculators. This represents a significant increase from 1992, when only 59 percent of fourth-grade students had teachers who had some training in the use of calculators. The growing emphasis on, and availability of, this technology may have had a positive influence on teachers' opportunities to learn how to use this instructional tool.

The *Professional Standards for Teaching Mathematics* (NCTM 1991) makes salient the need to reach all children, including students who are members of racial/ethnic minorities and students who are female. The data in table 5.7 show that approximately 50 percent of the students at grades 4 and 8 have teachers who have participated in a college or in-service course on gender issues or on teaching students from different cultural backgrounds. At grade 4, the percent of students whose teachers participated in college or in-service courses on gender issues is significantly higher than the 32 percent reported in 1992. This suggests an increased awareness of the need, particularly at the elementary school level, of providing opportunities for teachers to consider the importance of equitable treatment of female students in the classroom (American Association of University Women [AAUW] 1989).

Further examination of the data for different population subgroups show significant differences based on race/ethnicity in the percent of students whose teachers participated in a course or workshop that focused on teaching students from different cultural backgrounds. Almost 70 percent of Hispanic students in grade 8 had teachers who had taken a course in teaching students from different cultural backgrounds compared to about 45 percent of White students at grade 8 who had teachers with similar training. At grade 4, 60 percent of Hispanic students and 56 percent of Black students had teachers who had taken a course in teaching students from different cultural backgrounds, again compared to about 45 percent of White students who had teachers with similar training. This suggests that teachers of non-White students are receiving increased preparation for teaching in culturally diverse settings. This is encouraging because the majority of non-White students do not have teachers who are members of their racial/ethnic groups and because an increased understanding by teachers of the lives, interests, and cultures of students may have a positive effect on students' mathematics learning (Silver, Smith, and Nelson 1995).

Data from the NSSME (Weiss 1995) suggest that the majority of teachers feel well prepared to encourage all students—irrespective of gender or race/ethnicity—to participate in classroom instruction. Specifically, 95 percent of teachers in grades 1–8 reported feeling well prepared to encourage the participation of females in mathematics, 84 percent report feeling well prepared to encourage the participation of minorities in mathematics, and about 70 percent of the teachers in grades 1–8 consider themselves qualified to teach students from a variety of cultural backgrounds. Given that about one-half of the students at grades 4 and 8 have teachers who have taken courses or participated in in-service programs specifically addressing these issues, it is not clear what types of experiences contributed to feelings of preparedness or how feeling prepared influences teachers' actions and interactions in the classroom.

Participation in a Broader Set of Activities

Teachers responding to the NAEP questionnaire in 1996 were also asked about their participation in a wide range of professional development activities in the past five years, even if they did not deal exclusively with mathematics. As shown in table 5.8, students at both grades 4 and 8 had teachers who participated in many activities. Most notably, about three-fourths of the students at both grade levels had teachers who had participated in professional development related to technology and cooperative group instruction, and approximately 60 percent of students at grade 4 and 50 percent of the students at grade 8 had teachers who participated in activities related to teaching higher-order thinking skills. In addition, about 50 percent of students at grade 4 and 40 percent of students at grade 8 had teachers who participated in activities related to alternative assessment practices such as portfolios and performance-based assessments. Although these activities may not have focused on mathematics per se, these data indicate that a large percent of teachers at both grade levels had the opportunity to learn about pedagogical techniques that are receiving increased emphasis in mathematics instruction. However, the range of sessions in which teachers participated, coupled with the relatively limited amount of time in which teachers participated in professional development activities overall, raises questions about the depth of coverage in any one topic area and the lack of cohesion in the professional development experiences of teachers.

There were two places on the questionnaire where teachers were asked to indicate their participation in courses and other professional development activities related to teaching students from different cultural backgrounds. Table 5.7 indicates participation in such opportunities at any point in one's career, whereas table 5.8 indicates participation in the past five years. Therefore, the lower percents reported in table 5.8 suggest that although teachers participated in activities related to this topic, they had not done so recently (that is, within the past five years).

The Impact of the NCTM *Standards*

In 1996, for the first time, the NAEP questionnaire asked teachers about their knowledge of the NCTM *Curriculum and Evaluation Standards* (NCTM

Table 5.8
Teachers' Reports on Their Participation in Courses and Professional Development Activities during the Past Five Years

	Grade 4	Grade 8
	Percent of Students	
Use of telecommunications	23	21
Use of technology such as computers	81	76
Cooperative group instruction	76	71
Interdisciplinary instruction	46	50
Assessment by portfolios	56	39
Performance-based assessment	49	38
Teaching higher-order thinking skills	57	47
Teaching students from different cultural backgrounds	36	30
Teaching Limited English Proficient students	19	13
Teaching students with special needs (e.g., visually impaired, gifted and talented)	40	26
Classroom management and organization	55	43
Other professional issues	56	52
None of the above	2	2

1989) and about their opportunities to participate in professional development that was directly related to the *Standards*. As shown in table 5.9, only about one-fourth of the students at grade 4 and one-half of students at grade 8 had teachers who reported that they were either very knowledgeable or knowledgeable about the *Standards*. The teachers of fourth-grade students were less well informed about the *Standards*, with more than twice the number of teachers at grade 4 than at grade 8 reporting that they had little or no knowledge of the *Standards*.

The data in table 5.9 also suggest that teachers' knowledge of the *Standards* may have had some impact on students' performance at grade 8. Specifically, the students with teachers who were very knowledgeable about the *Standards* had significantly higher NAEP average scale scores than students who had teachers who were somewhat knowledgeable or had little or no knowledge of the *Standards*, and students with teachers who were knowledgeable about the *Standards* had significantly higher NAEP average scale scores than students whose teachers had little or no knowledge of the *Standards*.

Other studies have also reported on teachers' familiarity with the *Standards*. Weiss (1995) reported that relatively low percents of teachers (10 percent in grades 1–4 and 28 percent in grades 5–8) indicated that they were "well aware" of the *Standards*. These data seem to support NAEP findings that the majority of teachers are not very familiar with, or knowledgeable about, the *Standards*. Although knowledge alone does not indicate that

Table 5.9
Teachers' Reports on Extent of Knowledge about the National Council of Teachers of Mathematics Curriculum and Evaluation Standards

	Grade 4	Grade 8
Very Knowledgeable		
Percent of Students	5	16
Average Scale Score	236	282*+
Knowledgeable		
Percent of Students	17	32
Average Scale Score	223	276*
Somewhat Knowledgeable		
Percent of Students	32	33
Average Scale Score	224	270
Little or No Knowledge		
Percent of Students	46	19
Average Scale Score	223	267

* Indicates value is significantly higher than the value for Little or No Knowledge.
*+ Indicates value is significantly higher than the value for Somewhat Knowledgeable and Little or No Knowledge.
Source: Hawkins, Stancavage, and Dossey (1998).

teachers embrace the ideas on which the *Standards* are based, the relationship between students' performance and teachers' knowledge of the *Standards* raises interesting questions regarding the ways in which the practices of very knowledgeable teachers differ from those of teachers who report being somewhat knowledgeable or having little or no knowledge of the *Standards*.

Data from the TIMSS videotape analysis provides a somewhat different view of teachers' knowledge of the *Standards*. Stigler and Hiebert (1997) indicate that the majority of teachers at grade 8 reported having read the *Standards,* and 74 percent of the teachers claimed to be implementing ideas from the *Standards* in the lessons that were included in the videotape study. They go on to state that the analyses of these lessons, however, revealed an emphasis on the acquisition and application of skills without attention to underlying meaning or nontrivial application, thus calling into question the extent to which teachers had made the types of real changes in their practice being advocated by the *Standards*. Although TIMSS and NAEP data appear to provide conflicting accounts of teachers' knowledge of the *Standards*, the questions to which teachers responded may account for these differences. It may not be appropriate to interpret "Reading the *Standards*" and "Implementing Ideas from the *Standards*" as evidence that the teachers consider themselves to be very knowledgeable but, rather, as indicating that they have some knowledge and think they are acting on it.

Teachers were also asked about their opportunities to participate in professional development activities that provided them with strategies for implementing the *Curriculum and Evaluation Standards.* More than one-half of the students at grade 4 and about one-third of the students at grade 8 had teachers who responded that they did not participate in *any* professional development activities with this aim. The majority of students whose teachers have participated in *Standards*-related professional development learned about the *Standards* through attendance at local workshops.

When taken together, these data regarding the NCTM *Standards* point to the fact that teachers of fourth-grade students are both less well informed about the *Standards* and have fewer opportunities to learn about the *Standards* than teachers of eighth-grade students. It is worth noting that these data may reflect teachers' lack of familiarity with the *Standards* rather than with the ideas expressed therein. Teachers may have participated in sessions that were based on or consistent with the *Standards* but that did not make explicit connections or references to the document.

Preparedness for Teaching Mathematics

Although it is clear that teachers have participated in a number of courses and in-service programs on a range of topics considered to be important in the teaching and learning of mathematics, NAEP data provide no information regarding the quality of the experiences in which teachers engage, what they learn , and what impact these experiences have on classroom instruction. Additional questions, however, provide some information on the extent to which teachers feel prepared to teach concepts, procedures, and technology, and their perception of the importance of specific pedagogical techniques.

As shown in table 5.10, more than three-fourths of the students at grade 4 and nine-tenths of the students at grade 8 have teachers who report being "very well prepared" to teach both mathematical concepts and procedures. Although a smaller percent of students have teachers who report being very well prepared to teach calculators, about 90 percent of students at both grade levels have teachers who report being at least moderately well prepared. The students at both grade levels, however, have teachers who feel considerably less well prepared to teach computers, with fewer than three-fifths of students at each grade level having teachers who report being at least moderately well prepared to teach this technology.

These data also show a connection between a teacher's sense of preparedness for teaching and students' performance at grade 8. Students with teachers who indicated that they were not very well prepared to teach mathematical concepts and mathematical procedures had significantly lower NAEP average scale scores (238) than students whose teachers indicated

Table 5.10
Teachers' Reports on the Degree of Preparation for Teaching Mathematics Concepts and Procedures and for Using Technology in the Classroom.

	Very Well Prepared	Moderately Well Prepared	Not Very Well Prepared	Not at All Prepared
	Percent of Students			
Mathematical Concepts				
Grade 4	76	22	2	0
Grade 8	89	10	1	0
Mathematical Procedures				
Grade 4	78	21	1	0
Grade 8	91	8	1	0
Computers				
Grade 4	18	40	34	8
Grade 8	24	31	35	11
Calculators				
Grade 4	39	47	12	1
Grade 8	58	36	5	0

Note: Row percents may not add to 100 due to rounding.

that they were either very or moderately well prepared (270–275). This suggests that teachers who do not feel prepared to teach mathematics may have good reason to question their competence. This could result from the fact that some teachers of eighth-grade mathematics have limited preparation in mathematics and may not have elected to teach in this content area.

In 1996, for the first time, the NAEP questionnaire asked teachers to indicate the importance to their mathematics instruction of four pedagogical techniques: involving students in constructing and applying mathematical ideas, using problem solving both as a goal of instruction and as a means of investigating important mathematical concepts, using questions that promoted students' interaction and discussion, and using the results of classroom assessment to guide instructional decisions. Almost all students at both grades 4 and 8 had teachers who reported that the techniques were either somewhat or very important to their classroom instruction. These data suggest that teachers are favoring a form of instruction that is less didactic than traditional instruction and aligned with calls for reform (NCTM 1989, 1991, 1995). A question that naturally arises from these data is, "Do teachers incorporate their instructional beliefs into their classroom practices?" In the next section we examine the extent to which students had the opportunity to participate in classrooms where these practices were being enacted.

WHAT HAPPENS IN MATHEMATICS CLASSROOMS?

The mathematics learned by students is determined in large part by the interactions that take place within the classroom between teacher and students, among students themselves, and by the conditions affecting these interactions. In this section, we examine some of these conditions and interactions using teachers' responses to the NAEP questionnaire. We begin with the conditions under which teaching and learning take place and then focus on instructional activities and assessment practices.

Mathematics Instructional Time

The amount of class time allocated to mathematics is, to some extent, a measure of how much the subject is valued. Some research has shown that class time is positively related to how much students learn (Suarez et al. 1991). The strength of the relationship between class time and mathematics learning may be masked in some studies because extensive class time is often devoted to remedial activities. This could be true, for example, in studies that involve low-achieving schools. Dossey et al. (1994) suggest this explanation when discussing time allocation findings from the 1992 NAEP mathematics assessment.

As shown by the data in table 5.11, two-thirds of fourth-grade students had teachers who reported providing four or more hours of mathematics instruction for them each week, whereas only one-third of eighth-grade students had teachers who reported providing this much instruction.

This difference in instructional time between grade 4 and grade 8 seems problematic given the increased complexity of the mathematics curriculum that is taught at grade 8.

Table 5.11
Teachers' Reports on Amount of Time Spent on Mathematics Instruction Each Week

	Two and One-Half Hours or Less		More than Two and One-Half Hours but Less than Four Hours		Four Hours or More	
	Percent of Students	Average Scale Score	Percent of Students	Average Scale Score	Percent of Students	Average Scale Score
Grade 4						
1996	6	228	26	226	68	223†
1992	5	224	24	224	71	217
Grade 8						
1996	20	269	47	275	33	274
1992	13	270	55	270	32	268

† Indicates significant difference from 1992.

Also of concern is the data showing that 20 percent of students in grade 8 had teachers who allocated 2.5 hours or less for mathematics instruction a week; that is, 30 minutes or less for mathematics instruction each day. Data from the TIMSS (Beaton et al. 1996) also show that time allocated for mathematics for many students at grade 8 is quite low. In fact, 8 percent of the U.S. grade 8 teachers in TIMSS reported teaching mathematics to their students less than two hours a week. None of the other thirty-seven countries reporting TIMSS data had a higher percent of teachers reporting so little time spent on mathematics. This situation has not improved in recent years. In the 1992 NAEP assessment, for example, 13 percent of eighth-grade students had teachers who provided 30 minutes or less of mathematics instruction a day. Thus, there was a greater than 50 percent increase in the number of students who had teachers providing minimal instructional time for mathematics at grade 8 from 1992 to 1996. Mathematics class periods of 30 minutes and shorter may not reflect the importance mathematics should have in a student's education. Periods of this length may also not provide sufficient teaching opportunities to develop students' mathematical abilities sufficiently for them to have the mathematical power called for in the NCTM *Standards* (NCTM 1989). If school districts allocate more instructional time to alleviate the problem of very short mathematics lessons, it is essential that the additional time be accompanied by high-quality instruction.

Homework

Another important consideration in examining students' opportunity to learn mathematics is the amount of time students spend on homework assignments. Homework time is defined as the amount of time spent on mathematics assignments outside the regularly scheduled mathematics class time (Grouws, in press). As shown in table 5.12, 90 percent of students in grade 4 have teachers who expect them to spend approximately 15 minutes or 30 minutes daily on homework. Not surprisingly, students in grade 8 are expected to spend more time on homework than students in grade 4 are. More than two-thirds of students in grade 8 have teachers who expect them to spend 30 minutes or more on homework each day as compared to 46 percent of fourth-grade students.

The relationship between homework and students' performance is quite different at grade 8 from that at grade 4. At grade 8, students of teachers who assigned homework (15 minutes or more) performed significantly better on NAEP, with average scale scores ranging from 266 to 283, than students did whose teachers did not assign homework. In fact, at grade 8, increasing the amount of time on homework seems to be related to increases in students' performance, with a plateau reached at assignments that take approximately 45 minutes. At grade 4, however, there is a trend, although not statistically significant, toward time spent on homework and students' performance being inversely related. For example, students whose teachers assign 15 minutes of homework had a higher average scale score (226) than students did whose teachers assign 45 minutes of homework (214). Thus, when considering the value of homework at grade 4, the purposes for assigning homework other than increasing students' performance should be considered. These purposes include such things as conveying high academic expectations and developing good study habits. The increased value of homework at upper grade levels over lower grade levels was also found by Cooper (1989) in his meta-analysis of homework studies.

Table 5.12
Teachers' Reports on the Amount of Homework Assigned Daily

	Grade 4		Grade 8	
	Percent of Students	Average Scale Score	Percent of Students	Average Scale Score
None	4	232	2	241*
15 Minutes	50	226	30	266
30 Minutes	40	222	54	276†
45 Minutes	4	214	10	284
One Hour or More	2	207	4	283

† Indicates significant difference from 1992.
* Indicates average scale score is significantly different from average scale scores of students whose teachers assigned 15 or more minutes of homework daily.

Class Size

Class size and students' performance are issues of national interest, but the relationship between them remains unclear in spite of numerous investigations. NAEP data show no discernible pattern concerning the relationship. At grade 8, for example, 25 percent of students are in classes with 20 or fewer students, 28 percent are in classes with 21 to 25 students, 28 percent are in classes with 26 to 30 students, and 20 percent are in classes with 31 or more students. The corresponding performance levels of students in terms of average NAEP scale scores for these categories are: 273, 275, 279, and 268, respectively.

The relationship between class size and students' performance may, of course, be masked by other variables such as the disproportionate number of Title I students being taught in small classes. For instance, NAEP data in 1996 indicate that 44 percent of Title I students in grade 8 were in classes of 20 or fewer, but only 23 percent of non-Title I students in grade 8 were in classes of 20 or fewer. This contradicts the common presumption that small classes are not prevalent in schools whose students have low socioeconomic status. But the corresponding performance levels for the two groups were 246 and 279, respectively, suggesting that the potential benefits of small class size were overwhelmed by other factors.

Ability Grouping

Issues associated with ability grouping are not unique to mathematics education, but there is strong interest among many constituencies in the prevalence of this practice in mathematics. NAEP data shed some light on the questions of prevalence and change over time. At grade 4, 18 percent of students had teachers who reported that students were assigned to their class by ability; and at grade 8, 59 percent reported being assigned thus. Therefore, forming classes on the basis of ability is much more common at grade 8 than at grade 4. Average scale scores were significantly higher at both grade 4 and grade 8 when students were assigned to classes by ability. At grade 4, the average student scale score for classes formed by ability was 230 and for classes not formed on the basis of ability the average student scale score was 223. At grade 8, the average student scale score for classes formed by ability was 276, and for classes not formed on the basis of ability, the average student scale score was 270. There was a significant decrease in the forming of classes by ability from 1992 to 1996 at grade 4 (25 percent to 18 percent) and no significant change at grade 8 over this time period.

A related question of interest concerning ability grouping is the extent to which teachers create groups based on ability within mathematics classes. This practice was more common at grade 4 than at grade 8, with 39 percent of students at grade 4 having teachers who reported forming such groups within their classes as compared to 24 percent at grade 8. From 1992 to 1996 there was no significant change in this practice at grade 4 or at grade 8.

Availability of Resources

Teachers were asked about the availability of instructional materials and resources using the following response choices: all of what I need, most of what I need, some of what I need, and none of what I need. At grade 4, the percentage of students who had teachers who responded with "all" was 13 percent; "most," 53 percent; and "some" or "none," 34 percent. At grade 8, the comparable percentages were 21 percent, 58 percent, and 21 percent. Thus, the resources needed to teach mathematics seem to be more available at grade 4 than at grade 8. Also at grade 8, resources seem to be scarcer in Title I schools than in non-Title I schools. For example, 22 percent of non-Title I students had teachers who reported they had all the resources they needed, whereas only 10 percent of Title I students had teachers who reported having all the resources they needed. Teachers' reports of the adequacy of resources available in their classrooms were not associated with students' performance on the 1996 NAEP mathematics assessment.

Instructional Activities

Adequate instructional time is an important condition for students' high achievement, but how that time is used is equally important. In fact, the quality of mathematics instruction may be the single most important nondemographic factor in how much mathematics students learn. Currently, there is considerable support at the national level for changing the way mathematics is taught in schools. The NCTM *Standards,* for example, advocate increased involvement of students, more cooperative group work, a greater focus on communication, and more attention to such topics as algebra and probability and statistics. In the following paragraphs, we examine the NAEP data that pertain to teaching practice, beginning with the question of what mathematical topics are emphasized in instruction.

Attention to Mathematics Topics

One of the goals of reform in mathematics education has been to broaden the scope of the curriculum to include topics not historically or frequently taught in grades K–8. The NAEP provides information on the frequency with which mathematical topics are taught. As shown in table 5.13, Numbers and Operations is the topic addressed with greatest frequency at both grades 4 and 8, with approximately 90 percent of students having teachers who reported emphasizing this topic "a lot." All other content areas received considerably less attention, although the emphasis on a particular topic varies by grade level.

Almost three-fifths of students at grade 8 have teachers who report teaching algebra and functions "a lot." The NCTM *Curriculum and Evaluation Standards* (1989) has suggested increased emphasis on this topic at all grade levels, but this recommendation appears to have had a greater impact at grade 8 than at grade 4, since nearly 60 percent of students at grade 4 have teachers who report spending little or no time on the topic of algebra and functions. It is interesting to note that Weiss (1995, p. 6) reports that the majority of mathematics teachers at all levels (elementary, middle, high school) indicate that "students must master arithmetic computations before going on to algebra." Since such mastery is not likely to have occurred by fourth grade, this may explain the low emphasis on algebraic ideas at this grade level. Another factor that may influence this low level of attention is the lack of readily available algebraic activities in textbooks at grade 4, in contrast to textbooks at grade 8. Finally, the fact that more than 40 percent of teachers in

Table 5.13
Teachers' Reports on the Frequency with which They Address Specific Mathematical Topics

	None	A Little	Some	A Lot
		Percent of Students		
Numbers and Operations				
Grade 4	0	0	7	93
Grade 8	0	2	10	88
Measurement				
Grade 4	1	16	64	20
Grade 8	2	21	58	19
Geometry				
Grade 4	2	27	58	12
Grade 8	2	20	54	24
Data Analysis, Statistics, and Probability (informal introduction of concepts)				
Grade 4	9	41	42	8
Grade 8	7	30	47	15
Algebra and Functions (informal introduction of concepts)				
Grade 4	21	39	30	9
Grade 8	2	8	34	57

Note: Row percents may not add to 100 because of rounding.

grades 1–4 indicated they did not feel qualified to teach algebra (Weiss 1995) may also account for the relative lack of attention given to this topic.

Measurement, geometry, and data analysis, statistics, and probability are addressed with less frequency at both grades 4 and 8 than numbers and operations. Most students, however, are receiving some instruction in these content areas. Specifically, about 80 percent of students at grades 4 and 8 have teachers who report spending at least some time teaching measurement, and approximately 75 percent of students at these grades have teachers who report spending at least some time teaching geometry. Data analysis and related topics are addressed somewhat less frequently than other mathematical topics, with 50 percent of students at grade 4 and about 60 percent of students at grade 8 having teachers who report spending some time on these topics. Despite differences across content areas and grade levels, the NAEP data overall suggest that Numbers and Operations remains the central focus of instruction at grades 4 and 8, with students receiving some instructional exposure to other content areas.

Attention to Types of Learning

As part of the 1996 NAEP teachers were asked how often they addressed each of the following skills: learning mathematics facts and concepts, learning skills and procedures needed to solve routine problems, developing reasoning and analytic ability to solve unique problems, and learning how to communicate ideas in mathematics effectively. As shown in table 5.14, more than 90 percent of fourth-grade students and about 80 percent of eighth-grade students had teachers who reported that they addressed facts and concepts, as well as skills and procedures for routine problems, "a lot." In contrast, at both grade levels only 52 percent frequently addressed developing reasoning and analytic ability to solve unique problems. This teacher self-report data is consistent with the finding from the TIMSS videotape study that thinking and reasoning are seldom present in grade 8 mathematics lessons in U.S. classrooms (Stigler and Hiebert 1997). Thus there appears to be an implementation gap between recommendations to focus on developing students' higher-order thinking skills and what is currently happening in mathematics classrooms.

There is a trend (containing statistically significant differences) in the data in table 5.14, particularly at grade 8, toward higher proficiency scores for students (from 251 to 267 to 282) when teachers' attention to reasoning and analytic skills in unique problem situations increases. Interestingly, students at grade 8 whose teachers reported giving little or no attention to skills for routine problems, or to facts and concepts, had higher proficiency scores than those students whose teachers reported giving little or no attention to reasoning skills for unique problems or communicating mathematical ideas. Why teachers do not address reasoning and higher-order thinking skills more frequently is, therefore, an important question. Do the reasons involve teacher beliefs about the value of these skills? Probably not. At least NAEP data suggest that teachers almost uniformly rate current pedagogical recommendations as quite important. For example, over 75 percent of teachers at grades 4 and 8 rated as very important such recommended practices as having students construct and apply ideas, using problem solving as a means of investigating mathematical concepts, and using questioning techniques that promote student interaction and discussion. Do the reasons for not emphasizing higher-order thinking and reasoning then center on a lack of appropriate curriculum material, school district evaluation procedures, or what? There is currently insufficient data to resolve this issue, but teachers should be made aware that students of those teachers who report devoting "a lot" of attention to higher-order thinking and reasoning skills are performing well on national tests such as NAEP.

Students' Activities in Mathematics Class

There are many ways for students to be actively involved in learning mathematics. Teachers were asked about specific activities that have been proposed for increased use in national recommendations because they are believed to enhance students' conceptual understanding. Data from the

Table 5.14
Teachers' Reports on Attention to Skills

	A Little or None		Some		A Lot	
	Percent of Students	Average Scale Score	Percent of Students	Average Scale Score	Percent of Students	Average Scale Score
Learning mathematics facts and concepts						
Grade 4	<1	***	7	221	93	224
Grade 8	5	279	16	270	79	274
Learning skills and procedure needed to solve routine problems						
Grade 4	<1	***	8	221	91	225
Grade 8	3	282	18	273	79	273
Developing reasoning and analytical ability to solve unique problems						
Grade 4	8	224	41	221	52	227
Grade 8	8	251	40	267	52	282
Learning how to communicate ideas in mathematics effectively						
Grade 4	18	228	45	221	38	226
Grade 8	16	264	42	272	43	279

*** Sample size insufficient to permit a reliable estimate of average scale score
Note: Row percents may not add to 100 because of rounding.

teacher questionnaire concerning five such activities are displayed in table 5.15. There is much similarity between the distribution of responses made by fourth-grade students' teachers and eighth-grade students' teachers within each of the five activities. If classroom instruction provided by mathematics teachers is in a state of transition, then apparently the transition is at about the same stage at these two grade levels with respect to these five students' activities.

The NAEP data show that more than two-thirds of students in grade 4 and grade 8 have teachers who at least once or twice a week ask them to solve problems in a group or with a partner, ask them to talk to the class about their math work, discuss problem solutions with other students, and solve problems that reflect real-life situations. About 30 percent of fourth-grade and eighth-grade students have teachers who never, or hardly ever, require them

Table 5.15
Teachers' Reports on Frequency of Occurrence of Students' Activities

	Never or Hardly Ever	Once or Twice a Month	Once or Twice a Week	Almost Every Day
		Percent of Students		
Solve problems in small groups or with a partner				
Grade 4	7	18	50	25
Grade 8	7	26	40	27
Write a few sentences about how to solve a problem				
Grade 4	29†	36	26†	9†
Grade 8	33	37	25	5
Talk to the class about their work				
Grade 4	16	18	25	41
Grade 8	22	20	25	33
Discuss solutions to problems with other students				
Grade 4	6	22	37	35
Grade 8	2	12	37	49
Work and discuss problems that reflect real-life situations				
Grade 4	4	23	45	29
Grade 8	4	22	47	27

† Indicates significant difference from 1992.
Note: Row percents may not add to 100 because of rounding.

to write about how they solve a problem. Given this lack of attention to writing about mathematics in some classrooms, it probably should not be a surprise that students have great difficulty with constructed-response items on assessments such as NAEP (Lindquist 1997; see also chapter 11 by Silver, Alacaci, and Stylianou in this volume). There is an encouraging trend from 1992 to 1996 in the percent of students at grade 4 whose teachers have them write about how to solve a problem. Specifically, in the category "never or hardly ever" the percent of students significantly decreased from 45 percent in 1992 to 29 percent in 1996, and the categories "once or twice a week" and "every day" had significant increases. These NAEP results suggest a significant increase in students' writing about mathematics at grade 4.

For some teachers the textbook determines what mathematics is taught and how it is taught. For other teachers the textbook serves as but one resource as they plan and teach lessons. NAEP data in table 5.16 provide insight into the extent of the use of the textbooks and worksheets. The teachers of 61 percent of students in grade 4 and 72 percent of students in grade 8 indicate that the textbook is used on a daily basis, whereas worksheets are used most often on a "once or twice a week" basis. Interestingly, there has been a significant decrease from 1992 to 1996 in the use of textbooks on a daily basis for both students in grade 4 (from 76 percent to 61 percent) and students in grade 8 (from 83 percent to 72 percent). One might speculate that this trend may result from teachers attempting to increase the amount of active involvement of students through greater use of activity-based instruction, curriculum replacement units, and naturally arising real-life situations.

One technique for increasing students' opportunity to communicate about mathematics is for teachers to have students work on tasks in small groups or with a partner. At grade 4, 79 percent of students have teachers who provided one-half hour or more a week in these types of settings, and at grade 8, 68 percent of students. The extent of the popularity of these methods should be interpreted in light of the data on total time allocated to mathematics reported in table 5.11. There were no significant students' performance differences in the NAEP data at grade 4 or at grade 8 related to the

Table 5.16
Teachers' Reports on Frequency with which Students Use Textbooks and Worksheets

	Never or Hardly Ever	Once or Twice a Month	Once or Twice a Week	Almost Every Day
		Percent of Students		
Do problems from textbooks				
Grade 4				
1996	5†	10†	24	61†
1992	1	3	20	76
Grade 8				
1996	5	3	19	72†
1992	2	2	14	83
Do problems on worksheets				
Grade 4				
1996	7	15	51	27
1992	5	12	56	26
Grade 8				
1996	13	22	49	16
1992	8	28	52	12

† Indicates significant difference from 1992.
Note: Row percents may not add to 100 because of rounding.

amount of time students spent on work with partners and in small group activities.

Calculator Use

There is a substantial amount of research that demonstrates the value of using calculators in the teaching of mathematics at most grade levels (see for example the meta-analysis by Hembree and Dessart 1986). Thus, NAEP data on the extent of the use of calculators and information about how they are used are of interest. Table 5.17 shows data on the frequency of the use of calculators at grade 4 and grade 8. There was a noticeable shift toward increased use of calculators from 1992 to 1996 at both grade 4 and grade 8. Significantly more students had teachers who reported students' use of calculators as part of mathematics instruction in 1996 than in 1992. In 1992, 51 percent of students in grade 4 had teachers who reported students never, or hardly ever, used calculators as part of mathematics class, but in 1996 only 26 percent of students had teachers who reported such lack of use, a substantial and statistically significant decline. At grade 8 there was a significant increase in the daily use of calculators from 1992 to 1996 from 34 percent to 55 percent.

It is interesting to note in table 5.17 the data concerning daily use of calculators. There is a large difference here between grade 4 and grade 8. In

Table 5.17
Teachers' Reports on Frequency of the Use of Calculators and Computers

	Never or Hardly Ever	Once or Twice a Month	Once or Twice a Week	Almost Every Day
	Percent of Students			
Use a calculator				
Grade 4				
1996	26†	42	28	5†
1992	51	32	15	1
Grade 8				
1996	9†	14	21	55†
1992	23	20	22	34
Use a computer				
Grade 4				
1996	22	19	46	13
1992	24	20	45	10
Grade 8				
1996	69	20	9	2
1992	72	18	9	1

† Indicates significant difference from 1992.
Note: Row percents may not add to 100 because of rounding.

1996, 55 percent of students in grade 8 had teachers who reported students used calculators in mathematics class on a daily basis, whereas only 5 percent of students in grade 4 had teachers who reported similar use. The frequency of calculator use data not seem to be related to calculator availability, as teachers of 84 percent of fourth-grade students and 80 percent of eighth-grade students reported that their students had access to school-owned calculators.

Computer Use

NAEP teacher questionnaire data provide information on the frequency of computer use at grade 4 and grade 8. Computer use is much more prevalent at grade 4 than at grade 8. In fact, as shown in table 5.17, 69 percent of eighth-grade students have teachers who report that they never (or hardly ever) use a computer as part of mathematics instruction as opposed to 22 percent of fourth-grade students. This differential usage is noteworthy because recent in-depth analyses of NAEP data seem to show that the effects of technology appear to be much smaller in the fourth grade than the eighth grade (Wenglinsky 1998). A pertinent question, then, is the extent to which computer use is influenced by teacher philosophy, availability of computers, opportunities for staff development, or some other factor. We do know that only 6 percent of students in grade 4 have teachers who indicated that a computer was not available for use by their students, whereas 24 percent of students in grade 8 have teachers who responded similarly. Although computer availability may be one of the reasons for the lack of use of computers, it probably does not fully account for the low level of computer use at grade 8. Another factor that may be significant in the extent of use of computers is that the use of computers often requires changes in classroom organization, changes that are not required for calculator use.

Fourth-grade students have teachers who report that they most often use computers with students as follows: for playing mathematics or learning games, 41 percent; for drill, 27 percent. The comparable usage figures for grade 8 are playing mathematics or learning games, 14 percent; drill, 16 percent; and simulations, 12 percent. These usage figures take on special importance given the findings of Wenglinsky (1998) that when computers are used primarily for drill and practice in classrooms at either grade level, their value to learning is greatly reduced and may even be counterproductive.

Assessment

About 75 percent of fourth-grade and eighth-grade students in the 1996 NAEP sample have teachers who responded that it was very important to use the results of classroom assessments to guide instructional decisions. Tests are a regularly used assessment method in mathematics classrooms. About 64 percent of students in grade 4 and 55 percent of students in grade 8 have teachers who report giving one or two mathematics tests a month; almost all the remaining students at these grade levels (32 percent and 44 percent, respectively) have teachers who report giving tests one to two times a week. Thus, eighth-grade students are tested more frequently than fourth-grade students are.

Reformers are calling for fundamental changes in the content, format, and particularly the spirit of assessment in mathematics (Leder 1992; MSEB 1993; NCTM 1995). One theme of these recommendations is that assessment should be a continuous process that relies on a variety of sources of data. The data in table 5.18 show the frequency of use of five specific assessment methods. As shown in the table, the use of problem sets to assess students' progress continues to be prevalent at both grade 4 and at grade 8. At both grades 4 and 8, more than one-half the students have teachers who indicate that they use problem sets for this purpose once or twice a week.

There are differences in grade 4 and grade 8 classrooms in the frequency of use of such forms of assessment as short and long written responses to questions, individual and group projects or presentations, and portfolios.

Table 5.18
Teachers' Reports on Frequency and Type of Assessments

	Never or Hardly Ever	Once or Twice a Month	Once or Twice a Week	Almost Every Day
	\multicolumn{4}{c}{Percent of Students}			
Multiple-choice tests				
Grade 4	32	20	42	6
Grade 8	34	31	32	3
Problem sets				
Grade 4	6	4	36	54
Grade 8	4	7	36	53
Short (e.g., phrase or sentence) or long (e.g., several sentences or paragraphs) written responses				
Grade 4	18	18	38	26
Grade 8	21	21	41	17
Individual or group projects or presentations				
Grade 4	23	31	30	16
Grade 8	23	43	27	7
Portfolio collections of each student's work				
Grade 4	39	17	30	15
Grade 8	50	21	19	10

Note: Row percents may not add to 100 because of rounding.

Basically, students in grade 4 have teachers who use these methods of assessment more frequently than students in grade 8 do. For example, 15 percent of fourth-grade students have teachers who report using portfolios to assess students' progress once or twice a week, whereas 10 percent of eighth-grade students have teachers who report using this assessment technique that frequently. The reasons for the grade-level differences shown in table 5.18 in the use of these assessment methods are not clear. One part of the explanation may involve differences in participation in professional development activities. Data in table 5.8 show, for example, that 56 percent of teachers at grade 4 had recent course work or staff development experiences focused on assessment using portfolios, whereas only 39 percent of teachers at grade 8 reported similar experiences. Thus the prevalence with which a particular assessment method is used in the classroom seems to be related to teachers' recent in-service training in the use of the method.

CONCLUSIONS

Improving students' learning of mathematics in American schools depends on knowledgeable teachers conducting high-quality lessons focused on important mathematics under conditions that support students' opportunity to learn. In this chapter we have used NAEP data to paint a portrait of mathematics teachers. We have characterized their education, experience, beliefs, professional development, and the conditions under which they teach. We have described their teaching techniques, the topics they emphasize, and how they value a variety of educational practices and recommendations. The findings we have reported are important because the "bottom line is that there is just no way to create good schools without good teachers" (NCTAF 1996, p. 9).

Given the importance of the teacher and what happens in the classroom, we close with two suggestions. First, we must provide our nations' teachers with sustained support for implementing instructional practices that undergird current reform efforts. In designing such support, we should keep in mind the wealth of evidence documenting the lack of effectiveness of "one-shot" in-service programs (for example, Fullan 1991) and consider more comprehensive approaches to professional development that assemble a combination of learning activities intended to meet a specific goal in a particular context (Loucks-Horsley et al. 1998). Second, it is important that additional research be done to confirm and extend the findings we have reported in this chapter. In particular, we need to investigate further the linkages that appear to exist between teachers' characteristics and teaching practices and differential levels of students' performance. Such research should address the strength of the relationships in different classroom contexts and examine the extent to which cause-and-effect relationships exist. If cause-and-effect relationships are identified, then it will also be important to document the necessary supporting conditions surrounding the situation.

REFERENCES

American Association of University Women. *Equitable Treatment of Girls and Boys in the Classroom.* Washington, D.C.: American Association of University Women, 1989.

Armour-Thomas, Eleanor, Camille Clay, Raymond Domanico, K. Bruno, and Barbara Allen. *An Outlier Study of Elementary and Middle Schools in New York City: Final Report.* New York: Board of Education, 1989.

Beaton, Albert E., Ina V. S. Mullis, Michael O. Martin, Eugenio J. Gonzalez, Dana L. Kelly, and Teresa A. Smith. *Mathematics Achievement in the Middle School Years: IEA's Third International Mathematics and Science Study.* Chestnut Hill, Mass.: Center for the Study of Testing, Evaluation, and Educational Policy, Boston College, 1996.

Cooper, Harris. *Homework.* New York: Longman, 1989.

Dossey, John A., Ina V. S. Mullis, Steven Gorman, and Andrew S. Latham. *How School Mathematics Functions: Perspectives from the NAEP 1990 and 1992 Assessments.* 23-FR.-02. Washington, D.C.: National Center for Educational Statistics, 1994.

Fullan, Michael. *The New Meaning of Educational Change.* New York: Teachers College Press, 1991.

Grouws, Douglas A. "Homework." In *Encyclopedia of Mathematics Education*, edited by Louise S. Grinstein and Sally I. Lipsey. New York: Falmer, in press.

Hawkins, Evelyn F., Frances B. Stancavage, and John A. Dossey. *School Policies and Practices Affecting Instruction in Mathematics: Findings from the National Assessment of Educational Progress.* Washington, D.C.: National Center for Education Statistics, 1998.

Hembree, Ray, and Donald J. Dessart. "Effects of Hand-Held Calculators in Pre-College Mathematics Education: A Meta-Analysis." *Journal for Research in Mathematics Education* 17 (March 1986): 83-99.

Leder, Gilah C., ed. *Assessment and Learning of Mathematics.* Hawthorn, Victoria: Australian Council for Educational Research, 1992.

Lindquist, Mary Montgomery. "NAEP Findings Regarding the Preparation and Classroom Practices of Mathematics Teachers." In *Results from the Sixth Mathematics Assessment of the National Assessment of Educational Progress*, edited by Patricia Ann Kenney and Edward A. Silver, pp. 61–86. Reston, Va.: National Council of Teachers of Mathematics, 1997.

Loucks-Horsley, Susan, Peter W. Hewson, Nancy Love, and Katherine E. Stiles. *Designing Professional Development for Teachers of Science and Mathematics.* Thousand Oaks, Calif.: Corwin Press, Inc., 1998.

Mathematical Sciences Education Board. *Measuring Up: Prototypes for Mathematics Assessment.* Washington, D.C.: National Academy Press, 1993.

National Board for Professional Teaching Standards. *Middle Childhood and Early Adolescence/Mathematics: Standards for National Board Certification.* Washington, D.C.: National Board for Professional Teaching Standards, 1997.

———. *Doing What Matters Most: Investing in Quality Teaching*. New York: National Commission on Teaching and America's Future, 1996.

———. *What Matters Most: Teaching for America's Future*. New York: National Commission on Teaching and America's Future, 1997.

National Council of Teachers of Mathematics. *Assessment Standards for School Mathematics*. Reston, Va.: National Council of Teachers of Mathematics, 1995.

———. *Curriculum and Evaluation Standards for School Mathematics*. Reston, Va.: National Council of Teachers of Mathematics, 1989.

———. *Professional Standards for Teaching Mathematics*. Reston, Va.: National Council of Teachers of Mathematics, 1991.

National Science Foundation. *Indicators of Science and Mathematics Education 1995*. Arlington, Va.: National Science Foundation, 1996.

Silver, Edward A., Margaret S. Smith, and Barbara S. Nelson. "The QUASAR Project: Equity Concerns Meet Mathematics Education Reform in the Middle School." In *New Directions in Equity in Mathematics Education*, edited by Elizabeth Fennema, Walter Secada, and Lisa Byrd Adajian, pp. 9–56. New York: Cambridge University Press, 1995.

Simon, Martin A., and Deborah Schifter. "Towards a Constructivist Perspective: An Intervention Study of Mathematics Teachers." *Educational Studies in Mathematics* 22 (August 1991): 309–31.

Stigler, James W., and James Hiebert. "Understanding and Improving Classroom Mathematics Instruction: An Overview of the TIMSS Video Study." *Phi Delta Kappan* 79 (September 1997): 14–21.

Suarez, Tanya M., Daniel J. Torlone, Sue T. McGrath, and David L. Clark. *Enhancing Effective Instructional Time: A Review of Research*. Policy Brief, Vol. 1, no. 2. Chapel Hill, N.C.: North Carolina Educational Policy Research Center, 1991.

Weiss, Iris R. "Mathematics Teachers' Response to the Reform Agenda: Results of the 1993 National Survey of Science and Mathematics Education." Paper presented at the annual meeting of the American Education Research Association, San Francisco, Calif., April 1995.

Wenglinsky, Harold. *Does it Compute? The Relationship Between Educational Technology and Student Achievement in Mathematics*. Princeton, N.J.: Educational Testing Service, Policy Information Center, September 1998.

6
Whole Number Properties and Operations

Vicky L. Kouba and Diana Wearne

THE mathematics framework for the 1996 National Assessment of Educational Progress (National Assessment Governing Board 1994) continued the movement begun in 1990 from a dominance of items dealing with number properties and operations to a more balanced distribution of items across the five mathematics strands. According to the NAEP framework, 40 percent of the items given to fourth-grade students, 25 percent of the items given to eighth-grade students, and 20 percent of the items given to twelfth-grade students involved the NAEP content strand called Number Sense, Properties, and Operations (Reese et al. 1997). A total of 107 items in the 1996 NAEP addressed topics in the content strand Number Sense, Properties, and Operations. About half of items addressed students' facility with whole numbers and about half addressed students' facility with rational numbers. This chapter examines the extent to which students understand properties and operations when using whole numbers in varied contexts such as numeric, real-world application, and problem solving. In chapter 7 we discuss rational-number properties and operations.

RESULTS FROM STUDENTS' PERFORMANCE ON WHOLE NUMBER ITEMS

We know from prior research and from earlier NAEP assessments that students generally do well with whole-number concepts and operations in simple familiar contexts but have some difficulties with nonroutine and

Highlights

- About two-thirds of the fourth- and eighth-grade students and half of the twelfth-grade students have difficulty with number-line contexts that require some form of interpolation to determine the values represented by intermediate unlabeled marks on a number line.

- About 60 percent of the fourth- and eighth-grade students have at least a minimal understanding of place value.

- Fourth-grade students do better on addition than on subtraction and better on multiplication than on division.

- Students do well on the basic whole-number operations and concepts in numerical and simple applied contexts. However, students, especially those at the fourth- and eighth-grade level, continue to struggle with contexts that require more-complex reasoning.

- Students at all three grade levels have difficulty with division contexts involving interpreting remainders, problems that require explanations, nonroutine problems, and problems that are both nonroutine and multi-operational.

complex situations (Kouba, Zawojewski, and Strutchens 1997; Leinhardt, Putnam, and Hattrup 1992). The 1996 NAEP results were similar.

Although all whole-number items from the 1996 NAEP were examined as part of this interpretation, not all are described in detail in this chapter. Only when two or more items addressed a particular area was there sufficient data to warrant a detailed interpretation. For the purposes of this chapter, the items were classified into five groups: number lines, number theory, place value, addition and subtraction operations, and multiplication and division operations. The number theory group was subdivided into clusters: factors/multiples and even/odd numbers. The addition and subtraction and multiplication and division groups were also subdivided into clusters: numeric contexts, word problem applications, and either problem solving or number sense.

For some whole-number items, students were provided with, and permitted to use, calculators: four-function calculators at grade 4 and scientific calculators at grades 8 and 12. When information on calculator use is important to interpreting performance results (especially for released items), we note this in the text and the tables. Unfortunately, we did not have access to information about the percent of students reporting calculator use (or lack of use) on any particular item or about performance results according to reported calculator use.

Number Lines

A set of four secure whole-number items required students to interpret a number line. Descriptions and performance results for these items appear in table 6.1. Three of the four items were given only to fourth-grade students, and one was given to students at all three grade levels. All four items involved number lines whose successive markings indicated lengths greater than one unit (for example, two units or five units). Two of the items (items 2 and 3) were set in a measuring-instrument context, and the other two (items 1 and 4) were set in a traditional number-line context.

Table 6.1
Performance on Items Involving Number Lines

Item Description	Percent Correct		
	Grade 4	Grade 8	Grade 12
1. Given a number line, choose the correct sum after determining the values of particular unlabeled marks on a scale.	61	—	—
2. In the context of a measurement instrument, choose the correct result of calculations based on reading from a scale.	46	—	—
3. In the context of a measurement instrument, determine the value of unlabeled marks on a scale.	32	—	—
4. In a real-world context, given the distance between two locations on a number line, choose the correct distance between two other locations on the line.	22	38	50

Note: Item 3 was a short constructed-response item, and the rest were multiple-choice items. On all items, the scales were marked in lengths greater than 1 unit.

The three items given only to fourth-grade students differed in the numerical labels given to markings, the number of units between marked lengths, whether calculations were required, and whether the item was in multiple-choice or short constructed-response format. On the surface, item 3, which required students only to read and interpret a scale, might seem easier than items 1 and 2, which required students both to read and interpret a number line and to perform an operation. However, only 32 percent of the fourth-grade students responded correctly to item 3, whereas 61 percent and 46 percent, respectively, responded correctly to items 1 and 2. Item 3 may have been more difficult because it was a short constructed-response item; items 1 and 2 were multiple-choice items.

Another probable cause for differences with item 3 is that students had to determine the numerical value for unlabeled marks between two labeled marks on a number line. For item 1, students only needed to extend the numerical pattern established by the values given on labeled consecutive markings. For item 2, all the markings were labeled. In other words, item 3 called for interpolation, item 1 called for extrapolation, and item 2 required neither. Additionally, the difference between the percents of correct responses to items 1 and 2 may have been influenced by the shift from a traditional number context to a measuring-instrument context.

The item given to all three grade levels, item 4, was more complex than the items given only to fourth-grade students, in that no markings on the number line were labeled with numeric values. To complete the item successfully, students needed to glean the appropriate value labels for markings on the number line from the accompanying text and from calculation. Then they had to do a second calculation using their labels for the markings in order to arrive at a correct response. Twenty-two percent of the fourth-grade students responded correctly. Eighth- and twelfth-grade students' performance on item 4 was 38 percent and 50 percent correct, respectively. Thus, only about a fifth of the fourth-grade students and half the twelfth-grade students were able to solve this item.

Performance on these four items seems to indicate that students have difficulty with number line contexts that require some form of interpolation to determine the values represented by intermediate, unlabeled marks on a number line. The most common error for fourth-grade students was to treat all marked lengths on a number line as unit lengths rather than lengths that could be greater than one unit. Twenty-eight percent of the fourth-grade students made this type of error on item 2, and 47 percent made a similar error on item 3.

Research on representation in mathematics provides some insight into students' difficulties with number lines marked in increments other than one. We know from the work of Lesh, Behr, and Post (1987) that in matters of rational numbers and ratios the basis of comparison is usually taken to be one. Further, Dufour-Janvier, Bednarz, and Belanger (1987, p. 117) contend that the premature use of the number line as a representation in the learning of positive integers can lead children to "develop the notion of the number line as a series of 'stepping stones' ... [where] each step is conceived as a rock, and between two successive rocks there is a hole." Such a notion of a number line belies both the concept of the density of rational numbers and the concept of unmarked points existing between marked points and can contribute to the misconception of viewing all marks on a number line as being one unit apart.

Number Theory

The 1996 NAEP assessment included items that involved number theory concepts. Clusters of items from two areas often associated with number theory were examined: three items related to factors and multiples and five items related to the concepts of even and odd numbers.

Factors and Multiples

Three items assessed students' understanding of factors and multiples. Descriptions and performance results for these items appear in table 6.2. Two items were given to fourth-grade students, and the third item was given to fourth-, eighth- and twelfth-grade students. Fifty-four percent of the fourth-grade students responded correctly to item 1, and 45 percent responded correctly to item 2. Both items were multiple-choice items. Twelve percent of the fourth-grade students gave a correct response to item 3, which was a short constructed-response item. Eighth- and twelfth-grade students, 81 percent and 92 percent correct, respectively, did well on item 2, a multiple-choice item involving factors.

Table 6.2
Performance on Items Involving Factors and Multiples

Item Description	Percent Correct		
	Grade 4	Grade 8	Grade 12
1. Select the whole number that could result from multiplying by 5.	54	—	—
2. In a realistic context, choose which of several numbers can have a given number as a factor.	45	81	92
3. Given a number, generate a pair of factors.	12	—	—

Note: Item 3 was a short constructed-response item, and the rest were multiple-choice items.

Performance of fourth-grade students appeared to vary with different numerical contexts, which provides insight into a possible pattern of faulty reasoning that students use with factors and multiples. Item 1, a released multiple-choice item, follows:

A whole number is multiplied by 5. Which of these could be the result?

[Choices: 652; 562; 526; 265]

The correct response, 265, had as its units digit 5, which was the factor given in the item. In item 2, which was also in multiple-choice format, but is secure, the correct response did not have the given factor as its units digit. Fourth-grade students operating with an incorrect rule of "all multiples of a single-digit number must end in that single-digit number," (for example, *all* multiples of 5 must end in 5) could get item 1 correct because 5 was the factor given, and 265 was the correct response among the four choices. Students operating with the same rule would get item 2 incorrect. The distribution of the incorrect responses for items 1 and 2 provides further evidence for suspecting that many fourth-grade students operate with the faulty rule just described. For item 1, the incorrect responses were dispersed somewhat equally across the incorrect answer choices, 11 percent to 16 percent. For item 2, 37 percent of the fourth-grade students selected the incorrect response that had as its unit digit the factor given in the task. The other incorrect choices were made by 10 percent or fewer of the fourth-grade students.

Item 3 was more complex than items 1 or 2, both because it was a short constructed-response item and because it required students to find a pair of factors of a given number that satisfy certain conditions rather than to identify a single factor. Because 12 percent of the fourth-grade students responded correctly to item 3 and another 7 percent found a pair of factors that failed to satisfy the conditions of the problem, it appeared that about one-fifth of the fourth-grade students understood how to find a least one set of factors of a given number.

Even and Odd Numbers

Five items, whose descriptions appear in table 6.3, assessed concepts related to even and odd numbers. Three of the items were given to both eighth- and twelfth-grade students, one was given only to eighth-grade students, and one was given only to twelfth-grade students. All but the item given only to twelfth-grade students were multiple-choice items. In general, performance levels decreased as situations either increased in number of actions required or were in more of a problem-solving than a procedural context. Correct responses were given to the items involving even and odd numbers by 31 to 65 percent of the eighth-grade students and by 38 to 77 percent of the twelfth-grade students. On items administered to both grade levels, the percent of twelfth-grade students who correctly answered an item was usually about 10 percent higher than the percent of eighth-grade students.

Two items (items 1 and 5, described in table 6.3) were released items addressing concepts related to even and odd numbers; these items appear

Table 6.3
Performance on Items Involving Even and Odd Numbers

Item Description	Percent Correct	
	Grade 8	Grade 12
1. For a given set of whole numbers, identify the number of products that result in an even number.	65	77
2. Identify properties of the sum of prime numbers, relative to even and odd.	52	61
3. In a prealgebra context, identify an even number that does not have an odd number as a factor.	38	55
4. Choose an operation that always results in odd integers.	31	—
5. Evaluate an algebraic expression to determine when it produces an odd number and when an even number.	—	38

Note: Item 5 was a short constructed-response item, and the rest were multiple-choice items. Items 1 and 5, both released items, are shown in table 6.4.

in table 6.4. Their relative differences in complexity and difficulty demonstrate a facet of the design of the 1996 NAEP. The items administered at grade 8 were aimed at a more basic mathematical level than those administered at grade 12. Likewise, the items administered at grade 8 were aimed at more-advanced levels of mathematics than items administered at grade 4 (Reese et al. 1997). Performance on item 2 in table 6.4 may be more encouraging than the 38 percent correct value indicates because although only 38 percent of the twelfth-grade students had all three entries correct, another 54 percent had one or two entries correct. Unfortunately, NAEP's scoring guide did not allow us to determine the relative difficulty of the three parts of this question.

Place Value

The descriptions and performance results for the six items, five secure and one released, relating to aspects of understanding place value within whole numbers, appear in table 6.5. Three of the six items were given only to fourth-grade students, one was given to fourth- and eighth-grade students, one was given only to eighth-grade students, and one was given only to twelfth-grade students.

About 70 percent of the fourth-grade students appeared facile with basic notions of place value, as evidenced by their performance on the first three items described in table 6.5, all three of which were multiple-choice items. Performance was lower on item 4, a short constructed-response item that required students to apply place value concepts and to attend to more than

Table 6.4
Even and Odd Numbers: Sample Items

Item	Percent Responding	
	Grade 8	Grade 12
1. If each of the counting numbers from 1 through 10 is multiplied by 13, how many of the resulting numbers will be even?		
A. One	5	2
B. Four	14	10
C. Five*	65	77
D. Six	9	6
E. Ten	5	4
2. If x and y are integers, then the expression $4x + 5y$ has a value that is odd or even depending on the values of x and y. For example, if x and y are each even, $4x$ is even and $5y$ is even. Therefore, $4x + 5y$ is also even. Fill in each of the blank spaces in the following table with either "odd" or "even" for the value of $4x + 5y$.		

Value of x	Value of y	Value of $4x + 5y$
even	even	even
even	odd	
odd	even	
odd	odd	

All 3 entries correct [odd, even, even, in that order]	—	38
1 or 2 entries correct	—	54
No entries correct	—	3
Omitted	—	6

*Indicates correct response.
Note: Percents may not add to 100 because of rounding.

one condition. Students were to write a number given three digits (for example, 1, 2, and 9) and the place values for two of the digits (for example, the digit 9 means "nine tens," and the digit 2 means "two ones"). Only 58 percent of the fourth-grade students correctly produced a number that met the given conditions. It is somewhat surprising that eighth-grade students did only slightly better on this task, with 60 percent producing a correct response. Because we do not have detailed responses from students for this short constructed-response item, we cannot identify particular difficulties students had with this item. However, students' responses for an extended constructed-response task (item 5) were available, and this item is discussed in detail in chapter 11 by Silver, Alacaci, and Stylianou. Also, although only 15 percent of the eighth-grade students scored at a satisfactory or extended

WHOLE NUMBER PROPERTIES AND OPERATIONS

Table 6.5
Performance on Place-Value Items

Item Description	Percent Correct		
	Grade 4	Grade 8	Grade 12
1. Choose the number represented in a picture of base-10 blocks.	76	—	—
2. Given a set of four-digit numbers, choose which four-digit number is greatest.	70	—	—
3. Choose a number that is 10 more than another number.	72	—	—
4. Write a three-digit number given digits and conditions related to their place values.	58	60	—
5. Reason about how to maximize the difference in a subtraction problem based on place value.	—	15[a]	—
6. Choose correct description of a number resulting from changes in place value.	—	—	54

[a]Percent of students scoring at either the satisfactory or extended level.
Note: Item 5 was an extended constructed-response item, item 4 was a short constructed-response item, and the rest were multiple-choice items.

level on this task, a little over 60 percent scored at least the minimal level. This supports the conclusion that about 60 percent of the eighth-grade students who took the place value items on the 1996 NAEP had at least a minimal understanding of place value.

Item 6, given only to twelfth-grade students, required them to choose the number resulting when multiple changes are made to the digits in a given number. Fifty-four percent of the twelfth-grade students correctly identified the new number when one digit was increased and one digit was decreased. The most common error was to treat both changes either as increases or as decreases. Although this is only one item, it is discouraging that only a little over half of the twelfth-grade students responded correctly.

Addition and Subtraction Operations with Whole Numbers

The 1996 NAEP mathematics assessment included items involving addition and subtraction of whole numbers. Twelve such items were examined in three clusters: numeric contexts, word problems, and problem solving.

Addition and Subtraction in Numeric Contexts

The first three items described in table 6.6 assessed students' addition and subtraction in numeric contexts, which are the common number contexts we often associate with problems found on worksheets. Item 1, which was a multiple choice addition exercise presented in vertical format to fourth-, eighth-, and twelfth-grade students, did not appear to be problematic for students. Despite having to do two renamings (that is, regroupings, or "carries") to arrive at a correct response, 90 percent of the fourth-grade students, 91 percent of the eighth-grade students, and 93 percent of the twelfth-grade students were successful on this item.

Table 6.6
Performance on Whole-Number Addition and Subtraction Items: Numeric and Word Problems

Item Description	Percent Correct		
	Grade 4	Grade 8	Grade 12
Numeric			
1. Add two three-digit numbers with regrouping and choose the correct sum.	90	91	93
2. Subtract a one-digit from a two-digit number with regrouping.	79	—	—
3. Subtract a two-digit number from a three-digit number with regrouping.	73	86	—
Word Problems			
4. Subtract a three-digit number from a four-digit number with regrouping (extraneous information included).	64	—	—
5. Solve a one-step subtraction problem involving three-digit numbers.	57	—	—
6. Add three-digit numbers with a calculator and identify the cause of error.	51	—	—
7. Subtract five-digit numbers with regrouping (answer expressed as an estimate).	47	86	—

Note: Items 1, 4, and 7 were multiple-choice items, and the rest were short constructed-response items. Items 4 and 6, both released items, are shown in table 6.7.

In contrast, fewer fourth- and eighth-grade students could correctly solve the subtraction items, one given in a vertical format and one given in horizontal format. Seventy-nine percent of the fourth-grade students correctly solved item 2. Seventy-three percent of the fourth-grade students and 86 percent of the eighth-grade students correctly solved item 3. Both items were short constructed-response items rather than multiple choice items, which may account for some of the difference between performance on these items and performance on item 1.

Although the scoring guide for item 2 did not allow for an examination of types of errors, the guide for item 3 allowed for the reporting of the percentages of students who made particular errors. Thirteen percent of the fourth-grade and 6 percent of the eighth-grade students made an unidentified error. Eight percent of the fourth-grade and 1 percent of the eighth-grade students made the error of subtracting the smaller number from the larger number regardless of which was the minuend or subtrahend. Two percent of the fourth-grade students and 4 percent of the eighth-grade students made an error in renaming, and fewer than 1 percent of the students in grades 4 or 8 added instead of subtracting. Thus, the distributions and types of errors differ within and across grade levels, with eighth-grade students showing a somewhat greater understanding of the nature of subtraction because fewer students disregarded the roles of the subtrahend and minuend. One implication for the classroom is that with no single dominant error, instruction and remediation must allow for individual differences in what students know and can do.

Addition and Subtraction Word Problems

The four items from the 1996 NAEP requiring students to apply addition and subtraction concepts and procedures in word problems are described as items 4–7 in table 6.6. Three of the four were given only to fourth-grade students and one was given to fourth- and eighth-grade students. That 64 percent, 57 percent, and 51 percent of the fourth-grade students responded correctly to items 4, 5, and 6, respectively, indicates that more than half but fewer than two-thirds of the fourth-grade students appear able to solve one-step word problems involving whole numbers.

The percentage of fourth-grade students who solved such items dropped slightly when the context was somewhat nonroutine, as in item 6, where the computation was set in the context of identifying an error made with a calculator, and in item 7, where students were required to estimate a difference rather than calculate it. The change from working with 3- and 4-digit numbers (items 4–6) to working with 5-digit numbers may have accounted in part for the fact that only 47 percent of the fourth-grade students correctly solved item 7. However, 86 percent of the eighth-grade students chose the correct response for that item. Thus, although there is still room for improvement, many of the difficulties that item 7 presented for fourth-grade students appear to be resolved by eighth grade.

Two of the items (items 4 and 6 in table 6.6) were released and afforded a closer look at particulars of fourth-grade students' responses. These items appear in table 6.7 and illustrate some of the variation of format and context in the NAEP assessment. For these items, students were provided with, and permitted to use, four-function calculators.

The design of the choices offered for item 1 provided an opportunity to examine whether students realized that they were not to use extraneous information—the 300 miles a day. Once a student realized that only 849 and 1,723 were to be used, just one of the four choices presented was possible: B

Table 6.7
Whole-Number Addition and Subtraction Word Problems: Sample Items

Item	Percent Responding Grade 4
1. Kitty is taking a trip on which she plans to drive 300 miles each day. Her trip is 1,723 miles long. She has already driven 849 miles. How much farther must she drive?	
A. 574 miles	12
B. 874 miles*	64
C. 1,423 miles	10
D. 2,872 miles	12
2. Mark tried to add the numbers 489 and 263 on his calculator. What is the sum of these numbers? Answer: _____ The display on Mark's calculator showed his answer to be 128607. Mark had pressed a wrong key when trying to add. Which wrong key did he press? Answer: _____	
Correct response of 752 and multiplication (×) key (includes correct answers, but reversed)	51
One response correct	38
Incorrect response	9
Omitted	3

*Indicates correct response.
Note: Percents may not add to 100 because of rounding. Items are from item blocks for which students were provided with, and permitted to use, four-function calculators.

(874 miles). All three of the incorrect choices used 300 in some way. Thus, the primary challenge in this context was the identification of the numbers to use. Allowing for the possibility that some of the students chose the correct response by chance, it seems reasonable to think that the item provides evidence that about two-thirds of the fourth-grade students recognized 300 as unnecessary information.

Item 2 allows us to infer that about one-half of the fourth-grade students exhibited some understanding of the effect of operations on whole numbers. Fifty-one percent of the fourth-grade students responded with the correct sum of 752 *and* with the correct indication that Mark had pressed the multiplication key rather than the addition key on the calculator. Although students may have used trial and error to identify which operation key had been used in error, about half the fourth-grade students apparently realized that the single key error made had to be an operation error. No other single key could cause such a discrepancy in the size of the numbers.

Addition and Subtraction Problem Solving

Finally, five of the addition and subtraction items were word problems or abstract symbolic problems that involved nonroutine problem solving. All are described in table 6.8. None of the problem-solving items were released, and all were in short-constructed response format.

Table 6.8
Performance on Whole-Number Addition and Subtraction Items: Problem Solving

Item Description	Percent Correct		
	Grade 4	Grade 8	Grade 12
1. Construct a subtraction problem given digits and a target difference.	35	—	—
2. Solve a multistep word problem involving two- and three-digit numbers.	33	76	—
3. Find pairs of missing one-digit terms in an addition-and-subtraction equation.	27	—	—
4. Write calculator directions for solving a multistep word problem.	26	—	—
5. Solve a nonroutine, multistep problem and explain.	—	33	53

Note: All were short constructed-response items.

As may be expected, the percent of students who could correctly respond to these types of items was lower than for the numeric problems and the simpler word problems. Fourth-grade students' performance ranged from 26 to 35 percent correct on four items. Because these four items differed from one another in many ways, it is not possible to determine the probable causes for the somewhat better performance on items 1 and 2 than on items 3 and 4.

We can conclude that as students move up in grade they do improve on these types of problems. For example, students in grade 8 did better than students in grade 4 on item 2, and students in grade 12 did better than students in grade 8 on item 5. A portion of this improvement may be attributed to the students' gain in the ability to read and interpret word problems, as indicated by looking closely at the kinds of incorrect responses given for item 2. For that item, we know that 13 percent of the fourth-grade students, compared to only 4 percent of the eighth-grade students, used an incorrect operation—adding three numbers from the problem context rather than doing a subtraction and an addition or two successive subtractions. Unfortunately, the NAEP scoring guides did not allow for this level of response analysis on the other items. This may suggest to teachers that they should design scoring guides for their own assessments that will allow for identification and tracking of error patterns, not just the recording of overall performance.

Multiplication and Division Operations with Whole Numbers

Twenty-four items from the 1996 NAEP addressed students' use and understanding of multiplication and division. The items were examined in two clusters: (1) numeric contexts and number sentences and (2) problem solving.

Multiplication and Division in Numeric Contexts and Number Sentences

Table 6.9 presents descriptions of three items that assess students' ability to multiply and divide whole numbers and seven items that assess students' ability to write or recognize and identify number sentences involving multiplication or division. Eighty-six percent of the fourth-grade students chose the correct response for item 1, which required multiplying and dividing two-digit numbers. Students had access to a four-function calculator while working on this item. The high performance on item 1 stands in contrast to the 48 percent of the fourth-grade students who chose the correct response for item 2, which involved multiplying a series of single-digit numbers, including zero, without a calculator. It appears as if the availability of a calculator, as well as whether zero appears as a factor, may affect what fourth-grade students are able to do. Because 80 percent of the eighth-grade students provided a solution for item 3 when no calculator was available, it may be the case that the availability of calculators makes less of a difference by the time students have reached grade 8. However, because no other numerical multiplication and division items were given to students, these conclusions are tentative at best.

Of the remaining seven items described in table 6.9, five were given only to fourth-grade students (items 4–8), one was administered to all three grade levels (item 9), and one was given to both fourth- and eighth-grade students (item 10). Items 4–8 required that students either choose or write a number sentence. A released multiple-choice item (item 8 in the table) is typical of these four items. The first part of the item includes this scenario: "Martha planted 32 seeds. She put 8 seeds in each row. How many rows did she plant?" The question asked of the students was, "Which of the following could Martha use to solve the problem correctly?" The choices were $32 + 8$, $32 - 8$, 32×8, and $32 \div 8$. Items 9 and 10 required that students focus on other ideas, such as representations of multiplication and division with respect to appropriate procedures, and extraneous information.

The results in table 6.9 show that fourth-grade students did better on items 4 and 5, which involved multiplication sentences, than on item 8, which involved a division sentence. It appears that two-thirds to three-

WHOLE NUMBER PROPERTIES AND OPERATIONS 155

Table 6.9
Performance on Whole-Number Addition and Subtraction Items: Numeric Situations and Number Sentences

Item Description	Percent Correct		
	Grade 4	Grade 8	Grade 12
Numeric			
1. Multiply and divide two-digit numbers.	86	—	—
2. Multiply a series of single-digit numbers including zero.	48	—	—
3. Divide a three-digit number by a two-digit number with no remainder.	—	80	—
Number Sentences			
4. Choose the correct number sentence for a multiplication word problem involving one-digit numbers.	74	—	—
5. Translate a repeated-addition number sentence into a multiplication number sentence.	62	—	—
6. Choose the correct number sentence for a multiplication word problem with extraneous numbers.	52	—	—
7. Given an array of groups and number in each group, choose the correct number sentence for determining the total.	50	—	—
8. Choose the correct number sentence for a division word problem involving a two-digit number divided by a one-digit number.	48	—	—
9. Identify the correct procedure to solve a multiplication problem.	48	78	90
10. Choose extraneous information in a multistep word problem with one-digit numbers.	33	64	—

Note: Items 3 and 4 were short constructed-response items, and the rest were multiple-choice items. Items 1, 6, and 8 are from item blocks for which students were provided with, and permitted to use, four-function calculators.

fourths of the fourth-grade students can correctly match a multiplication sentence to a multiplication context. However, only half can do the matching when division is involved. The 52 percent of fourth-grade students responding correctly to item 6, which required identifying a multiplication sentence, may have been lower than the percents for items 4 and 5 because item 6 required students to determine which of the numbers given in the context were to be used in the number sentence. That is, item 6 contained extraneous information, which seemed to cause difficulties for some fourth-grade students. About 39 percent of the fourth-grade students chose incorrect responses that included the extraneous information in the number sentence. This difficulty with extraneous information is further evident in the responses to item 10, in which only 33 percent of the fourth-grade students could identify the extraneous information. Although on item 10 eighth-grade students did much better, 64 percent responding correctly, even at that grade level about a third of the students still have difficulty with contexts involving extraneous information.

Multiplication and Division Word Problems

Fourteen items addressed performance on whole-number multiplication and division word problems. Of those fourteen items, seven items whose descriptions appear in table 6.10 involved single-step or multistep contexts that required students to identify and perform appropriate operations. Answers for these seven items were all whole numbers, and for all but item 7 students had access to a calculator. About one-half of the fourth-grade students provided correct responses for items 1, 2, and 3, of which item 3 (a released multiple-choice item shown below) is typical:

> Every hour, a company makes 8,400 paper plates and puts them in packages of 15 plates each. How many packages are made in one hour?
>
> [Choices: 560; 8,385; 17,857; 126,000]

On this item, 50 percent of the fourth-grade students correctly divided 8,400 by 15 to get 560 packages. The most common error for this item was either to subtract and answer 8,385 plates, or to multiply rather than divide and answer 126,000 plates. For item 1 the most common error was to divide rather than to multiply, and for item 2 the most common error was to subtract rather than to divide. Thus, although wrong-operation errors are common, there does not seem to be a dominant pattern.

Item 4 differed from the first three items in table 6.10 in that it required students to perform more than one operation to obtain a correct answer. Thirty-four percent of the fourth-grade students completed all the required operations, and another 50 percent completed at least one operation. Thus, 84 percent of the fourth-grade students appeared able to do the multiplication and division calculations required for item 4, which parallels the 86 percent of fourth-grade students who could do simple numeric multiplication and division for item 1, table 6.9. For both item 1 in table 6.9 and item 4 in table 6.10, students had access to a calculator.

Eighth-grade and twelfth-grade students did better on item 5 (84 percent and 91 percent, respectively) than on item 6 (43 percent and 66 percent, respectively). The major difference between the items was that item 5 required only one calculation, whereas item 6 required more than one calculation. Item 7 also required multiple calculations. Because there were no discernible patterns in the incorrect responses and no other similar items

Table 6.10
Performance on Whole-Number Addition and Subtraction Items:
Word Problems with Integer Answers

Item Description	Percent Correct		
	Grade 4	Grade 8	Grade 12
1. Choose the correct answer for multiplication of a one-digit number by a one-digit number in a measurement context.	57	—	—
2. Choose the correct answer for multiplication word problem involving one- and two-digit numbers.	57	—	—
3. Choose the correct answer for the number of packages of plates produced per hour based on the number of plates in each package.	50	—	—
4. Choose the correct answer for a multioperational problem with one-digit and two-digit numbers.	34	—	—
5. Choose the correct answer for a measurement conversion involving two-digit to four-digit numbers.	—	84	91
6. Choose the correct answer to a multioperational problem involving one-digit to three-digit numbers.	—	43	66
7. Given conversion equivalencies, choose which measurement is largest.	—	—	58

Note: All were multiple-choice items. Items 1, 6, and 8 are from item blocks for which students were provided with, and permitted to use, calculators (four-function at grade 4; scientific at grades 8 and 12).

were included in the 1996 NAEP, it was not possible to determine why twelfth-grade students did slightly better on item 6 (66 percent responding correctly) than on item 7 (58 percent responding correctly). In general, it appears that for students in grades 4, 8, and 12, multistep items continue to be considerably more challenging than single-step items.

The other seven items involving whole-number multiplication and division, which are described in table 6.11, required students to use remainders in division contexts. Fourth-grade students did not do quite as well on these as on items with whole-number answers. About 10 percent fewer fourth-grade students solved the remainder items correctly than solved the integer-answer items. Eighth- and twelfth-grade performance varied across items.

Items 1, 2, 3, and 6 required students to provide the remainder of a division; items 4 and 5 required students to round the result of a division to the

Table 6.11
Performance on Division-with-Remainder Word Problems

Item Description	Percent Correct		
	Grade 4	Grade 8	Grade 12
1. Of three numbers, choose which has a remainder for division by a given one-digit number and explain the answer.	45	—	—
2. Choose the correct answer to a multioperational problem involving one-digit to two-digit numbers.	42	80	93
3. Determine the correct remainder for a two-digit number divided by a one-digit number.	39	—	—
4. Solve a multioperational problem involving division with remainders and two-digit numbers.	34	—	—
5. Solve a multioperational problem involving division with remainders and two- and three-digit numbers.	—	52	—
6. Select the correct remainder for division involving two-digit numbers.	—	35	—
7. Select the correct answer to a multioperational, rate-versus-time problem.	—	—	49

Note: Items 2, 6, and 7 were multiple-choice items, and the rest were short constructed-response items. Items 6 and 7, both released items, are shown in table 6.12.

next higher whole number; and item 7 required students to express both the whole-number quotient and the remainder. These classes of items appear to be the ones that continue to be difficult for most students, although item 2 for eighth- and twelfth-grade students appears to be an exception. Item 2 was also administered in the 1990 and 1992 NAEP. Performance on this item has stayed about the same for eighth-grade students, with 80 percent responding correctly in all three years; has been maintained for twelfth-grade students, with 90 percent responding correctly in 1990, 92 percent in 1992, and 93 percent in 1996; and has steadily increased for fourth-grade students, with 35 percent responding correctly in 1990, 38 percent in 1992, and 43 percent in 1996.

Two released items that proved the most difficult for eighth- and twelfth-grade students are presented in table 6.12 and revealed some interesting aspects of students' solutions. Item 1 required eighth-grade students to find the number of pieces of candy left over after equally distributing 65 pieces of candy among 15 bags. Students had a scientific calculator available for this item. Thirty-five percent of the eighth-grade students responded correctly: 5 pieces left over. Forty percent reported the

Table 6.12
Division with Remainders: Sample Items

Item	Percent Responding	
	Grade 8	Grade 12
1. Anita is making bags of treats for her sister's birthday party. She divides 65 pieces of candy equally among 15 bags so that each bag contains as many pieces as possible. How many pieces will she *have left*?		
A. 33	6	—
B. 5*	35	—
C. 4	40	—
D. 3	5	—
E. 0.33	12	—
2. A certain machine produces 300 nails per minute. At this rate, how long will it take the machine to produce enough nails to fill 5 boxes of nails if each box will contain 250 nails?		
A. 4 min	—	9
B. 4 min 6 sec	—	18
C. 4 min 10 sec*	—	49
D. 4 min 50 sec	—	15
E. 5 min	—	7

*Indicates correct response.
Note: Percents many not add to 100 because of rounding. Item 1 is from an item block for which students were provided with, and permitted to use, scientific calculators.

number of bags filled (4) rather than the number of pieces left (65/15 = 4 r 5, or 4.33). This error, focusing on the whole-number quotient rather than on the remainder, appeared to be the most common error on items requiring attention to the remainder. This item also brought to light another category of error. Six percent of the students incorrectly chose 33 as the answer, while another 4 percent incorrectly chose 0.33 as the answer. Students who gave these answers seemed not to have mastered the idea of a remainder.

Item 2 in table 6.12 required twelfth-grade students to find how long a certain machine took to produce enough nails to fill five boxes. The calculations involved multiplication, division, and a conversion of a remainder to a fraction of a minute expressed as seconds. Because there was no dominant wrong choice, it seemed that after the students did the division, they did not know what to do with the remainder. This suggests that students need more experience with these types of contexts and also need help developing meaningful connections among fractions, decimals, whole numbers, and units within a measurement system. (See chapter 7 by Wearne and Kouba for related comments concerning performance on rational-number concepts.)

CONCLUDING COMMENTS

From the teacher questionnaire data for the 1996 NAEP, we know that 93 percent of the fourth-grade students and 88 percent of the eighth-grade students had teachers who reported spending "a lot" of time (as opposed to "some," "little," or "none") on numbers and numeration in their classes. We also know that 43 percent of the fourth-grade students and 50 percent of the eighth-grade students had teachers who reported having taken one or more college or university courses devoted to number systems and numeration. Another 33 percent of the fourth-grade and 35 percent of the eighth-grade students had teachers who reported having taken a college or university course devoted in part to number systems and numeration. Additionally, 33 percent of the fourth-grade and 20 percent of the eighth-grade students had teachers who reported having had professional development seminars or workshops on number systems and numeration. It appeared that the majority of teachers had an educational background in, and placed emphasis on, number systems and numeration in classroom instruction. However, overall students' performance, which was lower on complex, advanced concepts and operations than on the simpler, more-basic concepts and operations, indicated that the professional development experiences of teachers need to be directed at helping students understand the numbers and operations beyond a basic level.

In summary, as on previous NAEP assessments (Kouba, Zawojewski, and Strutchens 1997; Kouba, Carpenter, and Swafford 1989) students did well on the basic whole-number operations and concepts in numerical and simple applied contexts. However, students, especially those at the fourth- and eighth-grade levels, continue to struggle with contexts that require more-complex reasoning such as the type of reasoning that Thompson (1993) identifies as quantitative reasoning. Quantitative reasoning, which is related to an analysis of the relationships among quantities rather than just with the numbers or operations themselves, is the type of reasoning called for in the multiplication and division items in table 6.12. Students also continue to have difficulties with whole number items that are multistep or that include more than one operation.

REFERENCES

Dufour-Janvier, Bernadette, Nadine Bednarz, and Maurice Belanger. "Pedagogical Considerations concerning the Problem of Representation." In *Problems of Representation in the Teaching and Learning of Mathematics*, edited by Claude Janvier, pp. 109–22. Hillsdale, N.J.: Lawrence Erlbaum Associates, 1987.

Kouba, Vicky L., Thomas C. Carpenter, and Jane O. Swafford. "Numbers and Operations." In *Results from the Fourth Mathematics National Assessment of Educational Progress*, edited by Mary M. Lindquist, pp. 64–93. Reston, Va.: National Council of Teachers of Mathematics, 1989.

Kouba, Vicky L., Judith S. Zawojewski, and Marilyn E. Strutchens. "What Do Students Know about Numbers and Operations?" In *Results from the Sixth Mathematics Assessment of the National Assessment of Educational Progress*, edited by Patricia A. Kenney and Edward A. Silver, pp. 33–60. Reston, Va.: National Council of Teachers of Mathematics, 1997.

Leinhardt, Gaea, Ralph Putnam, and Rosemary A. Hattrup. *Analysis of Arithmetic for Mathematics Teaching*. Hillsdale, N.J.: Lawrence Erlbaum Associates, 1992.

Lesh, Richard, Merlyn Behr, and Tom Post. "Rational Number Relations and Proportions." In *Problems of Representation in the Teaching and Learning of Mathematics*, edited by Claude Janvier, pp. 41–58. Hillsdale, N.J.: Lawrence Erlbaum Associates, 1987.

National Assessment Governing Board. *Mathematics Framework for the 1996 National Assessment of Educational Progress*. National Assessment Governing Board, 1994.

Reese, Clyde M., Karen E. Miller, John Mazzeo, and John A. Dossey. *NAEP 1996 Mathematics Report Card for the Nation and the States*. Washington, D.C.: National Center for Education Statistics, 1997.

Thompson, Patrick W. "Quantitative Reasoning, Complexity, and Additive Structures." *Educational Studies in Mathematics* 25 (1993): 165–208.

7
Rational Numbers

Diana Wearne and Vicky L. Kouba

THIS chapter continues the discussion begun in chapter 6, on whole-number properties and operations, with a discussion of the results for rational numbers. There were fifty-eight rational number items on the 1996 NAEP, including twenty-four fraction items, seven decimal items, thirteen money items, and fourteen items related to percent and proportions. There were twenty-six rational number items on the fourth-grade assessment, twenty-seven on the eighth-grade assessment, and thirty on the twelfth-grade assessment.

Some of the items measuring students' performance on the 1996 NAEP were also included on the 1990 NAEP and the 1992 NAEP assessments. The items showing a significant change from the 1990 or the 1992 assessment are noted in the accompanying text.

The students had the option of using a calculator for some of the rational number items, four-function calculators at grade 4 and scientific calculators at grades 8 and 12. If the availability of a calculator might affect the interpretation of results, this is noted in the text and in the tables. Unfortunately, as was the case with calculator use reported in chapter 6, information concerning performance results according to reported calculator use was not made available by NAEP and so cannot be a subject of this discussion.

FRACTIONS

There is ample evidence that students experience difficulty solving problems involving fractions and proportions, even simple ones (for example, Behr et al. 1984; Noelting 1980; Vergnaud 1983). These topics not only are central to the middle school mathematics program but also are the basis for

> **Highlights**
>
> - Rational numbers (fractions, decimals, proportions) continues to be a difficult topic for students.
>
> - Students at all three grade levels are more successful in solving routine one-step tasks than nonroutine multistep tasks.
>
> - In spite of the call for increased attention to student communication, many students find it difficult to explain their responses.
>
> - Students at all three grade levels performed at the same or at a higher level on the set of items included on previous NAEP assessments (1990 and 1992).

much of the mathematics in secondary school and beyond (Lesh, Post, and Behr 1988). Thus, information on students' performance related to these topics can assist in designing more effective instruction.

Overview of 1996 NAEP Fraction Results

The 1996 NAEP data provide evidence that fractions continue to be difficult for students, particularly for fourth-grade students. The results of the 1996 NAEP are quite similar to those reported for the 1990 and 1992 assessments, with older students more successful than younger students.

The fourth-grade students found this set of items to be quite difficult. There was only one domain in which at least half of the grade 4 students responded correctly: identifying the region or set shaded to represent a given fraction, where the region or set was divided into the number of parts indicated by the denominator of the fraction.

The eighth-grade students were more successful, with at least two-thirds of them able to identify a region representing a given fraction, represent a fraction on a rectangle divided into parts the number of which was a multiple of the denominator (for example, divided into 6 parts when the denominator of the fraction was 3), identify the fraction in reduced form corresponding to a given representation, and identify a diagram illustrating the equivalence of two fractions. However, ordering a set of fractions, using fractions to solve a verbal problem, and writing a story to go with a given division number sentence, all proved to be difficult for the eighth-grade students.

This set of items represented fewer than one-fourth of the items included on the twelfth-grade assessment. Twelfth-grade students were quite successful in responding correctly to most of these items, with the exception of those items assessing using fractions to solve verbal problems, the application items.

Representations of a Fraction

Developing meaning for mathematical symbols is essential for using these symbols effectively (Hiebert and Carpenter 1992). Students with an understanding of the written symbols for fractions are able to connect them with other representations, such as physical objects, pictorial representations, and spoken language. Creating meaning for symbols would seem to be necessary if students are to use procedures flexibly (Hiebert and Carpenter 1992). The 1996 NAEP contained items in which the students were asked either to write or to identify the fraction associated with a given pictorial representation or to represent the fraction on a region. Table 7.1 contains both a description of, and performance results for, those items.

Table 7.1
Performance on Items Involving Writing, Identifying, or Representing a Given Fraction

Item Description	Percent Correct		
	Grade 4	Grade 8	Grade 12
1. Select the diagram representing the fraction n/d (correct figure divided into d parts).	73	89	93
2. Identify a fraction n/d representing an indicated part of a set (n of d parts indicated in figure).	54	—	—
3. Indicate the number of fourths in a whole.	50	—	—
4. Indicate the placement of a fraction n/d on a unit number line divided into a multiple of d parts.	31	60	75
5. Represent the fraction $1/d$ on a figure divided into a multiple of d parts.	20	67	—

Note: Items 1 and 2 were multiple-choice items, and the rest were short constructed-response items. Item 3, a released item, is shown in table 7.2.

Fourth-grade students found representing a fraction or writing or identifying the associated fraction to be easier if the unit was a region (item 1, 73 percent correct) than if it was a set (item 2, 54 percent correct). This is consistent with earlier NAEP results (Post 1981; Carpenter et al. 1978). This suggests that students may have a fragile understanding of "unit" in their work with fractions. The meaning that some students have constructed for the unit when fractions are represented as regions is more robust than the meaning attached to the unit when fractions are represented as sets, as collections of objects.

Fourth-grade students encountered fewer problems representing a fraction on a region if the number of parts into which the region was divided was equal to the denominator of the fraction (item 1) than if the number of parts into which the region was divided was a multiple of the denominator (item 5). This was true even when one side of the rectangular region was divided into the number of parts equal to the denominator of the fraction. These additional divisions seemed to interfere with students' ability to shade a given fractional part of a region. This difficulty was noted in previous NAEP results (Carpenter et al. 1980; Kouba, Zawojewski, and Strutchens 1997) and in other studies (Larson 1980). The 20 percent of fourth-grade students who responded correctly to item 5, which involved representing a unit fraction ($1/d$) on a region divided into a number of parts equal to a multiple of the denominator, also included those students who chose to ignore the divisions in the given diagram and make their own divisions. As might be expected from the results on item 5, fourth-grade students also found it difficult to locate a fraction on a number line on which the unit was divided into a number of parts equal to a multiple of the denominator of the fraction (item 4), even though a unit fraction was placed on the number line as an aid to solving this task. Although only 31 percent of the fourth-grade students responded correctly to this item on the 1996 NAEP, it is a significantly greater percent than the 25 percent correct reported on the 1990 NAEP.

Approximately one-third of the students in grade 8 experienced some difficulty in responding correctly to item 5, which asked them to represent a unit fraction on a rectangular region divided into a number of parts equal to a multiple of the denominator. There was, however, a slight but significant increase in the percent of students in grade 8 responding correctly to this item on the 1996 assessment (67 percent) from the 1990 assessment (63 percent).

Essential to the notion of the fraction n/d is that the unit has been divided into d equal parts. It is difficult for students to understand equivalent fractions, and operations with fractions, without recognizing this. As shown in table 7.2, only 50 percent of students in grade 4 were able to state how many fourths were in a whole. Because this item assesses fundamental knowledge of fractions, a topic introduced in the first grade, the results are disturbing. Perhaps equally disturbing is the fact that one-sixth of the students omitted this item, the initial item in that item block. This may show, in another way, the difficulty students have in fully understanding the meaning of the unit in fraction tasks.

Table 7.2
Fractions: Sample Items

Item	Percent Responding Grade 4
How many fourths make a whole? Answer: _____	
Correct response of 4	50
Any incorrect response	35
Omitted	16

Note: Percents many not add to 100 because of rounding.

Equivalence and Ordering of Fractions

An understanding of equivalence of fractions is important in developing a sense of relative size of fractions and helping students connect their intuitive understandings and strategies to more general, formal methods. The results on NAEP items assessing the notion of equivalent fractions are contained in table 7.3.

Table 7.3
Performance on Items Involving Equivalence and Ordering of Fractions

Item Description	Percent Correct		
	Grade 4	Grade 8	Grade 12
1. Identify a diagram illustrating the equivalence of two fractions.	45	70	—
2. Given three fractions equivalent to a unit fraction, write two more equivalent fractions.	47	—	—
3. Identify the fraction (in reduced form) representing the shaded part of a region.	—	65	—
4. Use equivalent fraction notions to solve a verbal problem.	23	51	69
5. Explain why one unit fraction is larger or smaller than another (set in context).	16	—	—
6. Identify the correct ordering of three fractions, all in reduced form.	—	35	—

Note: Items 2 and 5 were short constructed-response items, and the rest were multiple-choice items. Item 3, a released item, is shown in table 7.4.

As shown in the table, 70 percent of the eighth-grade students identified the diagram illustrating the equivalence of two fractions (item 1). However, fewer than 50 percent of the fourth-grade students were successful on this item. Similar fourth-grade results, with fewer than half of the students responding correctly, occurred on item 2 in which the students were presented with three equivalent fractions and asked to write two more; the three given fractions all were equivalent to a unit fraction that was not included in the set (for example, if the unit fraction had been 1/3, fractions included in the item could have been 2/6 and 3/9).

The assessment included one question, item 4 in table 7.3, in which the concept of an equivalent fraction was involved in solving a problem. This is not a released item, but an example of a question assessing similar information is as follows:

> Josie has 2/3 cup of cocoa. She is making chocolate cookies. Each batch of cookies uses 1/6 cup of cocoa. How many batches can she bake?

The percent of students responding correctly to item 4 rises from almost 25 percent in grade 4, to 50 percent in grade 8, to almost 70 percent in grade 12. It is somewhat surprising that at least three out of every ten twelfth-grade students did not identify the correct solution, because once one sees that 2/3 is equivalent to 4/6 in the example above, the problem is solved. (The problem also can be solved by the use of the number sentence 2/3 ÷ 1/6 = 4.) The percent of correct responses to this NAEP item for all three grades has remained virtually unchanged over the last three NAEP assessments (1990, 1992, 1996).

Table 7.4 contains the performance results for a released item in which the eighth-grade students were asked to identify a fraction equivalent to 4/12, the fraction associated with the shaded portion of the rectangle. The percents of eighth-grade students selecting 1/3 and 1/4 have remained constant over the last three NAEP assessments, with 65 percent selecting 1/3 and 25 percent selecting 1/4.

The 1996 assessment included two items related to the ordering of fractions, items 5 and 6 described in table 7.3. Item 5 asked the fourth-grade students to tell which of two unit fractions was the larger and to explain their reasoning. This question was posed within a context to facilitate students being able to explain, in words or by a diagram, why one unit fraction was larger than another. Only about one-sixth of the fourth-grade students responded correctly. These results are similar to those noted in other studies (Post et al. 1985). The other ordering item (item 6) asked the eighth-grade students to identify which sets of fractions were ordered from least to greatest. Despite the fact that all the fractions were less than 1 and in reduced form, only 35 percent of the eighth-grade students chose the correct ordering.

Results on items 5 and 6 are troubling because it is difficult to imagine how students can work meaningfully with fractions if they do not have a sense of their relative size. What might students be thinking as they try to order fractions? The responses of the students in grade 8 on item 6 provide some clues. About one-third of the students selected the incorrect option in

Table 7.4
Identifying a Fraction Representing the Shaded Part of a Region: Sample Item

Item	Percent Responding Grade 8

[figure: rectangle ABCD divided into a 3×3 grid with two squares shaded in the left column, second and third rows from top; corners labeled B (top-left), C (top-right), A (bottom-left), D (bottom-right)]

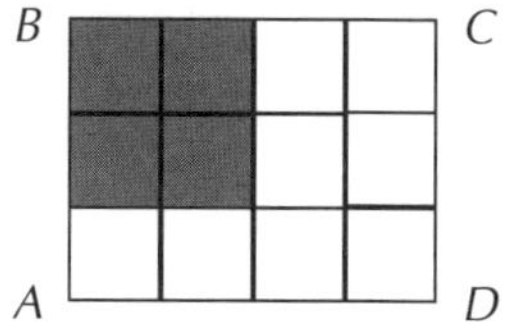	
In the figure above, what fraction of rectangle *ABCD* is shaded?	
A. 1/6	6
B. 1/5	2
C. 1/4	25
D. 1/3*	65
E. 1/2	2

*Indicates correct response.

which *both* the numerators and the denominators of the fractions were arranged in descending order, that is, from largest to smallest (for example: 6/7, 2/5, 1/2—fractions similar to those used in the item). One might guess that students selected this option because they thought that larger denominators represented smaller fractions, and so 6/7, for example, would be smaller than 2/5 because if you divide something into seven parts, each part is smaller than if you divide it into five parts. Approximately one-fifth of the students selected the incorrect option in which both the numerators and the denominators were arranged in ascending order, from smallest to largest. This would be the option selected by the students if they were employing whole-number ordering notions to ordering fractions. Research has shown that many students' misconceptions in tasks involving fractions are tied to whole-number ideas, and so it is not surprising that ordering fractions initially is dominated by the students' view of ordering whole numbers (Behr et al. 1984; Mack 1995). Thus, although it is disturbing that only about one-third of the eighth-grade students were successful on this item, it is encouraging that only one-fifth of the responses reflected domination by whole-number ideas. It also is encouraging that a significantly greater percent of students in grade 8 responded correctly to this item on the 1996 NAEP (35 percent) than on the 1990 NAEP (27 percent).

Other Fraction Items

Table 7.5 contains additional items assessing knowledge of fractions, including items that can be categorized as applications. The first three items asked students either to identify the size of the whole set given the size of a specified fractional part of the set or, given the size of the set, to find the size of the subset left after a specified fractional part of the original set had been removed. As indicated in the table, finding the size of the whole given an indicated fractional part (item 1) was the easier of the two types of tasks for the fourth-grade students; this multiple-choice task proved not to be difficult for the students in grades 8 and 12.

Table 7.5
Performance on Other Fraction Items Including Applications

Item Description	Percent Correct		
	Grade 4	Grade 8	Grade 12
Find the whole given a fractional part or find the size of a set after a fractional part has been removed.			
1. Given a region in which a value is assigned to a part of the region, determine the value of the whole region.	46	74	89
2. Find the size of a set after a fractional part has been removed.	11	—	—
3. Find the size of a set after a fractional part has been removed and explain your answer.	10	—	—
Applications involving fractions			
4. Write a story problem for a given division by a fraction number sentence. (set in context)	—	26	—
5. Given the scale and the scaled distance, find actual distances.	—	—	33
6. Solve a multistep problem involving finding a sum, finding a fractional part of a number, finding a product, and comparing two quantities.	—	8[a]	—
7. [Item similar to item 6, but used slightly different values and was scored differently].	—	—	24
8. Write a fraction resulting from dividing a fractional part of a unit into an integral number of parts.	—	—	25

[a]Percent of students scoring at the satisfactory or extended level.
Note: Item 1 was a multiple-choice item, item 6 was an extended constructed-response item, and the rest were short constructed-response items.

Items 2 and 3 contain descriptions of tasks that asked the students to find the size of a subset after a specified fractional part of the original set had been removed. These are not released items, but a similar task is the following:

There are 12 cookies on the plate. One-fourth of the cookies are chocolate chip cookies. The rest of the cookies are oatmeal cookies. How many cookies are oatmeal cookies?

Although one of the tasks only required the solution (item 2) and the other an explanation of the solution (item 3), the performance level is essentially the same. Both of these items are multistep problems. For example, in the similar problem stated above about cookies, the student could first find that one-fourth of 12 is 3, then subtract 3 from 12 to get 9, or the student could think that if one-fourth of the cookies are chocolate chip, then three-fourths must be oatmeal and then find that three-fourths of 12 is 9. One can hypothesize that the difficulty with the two NAEP items is that they involve two steps rather than a simple one-step procedure such as finding one-fourth of 12.

The remaining items, items 4–8 in table 7.5, were administered only at the upper two grades. This set included writing a story problem to go with a given division by a fraction number sentence (item 4) and solving four multistep problems (items 5–8) in which students had to apply basic fraction ideas. As reported in the table, at most one-third of the students in grade 12 responded correctly to these items. However, significantly more students in grade 12 responded correctly to item 8 on the 1996 NAEP (25 percent) than had on the 1992 NAEP (20 percent). Given the weak understanding younger students showed for some of these ideas, perhaps it is not surprising that students have trouble applying these ideas as they get older.

DECIMALS

As with fractions, an understanding of the symbolism for representing decimals is essential to developing understanding of symbolic operations with decimals. Students who do not understand decimal notation frequently resort to memorizing procedural rules (Bell, Swan, and Taylor 1981; Fischbein et al. 1985; Sackur-Grisvard and Leonard 1985; Resnick et al. 1989), rules that may be tied only to surface features of the tasks rather than to any underlying conceptual rationale (Hiebert and Wearne 1985; Wearne and Hiebert, 1988).

What kind of understanding of decimal symbols did students display on the 1996 NAEP? Table 7.6 contains descriptions of the decimal items and the percents of students responding correctly. In item 1, fourth-grade students were asked to select the number associated with a base-10 block representation in which the 1 and the 0.01 blocks were identified. Approximately one-half of the fourth-grade students chose the correct response. About one-third of the students responded either by selecting the option with all of the digits to the right or to the left of the decimal point, suggesting that they were probably using whole-number ideas to choose the number.

Table 7.6
Performance on Items Involving Decimals

Item Description	Percent Correct		
	Grade 4	Grade 8	Grade 12
1. Select the decimal number for a given representation of base-10 blocks.	53	—	—
2. Place a decimal number on a number line with divisions not 0.1 apart.	39	87	—
3. Select a set of whole numbers representing a set of decimal numbers rounded to the nearest whole number.	34	83	—
4. Write the common fraction equivalent (in reduced form) of a decimal number.	—	—	57
5. Given a scale drawing, the scale in common fractions, and a length, identify the scaled length in decimal form.	—	—	56

Note: Items 2 and 4 were short constructed-response items, and the rest were multiple-choice items.

Another way of assessing students' understanding of decimal symbols is to ask them to write the number associated with a given place on a number line. For item 2, the divisions on the given number line were a multiple of 0.1 apart, rather than 0.1. Almost 90 percent of the eighth-grade students but only about 40 percent of the fourth-grade students responded correctly to this item. This question is similar in some ways to the fraction items in which the region or number line was divided into a number of parts equal to a multiple of the denominator (see items 4 and 5 in table 7.1); similar in that the student has to attend to the number of divisions. The fourth-grade students found those fraction tasks to be difficult as well.

A third way to assess decimal number knowledge is to ask students to round decimals to the nearest whole number, as in item 3 in table 7.6. More of the fourth-grade students selected the response in which the decimal portion of the number was omitted (for example, selecting 56 for the number 56.8) than responded correctly; 39 percent of the fourth-grade students made this error. However, as in the previous example, the eighth-grade students did not find this task difficult.

A somewhat different way of assessing students' understanding of symbols is to ask them to translate between different systems of notation. Of course, students' difficulties may be caused by a lack of understanding of either one or both of the systems, so interpretations must be made with caution. There were two tasks which assessed knowledge of the relationship between decimals and common fractions, both of which were administered only to twelfth-grade students (items 4 and 5). The first task, item 4, asked the students to write the common fraction equivalent, in reduced form, of a given decimal. It would be of interest to know why so few of the twelfth-grade students responded correctly to this fairly straightforward question concerning the relationship between a decimal and a common fraction. It is possible that the students did not attend to the directions indicating they were to write the fraction in reduced form. The scoring of this short constructed-response item did not indicate what proportion of the students wrote an associated fraction that was not in reduced form. This item also was included on both the 1990 and 1992 assessments with approximately the same percent of students responding correctly.

The other question assessing the relationship between fractions and decimals is item 5. This task was the more demanding of the two items, in that the student had multiple steps to perform. In spite of this, the twelfth-grade students were as successful on this item as they were on the rather straightforward task of writing a fraction in reduced form that was equal to a given decimal. The item format may have played a role here; item 4 was a short constructed-response item and item 5 was multiple-choice.

MONEY

Money is a topic that is introduced in the early grades and continually referred to in the upper grades. Money is of interest because it is a concrete and very familiar referent for quantity and because it is a context for many practical applications of mathematical skills. Consequently, it is a potential source for both meaningful and practical experiences.

What does the evidence from NAEP say about the benefits students are deriving from these early experiences with money? There were a variety of items that assessed students' facility with money; the items are described and the results reported in table 7.7. As the results for item 1 illustrate, fourth-grade students can readily round an amount to the nearest dollar. Posing somewhat more of a challenge for them were tasks requiring them to identify an amount representing a multiple of a given amount, items 2 and 3. The cent symbol (for example, 65¢) was used in item 2 and the dollar symbol (for example, $0.65) in item 3. The symbols used in the tasks to denote monetary value did not affect performance for the fourth-grade students. At least 60 percent of the fourth-grade students responded correctly to each of these items. Over 80 percent of the eighth-grade students responded correctly to the item using the dollar symbol (item 3); the item involving the cent symbol (item 2) was not included on the eighth-grade assessment. Students had the option of using a calculator on both these items, an option which may have assisted some students in eliminating computational errors.

Table 7.7
Performance on Items Involving Money

Item Description	Percent Correct		
	Grade 4	Grade 8	Grade 12
1. Identify an amount rounded to the nearest dollar.	88	—	—
2. Identify multiples of an amount given in cents (for example, 65¢).	63	—	—
3. Identify multiples of an amount given in dollars (for example, $0.65).	60	83	—
4. Find 100 times an amount given in dollars.	—	71	—
5. Identify the change from a dollar amount when buying multiples of a given unit (in dollars).	30	68	88
6. Find multiples of a given amount (price given in dollars) and then identify how many one-dollar bills are necessary for the purchase.	26	—	—
7. Find the sum of amounts (given in cents), then state how many one-dollar bills are necessary for the purchase.	17	—	—
8. Given a table in which some items are to be added and others subtracted, identify a particular result.	—	60	88
9. Identify the number of a unit (given in cents) that could be purchased with a given number of dollars.	—	57	—
10. Given a problem involving multiple operations, select the correct response.	—	—	82
11. Select the amount saved when buying an item as part of a set as opposed to buying it alone.	—	—	40

Note: Items 4 and 7 were short constructed-response items, and the rest were multiple-choice items. Items 2, 3, and 7–9 were from item blocks for which students were provided with, and permitted to use, either four-function calculators at grade 4 or scientific calculators at grades 8 and 12. Item 7, a released item, is shown in table 7.8.

Two other items in this set, items 6 and 7 described in table 7.7, asked fourth-grade students to identify how many one-dollar bills would be necessary to make a purchase. Both of these tasks required the student to first find multiples of a given amount, as they had to do in items 2 and 3, and then round up to the next dollar, as opposed to rounding to the nearest dollar as in item 1. Both of these tasks proved to be difficult for the fourth-grade students and are described next.

Item 6 asked students to identify the number of one-dollar bills necessary to purchase more than one of an article, the price of which was given in dollars. Only about one-fourth of the students in grade 4 chose the correct answer. Forty-five percent of the students identified the number of one-dollar bills necessary to purchase *one* article. Because at least 60 percent of the students were able to find multiples of a given amount, based on the results for items 2 and 3, these results suggest the possibility that some students did not read the item very carefully.

The second of these two tasks is shown in table 7.8. The students had the option of using a four-function calculator on this item. This task is more complex than item 6 in that the amounts were all different. Instead of finding a multiple of the cost of one amount, the students had to find the amount needed for one day, then multiply by five, before determining the fewest number of dollar bills necessary for the week. As noted in table 7.8, in addition to the 17 percent who responded correctly, 20 percent of the students responded in ways indicating they either had completed the first step in solving the problem (for example, finding the sum of the amounts for all five days) or rounded the amounts. Unfortunately, we cannot tell from the NAEP scoring guide for this item what proportion of the students found the number of dollar bills necessary for just one day rather than for all five days, an error similar to that reported on item 6.

Table 7.8
Money: Sample Item

Item	Percent Responding Grade 4
Sam can purchase his lunch at school. Each day he wants to have juice that costs 50¢, a sandwich that costs 90¢, and fruit that costs 35¢. His mother has only $1.00 bills. What is the least number of $1.00 bills that his mother should give him so he will have enough money to buy lunch for 5 days?	
Correct response of 9 $1.00 bills	17
Specific incorrect responses: $8.75 (or 875) One day is $1.75, so $10 for the week Correct method but answer rounded to $8.00 Correct method with small error ($7 or $11)	20
Any other incorrect response	51
Omitted	10

Note: Percents may not add to 100 because of rounding or off-task responses. This item is from an item block for which students were provided with, and permitted to use, four-function calculators.

The eighth-grade students were relatively successful on these money items, with the possible exception of items 8 and 9. Both of these items were multistep items, and the students had access to a calculator for both items.

The only item that proved to be difficult for the twelfth-grade students was item 11, which asked the students to compare the price of a unit when purchasing it as part of a set or as a single piece. This is not a released item, but a similar item is the following:

> A set of 8 dinner plates cost $48. The plates also are sold individually for $7.50 per plate. If you bought the 8-plate set, how much money would you save *on each plate* over the individual plate price?

Forty percent of the twelfth-grade students responded correctly. It should be noted that the students had the option of using a calculator on this item, an option which may have eliminated computational errors for some students. Another 41 percent of the students selected the option indicating how much money you would save if you bought the entire set rather than individual units or, using the above example, how much money you would save by buying a set of eight plates rather than buying eight plates at the individual price. This response would seem to indicate that the students understood they were to compare the price under two different conditions, but did not read the item very carefully.

Only one item, item 5 in table 7.7, was administered at all three grade levels. This item asked the students to identify the change received when buying multiples of a given unit. Eighth-grade students were more successful than fourth-grade students, and the twelfth-grade students were more successful than the eighth-grade students. This item also appeared on the 1990 and 1992 NAEP assessments, with similar results at all three grade levels.

PROPORTION: PERCENT, RATIO, AND RATE

The topic of proportions occupies a central role in the middle school mathematics curriculum. Proportional reasoning has been described as the capstone of elementary school arithmetic and the cornerstone of all that is to follow (Lesh, Post, and Behr 1988). The 1996 NAEP results on this topic are described next. The discussion is divided into two sections. The first section focuses on the percent items, and the second looks at other proportional reasoning items.

Percent

Percent is an important mathematics topic. There is frequent use of percent in the popular press and in news broadcasts. Percents are used by stores to advertise sales, and salary increases in the workplace frequently are reported in terms of percents. Thus, percent is a topic the student has encountered prior to instruction and will continue to encounter after leaving school.

The topic of percent usually is introduced in grade 6 and continues to be part of the mathematics curriculum in grades 7 and 8. Previous NAEP reports (Carpenter et al. 1975; Carpenter et al. 1980; Kouba et al. 1988; Kouba, Zawojewski, and Strutchens 1997) as well as other studies (Edwards 1930; Kircher 1926; Risacher 1993) all indicate that students have difficulty solving problems involving percent. It has been suggested that one of the reasons percent is difficult is that it uses an extremely concise linguistic form, a form which leads students to "manipulate numbers based on learned procedures rather than on underlying relationships" (Parker and Leinhardt 1995, p. 446).

What do the results of the 1996 NAEP tell us about eighth- and twelfth-grade students' understanding of this important topic? The results, as shown in table 7.9, tell us that solving problems involving percent continues to be difficult for students and that students in grade 12 are more successful at solving these tasks than students in grade 8, the only two grade levels assessed on this topic. Students had the option of using a calculator on seven of the ten items assessing percent. The availability of the calculator may have encouraged some students to attempt the item and may have eliminated computational errors. These seven items are indicated in the text and in table 7.9.

Instruction frequently divides percent tasks into three categories. The first category, denoted here as Type A, involves finding a percent of a quantity. Eight of the ten items on the assessment were of this type and are the first eight items described in the table. Some of the items in this category are one-step problems; others involve more than one step, but one of these steps requires finding a specified percent of a quantity.

The eighth- and twelfth-grade students were reasonably successful in solving fairly simple one-step problems of Type A (for example, finding 40% of 130, a format similar to that used in items 1 and 3, table 7.10); at least half of the eighth-grade students and about 80 percent of the twelfth-grade students responded correctly. The eighth-grade students were not as successful on item 3, a released item shown in table 7.10, even though this item appears to be quite similar to items 1 and 2. Many students may have computed 15 percent of $24.99 ($3.7485) and then looked for the closest option ($3.75). Thus, although the item is at the same conceptual level as items 1 and 2 and involves the same size numbers (four-digit quantities, two-digit percent) as item 2, only 38 percent of the eighth-grade students responded correctly, as compared to 50 percent on item 2. Perhaps because the item asked for the closest approximation to a 15 percent tip, the students were encouraged to guess at an amount; 41 percent of the students selected either $2.50 or $3.00

Table 7.9
Performance on Items Involving Percent by Type

Item Description	Percent Correct	
	Grade 8	Grade 12
Type A—Finding a given percent of a quantity		
1. Find the percent of a quantity.	59	84
2. Find the percent of a given (percents shown on circle graph).	50	80
3. Identify the approximate percent of a quantity.	38	—
4. Given the percent of increase, identify the new amount.	35	—
5. Solve a multistep problem involving finding a sum, then computing a percent of that sum.	—	67
6. Compare two percent amounts computed under two different conditions; explain reasoning.	7	23
7. Given the percent of increase, identify the new amount written as a literal expression.	—	23
8. Compare the effect of interest rates over a two-year period; justify answer.	—	11[a]
Type B—Given the value of a specified percent, find the quantity.		
9. Given the amount of a percent of a quantity, find a different percent of the same quantity.	—	66
Type C—Determine what percent one number is of another.		
10. Solve a multistep problem involving how to construct a percent higher than a given percent at one point in time, yet lower overall.	—	23

[a]Percent of students scoring at the satisfactory or extended level.
Note: Item 8 was an extended constructed-response item, items 5, 6, and 10 were short constructed-response items, and the rest were multiple-choice items. Items 1, 2, and 5–10 were in item blocks for which students were provided with, and permitted to use, scientific calculators. Item 3, a released item, is shown in table 7.10.

as the correct response. Another contributing factor may have been that students had access to a calculator for items 1 and 2 but not for item 3.

Item 4 in table 7.9 is another example of a Type A task the grade 8 students found difficult; only 35 percent chose the correct response. This task asked the students to find a new amount, given the percent of increase from the original amount. This is not a released item, but a similar task is the following:

> Heidi works part-time at the Malt Shop. Her boss decided to give her a 7 percent raise in pay. Before the raise, she earned $50 a week. Now how much will she earn each week?

For many students this probably was a two-step problem. In the example above, the students might first find 7 percent of $50, and then add that

Table 7.10
Percent: Sample Item

Item	Percent Responding Grade 8
Of the following, which is the closest approximation of a 15 percent tip on a restaurant check of $24.99?	
A. $2.50	21
B. $3.00	20
C. $3.75*	38
D. $4.50	9
E. $5.00	11

*Indicates correct response.
Note: Percents may not add to 100 because of rounding. While working on this item, students did *not* have access to a calculator.

amount to $50 to find the new weekly wage. It is probable that many students proceeded in this manner rather than solving the problem in a single step (that is, finding 107% of $50). In the NAEP item, almost half of the students selected the amount obtained by adding the percent itself to the original quantity (that is, in the above example, adding 7 to $50) rather than adding 7 percent of the quantity. This error, adding the percent itself to the quantity, is similar to an error students made on item 1 in table 7.9, where 24 percent of the eighth-grade students subtracted the percent from the quantity (for example, subtracting 40 from 130 when asked to find 40 percent of 130).

The twelfth-grade students were presented with a similar but more challenging task, item 7 in the table. This item is more challenging in that the original quantity was identified by a letter (for example, t) rather than a numeral. Students in grade 12 found this item to be quite difficult, with fewer than one-fourth responding correctly. Forty-eight percent of the students selected a response that represented the amount of the increase, written as a decimal number, instead of the new quantity. This response indicates the students recognized how the percent of a literal quantity would be written (for example, $0.07t$), but did not recognize how to write the new amount (for example, $1.07t$). This may be the result of an emphasis in instruction on always solving this type of problem in two steps (that is, adding $.07t$ to the original amount t), or it may be the result of emphasizing finding percents less than 100 percent. The results on this item are similar to those reported on the 1990 and 1992 assessments.

The remaining Type A items are items 5, 6, and 8 in table 7.9. It should be noted that 26 percent of the students in grade 12 omitted item 8, an extended constructed-response question. (The students had the option of using a calculator on this item.) Perhaps some students chose not to attempt the item because they did not understand what is meant by "interest rate." However, a high omission rate is typical for NAEP extended constructed-response questions (see chapter 11 by Silver, Alacaci, and Stylianou), which are always the last items in an item block.

The second category of percent item, denoted here as Type B, are items in which the students are given the value of a specified percent of a quantity and asked to find the quantity. For example, if 40 percent of a number is 6, what is the number? The 1996 NAEP did not include an item of this exact form, but some students may have solved item 9 in table 7.9 in two steps, one of which was a Type B step; the students had the option of using a calculator on this item. The item included on the assessment is similar to the following example: "If 40 percent of a number is 6, what is 60 percent of the number?" It is possible that some students first found the size of the quantity (15 in the example), a Type B task, and then found 60 percent of 15, or 9. Other students, however, may have solved this task by writing a proportion (for example, $40/60 = 6/x$). Whatever method was used, about two-thirds of the students responded correctly to this item.

The remaining category of items, denoted here as Type C, are items in which the student has to determine what percent one number is of another; for example, what percent of 36 is 24? There was only one Type C item included on the assessment, item 10 in table 7.9; the students had the option of using a calculator on this item. This is an interesting item because it presents a situation which may be contrary to the student's intuition—that one set of percents could be lower than another set at each point in time yet higher overall. Twenty-three percent of the students responded correctly to this item. However, one-sixth of the twelfth-grade students omitted the item.

Other Proportional-Reasoning Items

The remaining proportional reasoning tasks included in the 1996 NAEP are summarized in table 7.11. Item 1 asks the students to identify the proportion satisfying a given criterion from a line graph showing percents. As indicated in table 7.11, approximately two-thirds of the students in grade 12 responded correctly to this item. The other three proportional-reasoning items are described in greater detail in tables 7.12–7.14. All three of these items are released items, and the students had the option of using the calculator on all three items. Accompanying these items are sample student responses that provide some insight into how students solved the problems. The responses for each task were from a sample of responses to be used for illustrative purposes; they do not necessarily represent the full range of responses to any particular item.

The task hereafter called "Cherry Drink" and shown in table 7.12 asked the twelfth-grade students to indicate who made the stronger drink and to

Table 7.11
Performance on Other Proportional Reasoning Items

Item Description	Percent Correct	
	Grade 8	Grade 12
1. Select approximate ratio from a graph listing percents on one axis.	—	63
2. Compare proportions from a situation; justify answer.	—	23
3. Use proportions (or rates) to solve a problem.	12	—
4. Compare amounts using percents or proportions; justify answer.	1	3

Note: Item 1 was a multiple-choice item. Items 2–4 were short constructed-response items from item blocks for which students were provided with, and permitted to use, scientific calculators. Items 2–4, all of which are released items, are shown in tables 7.12, 7.13, and 7.14, respectively.

explain their answer. Figure 7.1 shows some sample responses for different types of reasoning that students used when solving this task. On the basis of the set of sample responses, students used two different approaches to solve the problem, comparing syrup to water (or water to syrup) and comparing syrup to total liquid. Students who compared syrup-to-water ratios, 6/53 and 5/42, either found the decimals equivalent to each of the two fractions and then compared the decimals, as in response 1, or they compared the two fractions using a common denominator, as in response 2. Students also compared water-to-syrup ratios, sometimes reaching the correct conclusion, as in response 3, and other times arriving at an incorrect conclusion

Table 7.12
Cherry Drink

Item	Percent Responding
	Grade 12
Luis mixed 6 ounces of cherry syrup with 53 ounces of water to make a cherry-flavored drink. Martin mixed 5 ounces of the same cherry syrup with 42 ounces of water. Who made the drink with the stronger cherry flavor?	
Give mathematical evidence to justify your answer.	
Correct response ("Martin" with correct explanation)	23
Partial response (For example: compares correct ratios, but chooses "Luis" or makes a calculation error)	26
Incorrect response	42
Omitted	9

Note: This item is from an item block for which students were provided with, and permitted to use, scientific calculators.

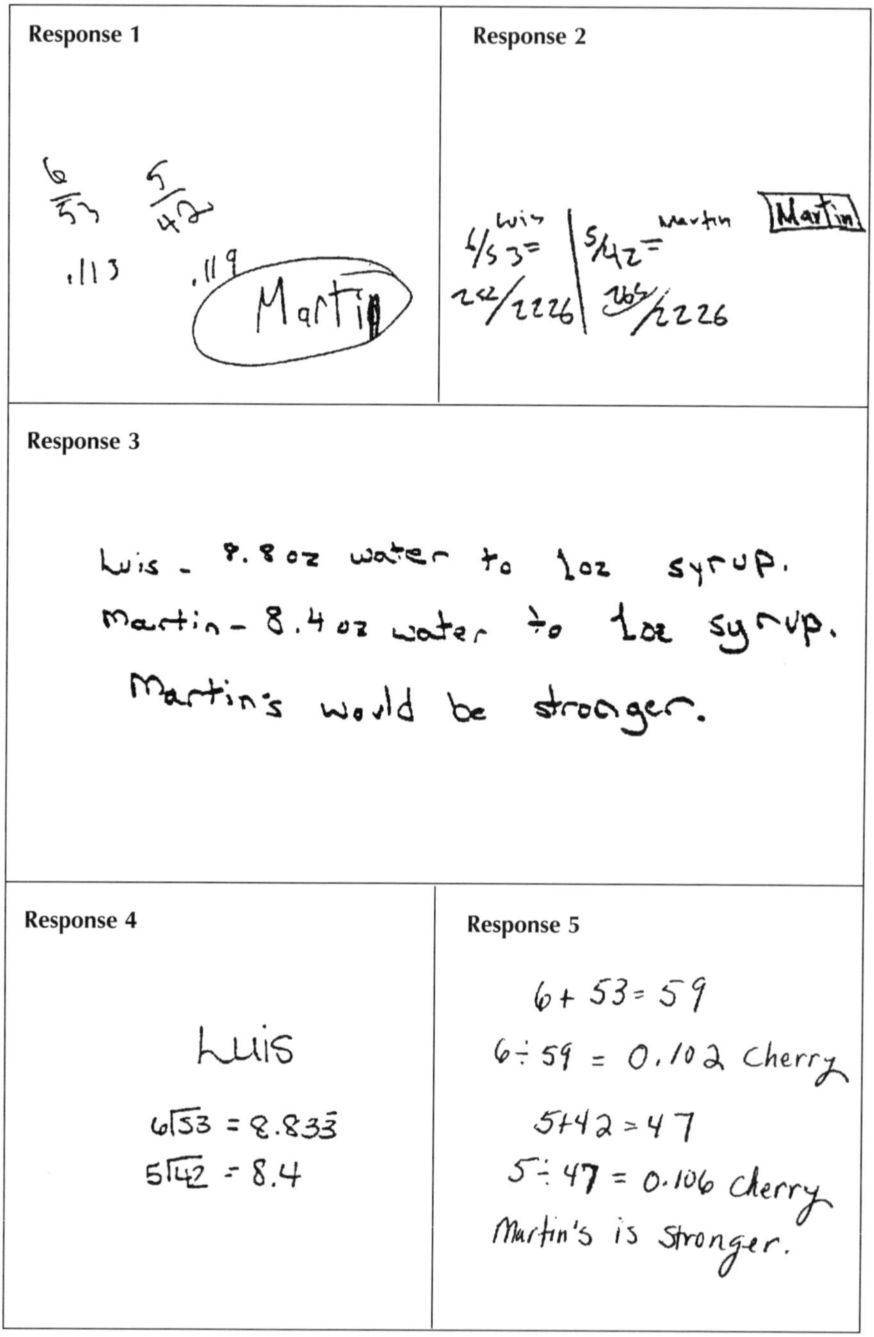

Response 1	Response 2
$\overset{6}{5\overline{\smash{)}3}}$ $\overset{5}{4\overline{\smash{)}2}}$ .113 .119 (Martin)	$\overset{\text{Luis}}{4/53=}$ $\overset{\text{Martin}}{5/42=}$ [Martin] $2\text{??}/2226$ $265/2226$

Response 3

Luis – 8.8 oz water to 1 oz syrup.
Martin – 8.4 oz water to 1 oz syrup.
Martin's would be stronger.

Response 4

Luis

$6\overline{\smash{)}53} = 8.83\overline{3}$

$5\overline{\smash{)}42} = 8.4$

Response 5

$6 + 53 = 59$
$6 \div 59 = 0.102$ Cherry
$5 + 42 = 47$
$5 \div 47 = 0.106$ Cherry
Martin's is Stronger.

Fig. 7.1. Selected responses to Cherry Drink

by neglecting to consider what the ratios represented, as in response 4. Another approach used by some students, illustrated in response 5, was to compare the ounces of syrup to the total ounces of liquid, water plus syrup. Unfortunately, the NAEP scoring guide provided no information about what percentage of the students selected each of these approaches, but we do know that 23 percent of the students responded correctly. Another 26 percent of the students compared ratios but either did not state what the comparison indicated or arrived at an incorrect solution.

Another released item, involving a comparison of rates for two vehicles, is described in table 7.13 and hereafter referred to as "Victor's Van and Sharon's Sedan." Designed as a two-part problem, eighth-grade students were asked to decide which of the two vehicles would first reach each of the two points and to explain their reasoning. As indicated in the table, only one-eighth of the students gave the correct response and provided a complete explanation for both parts of the question. An additional one-eighth either provided a correct answer and explanation for one of the parts or responded correctly to both questions but did not include an explanation.

Figure 7.2 contains two sample responses to this task. In responding, some of the students computed *miles per minute,* as in response 1; others computed *minutes per mile*, although not always appropriately labeled, as in response 2. Either approach to solving this item (miles per minute or minutes per mile) leads to a correct response; however, if the student is mislabeling the ratio, as the student did in response 2, this will lead to an incorrect response if the two are traveling at different speeds. For example, had the second vehicle (Sharon's sedan) been traveling at the rate of 20 miles in 30 minutes, the student who produced response 2 would have decided that the sedan was the faster vehicle because it was traveling at 1.5 miles per minute (when it really was traveling at the rate of 1.5 minutes per mile, making it the slower of the two vehicles).

On the basis of the sample set of responses, it appears that some students responded correctly to the first question by determining that the two vehicles traveled 8 miles in the same amount of time but were unable to respond to the second question. These students did not seem to recognize that the two vehicles were traveling at the same rate and so also would reach point B at the same time. The students may have thought they needed to know the distance in order to determine whether the vans would reach point B at the same time. There also were students who recognized that the van and the sedan were traveling at the same speed over the initial distance to point A but thought that was not the case over the second distance to point B.

The item shown in table 7.14 and referred to as "Population of Town A and Town B" asked the students to defend two different interpretations of a situation. Students received partial credit on this item if they defended one of the positions. Only 22 percent of the eighth-grade students and 27 percent

Table 7.13
Victor's Van and Sharon's Sedan

Item	Percent Responding Grade 8

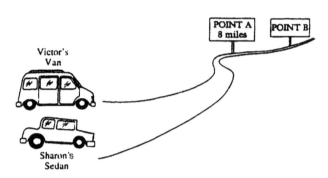

Victor's van travels at a rate of 8 miles every 10 minutes.
Sharon's sedan travels at a rate of 20 miles every 25 minutes.

If both cars start at the same time, will Sharon's sedan reach point A, 8 miles away, before, at the same time, or after Victor's van?

Explain your reasoning.

_____
_____
_____

If both cars start at the same time, will Sharon's sedan reach point B (at a distance further down the road) before, at the same time, or after Victor's van?

Explain your reasoning.

_____
_____

Correct response on both parts	12
Partial response: Answers either part correctly or answers both parts correctly with no justification	12
Incorrect response	67
Omitted	7

Note: Percents may not add to 100 because of rounding or off-task responses. This item is from an item block for which students were provided with, and permitted to use, scientific calculators.

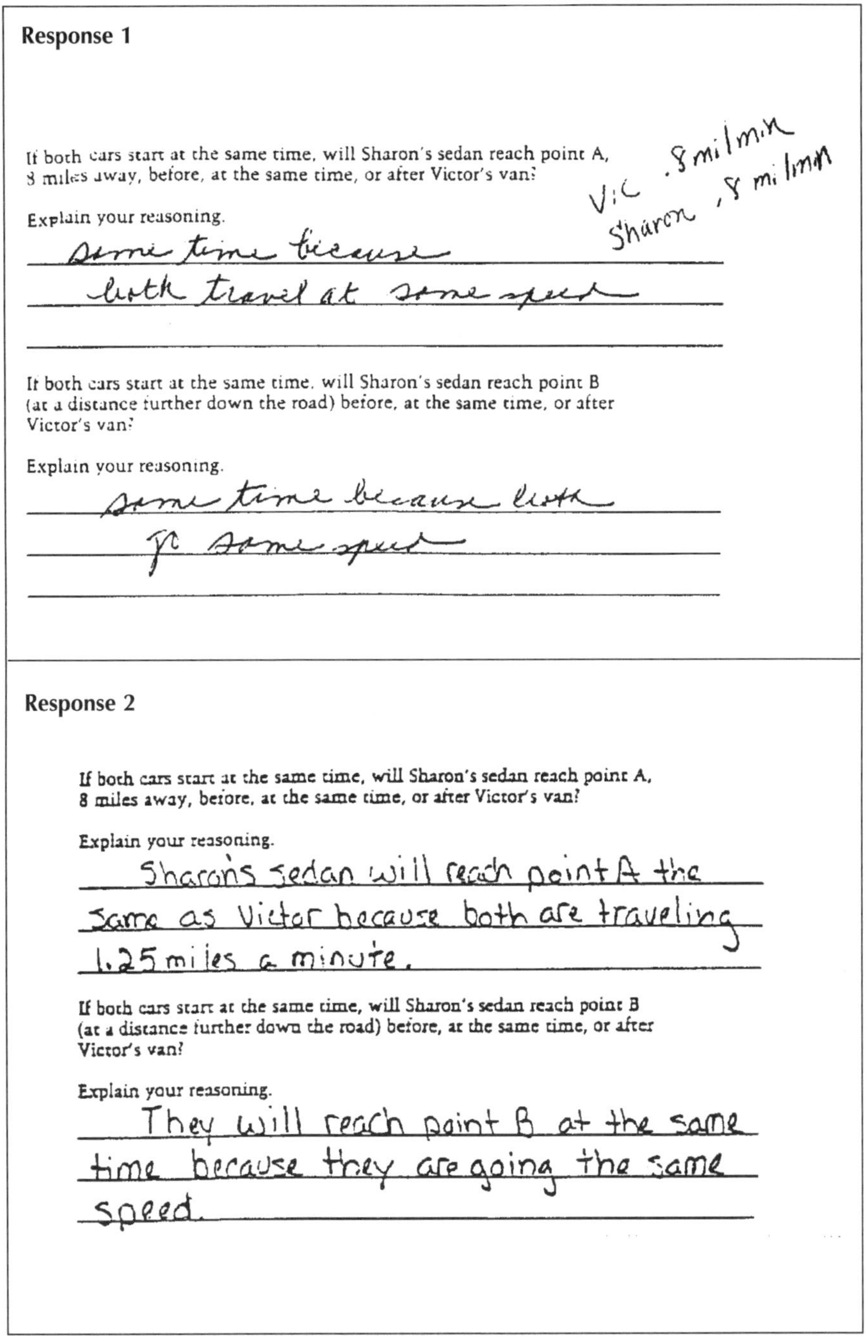

Response 1

If both cars start at the same time, will Sharon's sedan reach point A, 8 miles away, before, at the same time, or after Victor's van?

Explain your reasoning.

Same time because both travel at same speed

Vic .8 mi/min
Sharon .8 mi/min

If both cars start at the same time, will Sharon's sedan reach point B (at a distance further down the road) before, at the same time, or after Victor's van?

Explain your reasoning.

Same time because both go same speed

Response 2

If both cars start at the same time, will Sharon's sedan reach point A, 8 miles away, before, at the same time, or after Victor's van?

Explain your reasoning.

Sharon's sedan will reach point A the same as Victor because both are traveling 1.25 miles a minute.

If both cars start at the same time, will Sharon's sedan reach point B (at a distance further down the road) before, at the same time, or after Victor's van?

Explain your reasoning.

They will reach point B at the same time because they are going the same speed.

Fig. 7.2. Selected responses to Victor's Van and Sharon's Sedan

Table 7.14
Populations of Town A and Town B

Item	Percent Responding	
	Grade 8	Grade 12

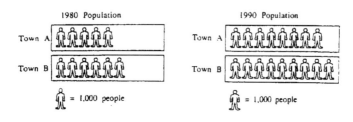

In 1980, the populations of Town A and Town B were 5,000 and 6,000, respectively. The 1990 populations of Town A and Town B were 8,000 and 9,000, respectively.

Brian claims that from 1980 to 1990 the populations of the two towns grew by the same amount. Use mathematics to explain how Brian might have justified his claim.

Darlene claims that from 1980 to 1990 the population of Town A had grown more. Use mathematics to explain how Darlene might have justified her claim.

Correct response for both Brian's and Darlene's claims	1	3
Partial response (for example: correct response for Brian or Darlene but not both)	21	24
Incorrect response	60	56
Omitted	16	16

Note: Percents may not add to 100 because of rounding or off-task responses. This item is from an item block for which students were provided with, and permitted to use, scientific calculators.

of the twelfth-grade students could adequately defend at least one of the positions, Brian's or Darlene's.

The responses in figure 7.3 provide some insight into how students solved this task. Defending Brian's position meant stating that the population increased by 3,000 in both towns. Students defended Darlene's position either by showing the percent of increase was more, as in response 1, or by showing the increase was proportionally more by comparing fractions, as in response 2. A common error students made in defending Darlene's position, as illustrated by response 3, was to compare fractions obtained by

RATIONAL NUMBERS

Response 1

B/c in 1980 the populations of A & B were 5000 & 6000 respectively and in 1990 the populations of A & B were 8000 & 9000 respectively. So they both increased by 3000 people
A 8000 - 5000 = 3000   &   9000 - 6000 = 3000

Town A increased by a greater percentage than Town B   3000 = __% of 5000 = 60%
3000 = __% of 6000 = 50%

Response 2

The population of town A & B both grew 3000 from 1980 to 1990, so they grew the same amount.

Town A had grown more than Town B percentage wise. Town A grew from 5,000 to 8,000 which was 3/5 of its population. Town B grew 6,000 to 9,000 which was 1/2.

Response 3

Town B -
grew 3,000
6/9 % = 67%

Town A -
grew 3,000
5/8 % = 63%

Fig. 7.3. Selected responses to Populations of Town A and Town B

comparing the 1980 population to the 1990 population in each town, 5/8 for Town A and 6/9 in Town B, rather than comparing the increase to the 1980 population. This error possibly led students to decide that because 6/9 was greater than 5/8, Darlene was wrong.

CONCLUDING COMMENTS

The NAEP assessments provide us with periodic pictures of student achievement in mathematics. From the students' responses, we infer the understandings they have of various topics in mathematics. Not only do we have these periodic pictures, but we also have the opportunity to look for change over time, to look for "progress." Approximately one-third of the rational number items on the fourth- and twelfth-grade 1996 NAEP and almost one-half of the rational number items on the eighth-grade assessment were also included on the 1990 and the 1992 assessments. It is on this set of items that we can look for change over time.

Due to the large sample size, a difference of 4 percent often represents a statistically significant difference; of the ten items referred to in this paragraph, eight involved more than a 4 percent change. Although a similar percentage of common items (items included on the 1990, 1992, and 1996 assessments) occurred at all three grade levels, the proportion of significant change was not uniform across grade levels. Fourth-grade students improved significantly in their performance on one of the eight items (12 percent) common to the last three NAEP assessments; students in grade 8 showed significant improvement on four of the eleven common items (36 percent); students in grade 12 showed significant gains on five of the nine common items (55 percent). Although most of the improvement occurred on fairly routine tasks, significant improvement is encouraging.

Several conclusions emerge from reviewing the results on the rational number portion of the 1996 NAEP. The first conclusion is that when students at multiple grade levels are presented with the same task, older students are more successful than younger students. This is not a surprising conclusion. However, the import of this conclusion is that eventually many students seem to acquire some understanding of the basic ideas of rational numbers. The second conclusion is that students at all grade levels are more successful in solving routine one-step tasks than nonroutine and multistep tasks. The rate of success on the items appears to decrease with the increasing complexity of the task.

The third conclusion that can be reached is that students appear to have great difficulty in justifying their responses, in explaining how they arrived at a particular conclusion. The results on the tasks requesting explanations do, of course, vary with the complexity of the tasks. However, even when the tasks are fairly straightforward, students appear to find it difficult to explain their responses. The role of communication in mathematics has received increased attention since the publication of the NCTM *Curriculum and Evaluation Standards for School Mathematics* (1989). There is increasing evidence of the importance of students being able to explain their solution strategies. Communication is included in virtually all state standards, and textbooks refer to its importance. Consequently, the results on this assessment are quite disappointing.

The fourth and final conclusion is that rational numbers continues to be a difficult topic for students; this is particularly evident at the fourth-grade level. The notion of what constitutes the unit appears to be at the root of many of their difficulties. It is difficult to think how fourth-grade students can make sense out of the various operations with fractions when they seemingly are unable to explain why one unit fraction is larger than another and have little understanding of equivalent fractions. In the same vein, one might question eighth-grade students' ability to judge reasonableness of answers if they have difficulty in judging if one fraction is larger than another. The topics of percent and decimals are built upon an understanding of rational numbers. It is thus not surprising that the topic of percent poses great problems for students in grade 8.

A number of implications for instruction can be drawn from these results. The first is that there is a need to develop more fully the basic notions of rational numbers. A second is that there is a need to provide students more opportunities to explain their reasoning and to hear the explanations of other students. A third implication of the results is that there is a need to furnish opportunities for students not only to solve number sentences but also to propose situations that would give rise to the number sentence. Finally, students need to have opportunities to solve more complex problems.

REFERENCES

Behr, Merlyn, Ipke Wachsmuth, Thomas R. Post, and Richard Lesh. "Order and Equivalence of Rational Numbers: A Clinical Teaching Experiment." *Journal for Research in Mathematics Education* 15 (November 1984): 323–41.

Bell, Alan, Malcolm Swan, and G. Taylor. "Choice of Operation in Verbal Problems with Decimal Numbers." *Educational Studies in Mathematics* 12 (November 1981): 399–420.

Carpenter, Thomas P., Terrance G. Coburn, Robert E. Reys, and James W. Wilson. "Results and Implications of the NAEP Mathematics Assessment: Elementary School." *Arithmetic, Teacher* 22 (October 1975): 438-50.

Carpenter, Thomas P., Terrance G. Coburn, Robert E. Reys, James W. Wilson, and Mary K. Corbitt. *Results from the First Mathematics Assessment of the National Assessment of Educational Progress*. Reston, Va.: National Council of Teachers of Mathematics, 1978.

Carpenter, Thomas P., Henry S. Kepner, Mary K. Corbitt, Mary Montgomery Lindquist, and Robert E. Reys. "Results of the NAEP Mathematics Assessment: Elementary School." *Arithmetic Teacher* 27 (September 1980): 8–12.

Edwards, Arthur. "A Study of Errors in Percentage." In *Twenty-ninth Yearbook of the National Society for the Study of Education*, edited by Guy M. Whipple, pp. 621–40. Bloomington, Ill.: Public Schools Publishing Company, 1930.

Fischbein, Efraim, Maria Deri, Maria S. Nello, and Maria S. Marino. "The Role of Implicit Models in Solving Verbal Problems in Multiplication and Division." *Journal for Research in Mathematics Education* 16 (January 1985): 3–17.

Hiebert, James, and Thomas P. Carpenter. "Learning and Teaching with Understanding. In *Handbook of Research on Mathematics Teaching and Learning*, edited by Douglas A. Grouws, pp. 65–97. New York: Macmillan, 1992.

Hiebert, James, and Diana Wearne. "A Model of Students' Decimal Computation Procedures." *Cognition & Instruction* 2 (1985): 175–205.

Kircher, H. W. "Study of Percentage in Grade VIII A." *Elementary School Journal* 26 (December 1926): 281–89.

Kouba, Vicky L., Catherine A. Brown, Thomas P. Carpenter, Mary M. Lindquist, Edward A. Silver, and Jane O. Swafford. "Results of the Fourth NAEP Assessment of Mathematics: Number, Operations, and Word Problems." *Arithmetic Teacher* 35 (April 1988): 14–19.

Kouba, Vicky L., Judith S. Zawojewski, and Marilyn E. Strutchens. "What Do Students Know about Numbers and Operations?" In *Results from the Sixth Mathematics Assessment of the National Assessment of Educational Progress*, edited by Patricia Ann Kenney and Edward A. Silver, pp. 33–60. Reston, Va.: National Council of Teachers of Mathematics, 1997.

Larson, Carol N. "Locating Proper Fractions on Number Lines: Effect of Length and Equivalence." *School Science and Mathematics* 53 (May-June 1980): 423–28.

Lesh, Richard, Thomas Post, and Merlyn Behr. "Proportional Reasoning." In *Number Concepts and Operations in the Middle Grades*, edited by James Hiebert and Merlyn Behr, pp. 93–118. Reston, Va.: Lawrence Erlbaum Associates and National Council of Teachers of Mathematics, 1988.

Mack, Nancy K. "Confounding Whole-Number and Fraction Concepts When Building on Informal Knowledge." *Journal for Research in Mathematics Education* 26 (November 1995): 422–41.

National Council of Teachers of Mathematics. *Curriculum and Evaluation Standards for School Mathematics*. Reston, Va.: National Council of Teachers of Mathematics, 1989.

Noelting, Gerald. "The Development of Proportional Reasoning and the Ratio Concept: Part 1-Differentiation of Stages." *Educational Studies in Mathematics* 11 (May 1980): 217–53.

Parker, Melanie, and Gaea Leinhardt. " Percent: A Privileged Proportion." *Review of Educational Research* 65 (Winter 1995): 421–81.

Post, Thomas R. "Fractions: Results and Implications from National Assessment." *Arithmetic Teacher* 28 (May 1981): 26–31.

Post, Thomas R., Ipke Wachsmuth, Richard Lesh, and Merlyn Behr. "Order and Equivalence of Rational Numbers: A Cognitive Analysis." *Journal for Research in Mathematics Education* 16 (January 1985): 18–36.

Risacher, Bille F. "Students' Reasoning about Ratio and Percent." In *Proceedings of the Fifteenth Annual Meeting of the North American Chapter of the International Group for the Psychology of Mathematics Education*, edited by Joanne R. Becker and Barbara J. Pence, pp. 261–67. San Jose, Calif.: San Jose State University, 1993.

Resnick, Lauren B., Pearla Nesher, Francois Leonard, Maria Magone, Susan Omanson, and Irit Peled. "Conceptual Bases of Arithmetic Errors: The Case of Decimal Fractions." *Journal for Research in Mathematics Education* 20 (January 1989): 8–27.

Sackur-Grisvard, Catherine, and Francois Leonard. "Intermediate Cognitive Organizations in the Process of Learning a Mathematical Concept: The Order of Positive Decimal Numbers." *Cognition and Instruction* 2 (1985): 157–74.

Vergnaud, Gerard. "Multiplicative Structures." In *Acquisition of Mathematics Concepts and Processes*, edited by Richard Lesh and Marsha Landau, pp. 127–74. New York: Academic Press, 1983.

Wearne, Diana, and James Hiebert. "Constructing and Using Meaning for Mathematical Symbols: The Case of Decimal Numbers." In *Research Agenda in Mathematics Education: Number Concepts and Operations in the Middle Grades*, edited by James Hiebert and Merlyn Behr, pp. 220–35. Reston, Va.: Lawrence Erlbaum Associates and National Council of Teachers of Mathematics, 1988.

8

Geometry and Measurement

W. Gary Martin and Marilyn E. Strutchens

GEOMETRY has long been an important part of the school mathematics curriculum because of its usefulness in relating spatial ability and problem-solving skills, in illustrating applications to real-life problems, and in developing both basic skills and advanced concepts (Sherard 1981). Similarly, measurement provides important connections between school mathematics and everyday life. In this chapter, we explore the current status of students' achievement in geometry and measurement based on the findings from the 1996 NAEP assessment. Although measurement extends beyond geometry to areas such as time and temperature, we limit our discussion to measurement within the context of geometry (for example, linear measurement, area, and perimeter). When appropriate, we compare results to items common to previous NAEP assessments in order to understand better the degree of progress that is being made.

In addition to reporting and interpreting performance on the basis of NAEP results, for some short and extended constructed-response items we also examined a set of sample responses. This sample was not a representative sample of all the responses; rather, it was a convenience sample of nonblank responses. Some of the items described in this chapter were extended constructed-response tasks, a type of NAEP item that is discussed extensively in chapter 11 by Silver, Alacaci, and Stylianou.

The 1996 NAEP assessment included about eighty items in the content strands Geometry and Spatial Sense and Measurement, and a subset of these items is discussed in this chapter. The items tested a variety of students' abilities using multiple-choice, short constructed-response, and extended constructed-response formats. On certain items students were permitted to use calculators, paper rulers or protractors, and geometric shapes.

Highlights

Geometry

- Fourth-grade students performed better on items that were accompanied by manipulatives than on items that asked them to outline figures on a grid. Similarly, eighth-grade students performed better on such items than on those that used pictures of geometric shapes. For both grades, students' performance on the manipulatives items increased significantly from 1992 to 1996.

- Fourth- and eighth-grade students performed better on items that asked for examples of geometric figures than on those that asked them to produce or identify nonexamples of figures.

- Students at all three grade levels seemed to be generally successful in identifying and naming two-dimensional and three-dimensional figures and their properties. However, when asked to identify figures in less familiar contexts or to recognize more-complex properties, students' performance dropped.

- Performance levels by eighth- and twelfth-grade students for even direct application of the Pythagorean theorem were quite low, and the rates were even lower in situations that asked students to use the theorem in a problem-solving situation.

- According to the average NAEP scale scores, performance in the NAEP content strand Geometry and Spatial Sense has steadily increased from 1990 and 1992 to 1996. The greatest gain was at grade 4, where the average scale score rose from 213 in 1990 to 225 in 1996.

Measurement

- Students at all three grade levels generally had a sense of appropriate units of measurement for particular situations but were more familiar with the customary system of measure than the metric system.

- Fourth-grade students had difficulty with all items that asked them to use a ruler to measure an object or to draw a shape with particular dimensions.

- Students across all three grade levels had difficulty with perimeter and area concepts, especially situations in which they had to explain or justify answers.

- Items assessing familiarity with volume and surface area concepts were difficult for students at all three grade levels.

- According to the average NAEP scale scores from 1990, 1992, and 1996, performance in the NAEP content strand Measurement has increased only modestly, with a difference of less than 10 scale points between 1990 and 1992 and 1992 and 1996.

The sections that follow (1) discuss students' performance on clusters of geometry items related to recognizing and drawing geometric figures and to properties of geometric figures and (2) describe students' performance on measurement of geometric figures.

RECOGNIZING AND DRAWING GEOMETRIC FIGURES

The ability to visualize figures and operations on them has been recognized as an important component of mathematical thinking (Wheatley 1990) and provides a basis for the development of advanced geometric thinking. The following description of NAEP items relating to recognizing and drawing figures, and students' performance on them, is separated into two sections, the first dealing with two-dimensional figures and the second with three-dimensional figures and their nets.

Two-Dimensional Figures

Eight NAEP items assessing students' ability to recognize and draw two-dimensional figures are described in table 8.1. Of these, one item was given to all three grade levels, two items were given to both grades 4 and 8, and the remaining five items were each given to only one grade level. Because twelfth-grade students answered only one item, the focus of this section is on those items administered at the other two grades.

Students in grade 4 were given six items related to two-dimensional figures and were asked to draw either examples or nonexamples of figures. These students performed better on the items that asked them to draw or recognize examples of figures than on items that asked for nonexamples. In particular, fewer than one-fifth of fourth-grade students answered items 7 and 8, the nonexample items correctly, whereas on the first four items in the table, which asked for examples of geometric figures, performance ranged from about 75 percent to about 35 percent correct.

The lower performance on items that required the drawing of nonexamples suggests that this activity is far more challenging. As suggested by van Hiele (1986), the drawing of an example of a familiar geometric figure (for example, a square, rectangle, or triangle) can be done using "visual prototypes;" that is, the students may keep in mind a typical square or triangle that they easily identify or reproduce when asked to do so. However, the drawing of a nonexample requires students to know the properties of a square—that is, what makes a figure a square (besides visual similarity to their prototype square)—and thus what properties are missing from a nonsquare. NAEP performance results for two released items (items 1 and 7 in table 8.1) can illustrate the difference between producing examples and

Table 8.1
Performance on Items Involving Recognizing or Drawing Examples and Nonexamples of Two-Dimensional Figures

Item Description	Percent Correct		
	Grade 4	Grade 8	Grade 12
Examples of figures			
1. Trace a square using manipulatives in the form of an isosceles right triangles.	73	89	—
2. Trace a hexagon using manipulatives in the form of various geometric shapes.	63	82	—
3. Identify squares in a given figure depicted on a grid.	46	—	—
4. Identify triangles in a given figure depicted on a grid.	38	—	—
5. Show how a rectangle can be formed from pictures of geometric shapes.	—	65	—
6. Show how a quadrilateral can be formed from pictures of geometric shapes.	—	62	—
Nonexamples of figures			
7. Trace a nonsquare using manipulatives in the form of an isosceles right triangles.	16	48	58
8. Show how a nonrectangle can be formed from drawn geometric shapes.	8	—	—

Note: All were short constructed-response items. Item 7, a released item, is shown in figure 8.1.

nonexamples of squares. Results for item 1 show the relative ease with which fourth-grade students could produce a square (73 percent correct) using manipulatives in the form of isosceles right triangles. However on item 7, which required students to use the same manipulatives to "make a 4-sided shape that is not a square," performance dropped to 16 percent correct.

Of particular note was a class of responses to item 7 in which students produced shapes like the one shown in figure 8.1. In these shapes, the base was not horizontal, and perhaps students thought that the figure was a diamond instead of a square. This class of responses is consistent with research suggesting that students may regard orientation as a salient feature of a figure, since they tend only to see figures in a prototypical orientation—for example, with horizontal bases (Vinner and Hershkowitz 1980). Students need to experience shapes in a variety of orientations so that they can recognize geometric figures regardless of how they appear.

Another finding from fourth-grade students' performance on the items involving two-dimensional figures was that, in general, they performed

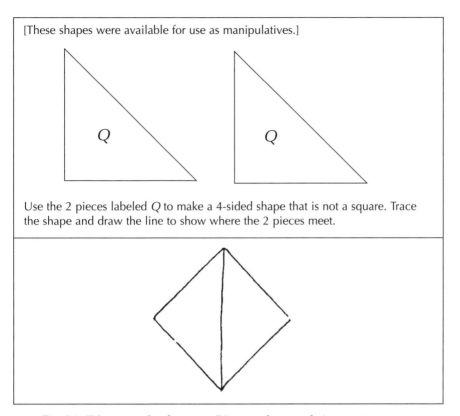

Fig. 8.1. "Nonexample of a square" item and a sample incorrect response

better on items that were accompanied by manipulatives (items 1 and 2) than on items that asked them to outline figures on a grid (items 3 and 4). This performance of students may suggest an increase in the use of manipulatives in elementary school mathematics classes, and thus greater familiarity with manipulatives.

Not surprisingly, the eighth-grade students were more successful than the fourth-grade students on the three common items, with correct response rates for items 1 and 2 exceeding 80 percent. In items 5 and 6 students were asked to use pictures of geometric shapes to form other shapes; as was the case with the fourth-grade students, the correct response rate on these items was somewhat lower than on items 1 and 2, where they used manipulative shapes. And as for the younger students, item 7, involving a nonexample, proved hardest for students in grade 8.

Another finding from performance on this set of items was that for the items that involved the use of geometric shapes as manipulatives, performance increased significantly from 1992 to 1996 for items 1 and 2 at grade 4 and for those two items and also item 7 at grade 8. This suggests that students may have experienced an increased familiarity with the use of manipulatives, a conjecture that is supported by data from the NAEP student questionnaires. On the questionnaires for both grades, students were asked about the frequency with which they worked with "objects like rulers, counting blocks, and geometric shapes or solids." From 1992 to 1996, the percent of students who responded that they worked with manipulatives on a weekly basis increased modestly but significantly from 34 percent to 41 percent at grade 4 and from 20 percent to 25 percent at grade 8.

Three-Dimensional Figures and Their Nets

Four NAEP items assessed students' ability to work with nets, which are two-dimensional patterns that can be folded to form hollow three-dimensional solids. Some examples of possible nets for geometric solids appear in figure 8.2. (The nets in the figure are not necessarily associated with any of the four NAEP items.)

Results for the NAEP items are given in table 8.2. For item 1, students in grade 4 were almost twice as successful in drawing a two-dimensional decomposition of a common three-dimensional object, with about two thirds answering correctly, than they were in identifying a net that could *not* be used to form a simple three-dimensional figure in item 2. This difference in performance is even more striking when considering that item 1 was a

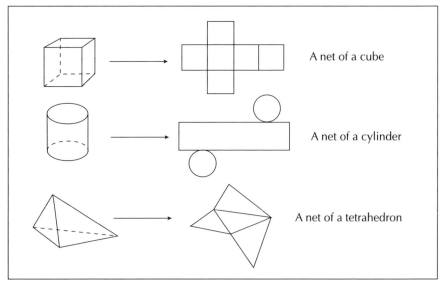

Fig. 8.2. Examples of nets that can be used to form solids

constructed-response item and item 2 a multiple-choice item. As noted in other chapters (for example, chapter 2 by Dossey; chapter 11 by Silver, Alacaci, and Stylianou), performance on NAEP constructed-response items tends to be lower than that on multiple-choice items. The difference may be explained in part by differences in the requirements of the items. The first item asked students to describe the two-dimensional shape formed when the lateral face of an object is decomposed; the second item asked students to select a nonexample of a net. As noted in the previous section, nonexamples of two-dimensional figures also proved to be of particular difficulty.

Table 8.2
Performance on Items Involving Three-Dimensional Figures and Their Nets

Item Description	Percent Correct		
	Grade 4	Grade 8	Grade 12
1. Identify the shape formed by a two-dimensional decomposition of a given three-dimensional figure.	67	83	—
2. Select the figure that is *not* a net that could be used to form a given three-dimensional figure.	34	65	75
3. Draw nets to form a rectangular solid of a given volume.	—	8[a]	—
4. Identify the three-dimensional figure formed by a given net.	—	—	51

[a]Percent of students scoring at either the satisfactory or extended level.
Note: Items 1 and 4 were short constructed-response items, item 2 was a multiple-choice item, and item 3 was an extended constructed-response item.

The same pattern of responses for items 1 and 2 held for students in grade 8. They were also more successful with item 1, where more than 80 percent responded correctly, than with item 2, where about two thirds were successful. For item 3, an extended constructed-response task, eighth-grade students' performance was dramatically lower. The low correct response rate on this item can be explained in part by the additional requirement that the net be accurately produced so that the solid formed has a given volume.

Students in grade 12 were somewhat more successful on item 2 than on item 4, in which they were asked to identify the figure formed by a net. It is noteworthy that the rates of correct responses on both items improved significantly from previous assessments for students in grade 12—up from 70 percent in 1990 on item 2 and up from 42 percent in 1992 on item 4. Eighth-grade students also improved significantly on item 2, up from 59 percent in 1990. These performance increases may suggest increased curricular attention given to three-dimensional geometry in general and to nets in particular. Again, there is evidence from the NAEP student questionnaire that supports this conjecture. As reported previously, the difference between the percentage of students in grade 8 who reported working with objects such as geometric solids on a weekly basis increased significantly from 1992 to 1996. A similarly modest but significant increase appeared at grade 12, with the percent of students reporting weekly use of geometric solids up from 20 percent in 1992 to 25 percent in 1996.

ANALYSIS OF PROPERTIES OF GEOMETRIC FIGURES

The description, comparison, and classification of geometric figures and their properties are important skills in the content area of geometry (NCTM 1989). Classifying geometric figures and understanding properties of geometric figures, such as parallelism and congruence, are necessary both for students' use of geometric models to solve problems and for students' development of an appreciation of the geometric nature of their physical world. In the next sections, we discuss items related to classification of geometric shapes using their properties, reflections and symmetry, the Pythagorean theorem, similarity, and properties of particular geometric figures such as triangles and circles.

Classification of Geometric Shapes Using Their Properties

Ten NAEP items, which are described in table 8.3, assessed students' ability to classify geometric figures using their properties. Four items were administered to only fourth-grade students, one item to both fourth- and eighth-grade students, two items to both eighth- and twelfth-grade students, one item to only eighth-grade students, one item to only twelfth-grade students, and the remaining item was administered to students at all three grade levels.

Students in grades 4 and 8 were generally successful in classifying and identifying simple two-dimensional figures such as triangles and familiar three-dimensional figures such as cubes, with percent-correct values ranging from 46 percent in grade 4 to 77 percent in grade 8. Results from items administered to eighth- and twelfth-grade students show that, as the geometric figures and the situations in the items became more complex (for example, parallelograms other than squares and rectangles; drawing a parallelogram with perpendicular diagonals), performance levels decreased.

Students in grade 4 were generally successful in classifying simple figures using their associated properties—for example, in item 1 more than 90 percent correctly identified a figure on the basis of a given property. However, performance levels dropped below 30 percent on items in which students

Table 8.3
Performance on Items Involving Properties of Geometric Figures

Item Description	Percent Correct		
	Grade 4	Grade 8	Grade 12
1. Choose the shape that satisfies a given property.	92	—	—
2. Explain how a square is different from two given triangles.	59	75	83
3. Identify a property of a cube and apply it to a given situation.	46	77	—
4. Draw a figure with a specific number of sides and angles.	29	—	—
5. Determine the shape formed under certain conditions.	28	—	—
6. Choose a property that is not true for parallelograms.	—	43	68
7. Identify a type of triangle on the basis of given properties.	—	32	59
8. Given several properties that define a particular geometric figure, select a deduction that must be true.	—	21	—
9. Use a ruler to draw a parallelogram with perpendicular diagonals.	—	—	19
10. Compare a rectangle and a parallelogram on the basis of features that are alike and different.	11[a]	—	—

[a]Percent of students scoring at either the satisfactory or extended level.
Note: Items 2, 4, and 9 were short constructed-response items, item 10 was an extended constructed-response item, and the rest were multiple-choice items. Items 2 and 10, both released items, are shown in table 8.4.

were asked to draw a figure with a given number of right angles (item 4) and were required to know that a particular geometric figure had been formed under specific conditions. Also, only about 46 percent of the fourth-grade students were able to answer item 3 correctly, which required them to know the properties of a cube; a little over 75 percent of the eighth-grade students answered the same item correctly.

As noted above, performance for eighth- and twelfth- grade students dropped when the geometric figures became more complex and when the task in the item was of a less-routine nature. A close look at performance on items 6 through 9 in table 8.3 illustrates this claim. A little more than two-fifths of the eighth-grade students and a little more than two-thirds of the twelfth-grade students were able to identify a false statement about a parallelogram in item 6. In item 7, only 32 percent of eighth-grade students and

59 percent of twelfth-grade students were able to identify a particular type of triangle; and in item 8, only about 20 percent of eighth-grade students were able to make a simple deduction about a figure, given a list of its properties.

Item 9, a released item, asked twelfth-grade students to use a ruler to draw a parallelogram with perpendicular diagonals and to show the diagonals. Fewer than one-fifth were able to produce a correct drawing of the parallelogram. An examination of a small set of students' responses revealed some misconceptions about the task. For example, some students drew parallel lines instead of a parallelogram. Other students drew parallelograms with no diagonals or with only one diagonal. Still others drew parallelograms with both diagonals shown but not perpendicular.

Two of the items in table 8.3 (items 2 and 10), involving properties of geometric figures, were released constructed-response items that required students to provide verbal responses. These items, which are shown in table 8.4, provide further insights regarding students' understanding of geometric properties and their level of geometric thinking.

The first item in table 8.4 asked fourth-grade students to compare a rectangle and a nonrectangular parallelogram. The results show that these students had a difficult time describing both the similarities and differences of a rectangle and a parallelogram drawn on a grid, as less than 1 percent of the students scored at the extended level. Descriptions of the criteria for various performance categories and additional discussion of performance on this task, called Compare Geometric Figures, can be found in chapter 11 by Silver, Alacaci, and Stylianou.

An analysis of a convenience sample of nonblank students' responses to the Compare Geometric Figures task revealed that many students identified similarities and differences that focused on the *appearances* of the two shapes. For example, students frequently described the similarities in terms of both shapes having four sides or edges or both shapes containing "little squares" (the grid marks). With respect to differences, the responses contained references to one shape being "slantier" than the other or one shape having only straight lines. Fewer students noticed that the shapes were the same height, and almost none used geometric terms like *parallel lines, perpendicular lines, right angles,* or *area.* At best, some students talked about four *equal* sides or corners. Some students noted that both figures contained the same number of squares if you counted them, a response that suggests some knowledge of area.

On the basis of this set of responses, there is evidence that as suggested by van Hiele (1986), most students were operating on the visual features of the geometric shapes rather than their geometric properties. This suggests that in responding to the Compare Geometric Figures task, students were performing at the first level of geometric thinking, often referred to as the

GEOMETRY AND MEASUREMENT

Table 8.4
Comparison of Geometric Figures: Sample Items

Item	Percent Responding		
	Grade 4	Grade 8	Grade 12
1. [General directions]			

Think carefully about the following question. Write a complete answer. You may use drawings, words, and numbers to explain your answer. Be sure to show all of your work.

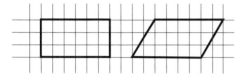

In what ways are the figures above alike? List as many ways as you can.

In what ways are the figures different? List as many ways as you can.

Extended response	<1	—	—
Satisfactory response	11	—	—
Partial response	29	—	—
Minimal response	31	—	—
Incorrect	23	—	—
Omitted	5	—	—

[These shapes were available for use as manipulatives.]

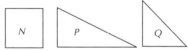

2. Laura was asked to choose 1 of the 3 shapes N, P, and Q that is different from the other 2. Laura chose shape N. Explain how shape N is different from shapes P and Q.

Correct response (for example, N is a square and P and Q are triangles; N has 4 sides, but the other two figures have 3 sides.)	59	75	83
Incorrect response based on shapes having different letters with no reference to geometric properties	9	4	3
Any other incorrect response	25	16	13
Omitted	6	5	2

Note: Percents may not add to 100 because of rounding.

visual level, in which the focus is on shape and other geometric configurations based on appearance.

The second item in table 8.4, which was administered at all three grade levels, asked students to compare the geometric properties of figures and provided another opportunity to examine students' geometric thinking. While working on this item, students had manipulatives in the form of geometric shapes: a square labeled N; an isosceles right triangle labeled Q; and a second right triangle, the same height as Q but with a longer base, labeled P. About 60 percent of the fourth-grade students, 75 percent of the eighth-grade students, and 83 percent of the twelfth-grade students correctly explained how shape N was different from shapes P and Q.

A convenience sample of fourth-grade students' responses provided additional information on how students explained the difference between the shapes. The majority of the responses stated that the shape labeled N was a square and the other two shapes were triangles, and thus, the focus was on the geometric properties of the three figures and the use of those properties as the criteria for recognition and characterization of the shapes. This stands in contrast to the focus of responses to item 1 (Compare Geometric Figures) in which the focus was on appearance. Thus, in item 2, students' explanations tended to be more descriptive and analytical rather than visual (van Hiele 1986), and thus, students were performing at the second level of the van Hiele model, often referred to as the *descriptive/analytic* level (Clements and Battista 1992).

From the set of students' responses, it became evident that students were often able to go beyond the surface features of the shapes in noting the differences between the shapes. In fact, some of the fourth-grade students were able to classify shape N in a number of ways (a square, a rectangle, and even a quadrilateral). This type of response was illustrated by one student who wrote, "Because shape N is a square, that makes it a quadrilateral. The rest of the shapes are triangles." Other responses included comments regarding the properties of the figures, such as "all of the sides of N are equal" and "all of the sides of N are parallel."

However, not all students focused on the geometric properties of the three shapes. Some of the responses indicated that students were still focusing on features of the geometric shapes other than their geometric properties. These students were not using geometric vocabulary; they were describing what they saw and not making any connections to formal mathematics. This item not only showed fourth-grade students' ability to tell the difference between a square and a triangle, it also provided examples of those students' perceptions of a triangle. A common type of response implied that shape P is not a triangle, as illustrated by one student who responded, "Shape N is a square, and shape Q is a triangle, and the shape P is like a flag on a stick." As we noted earlier, these students may still be operating on the visual features of the shapes, comparing each shape to visual prototypes (van Hiele 1986) of triangles and rejecting those shapes that do not appear to fit these prototypes.

Results from the NAEP assessment show that students at all three grade levels seemed to be quite successful in identifying and naming familiar two-dimensional and three-dimensional figures and their properties, usually operating at a visual level. However, when asked to identify figures in less familiar contexts or to recognize more-complex properties, students' performance dropped, which suggests that students should have experience with a variety of geometric figures in more complex contexts. Students' performance on this set of NAEP items indicates that students need more experience in examining geometric shapes and their properties. They need to experience exploring shapes at different cognitive levels; that is, students need to be able to identify shapes on the basis of their appearance, and they also need to understand the properties of shapes and what those properties mean. For example, students need to experience activities in which they are engaged in sorting shapes based on their geometric attributes, and at a higher cognitive level they should engage in activities that require them to identify a shape on the basis of a list of its properties or determine the least number of properties one would need to describe a particular shape (Van de Walle 1997).

Reflections and Symmetry

The NCTM *Curriculum and Evaluation Standards* (1989) recommends increased attention to topics associated with transformational geometry, including reflections and symmetry. Students encounter reflections and symmetry as properties of figures in everyday life; these properties of figures also provide a useful way to categorize figures. We examined seven NAEP items that assessed students' ability to recognize or produce reflections or lines of symmetry. One item was administered to all three grade levels, one item to both grades 4 and 8, two items to grades 8 and 12, and the remaining three items were given to one grade level each. Descriptions of, and performance results for, these items appear in table 8.5. All of these items are secure, which limits a detailed discussion of particular items.

Students' ability to answer items about reflections correctly varied by grade level and by the complexity of the situation in the item. In item 1, more than 70 percent of students in grade 4, more than 80 percent in grade 8, and more than 90 percent in grade 12 were able to identify a "flipped" figure. However, only about one-fourth of the students in grade 4 and slightly more than half of students at grade 8 were able to produce a figure folded over a vertical line as in item 3. Likewise, in item 4, in which eighth-grade students were asked to identify the image of a point when "folded" over an oblique line of reflec-

Table 8.5
Performance on Items Involving Reflections and Symmetry

Item Description	Percent Correct		
	Grade 4	Grade 8	Grade 12
Reflections			
1. Choose the correct representation for a figure when it is flipped over.	71	82	91
2. Indicate the location of several figures when folded over a vertical line.	—	59	—
3. Indicate the location of one figure when folded over a vertical line.	26	55	—
4. Identify the correct location of a point when folded over an oblique line.	—	48	62
Symmetry			
5. Select the figure that is not symmetrical.	91	—	—
6. Choose the figure that does not have a line of symmetry.	—	—	84
7. Draw lines of symmetry.	—	44	54

Note: Items 2, 3, and 7 were short constructed-response items, and the rest were multiple-choice items.

tion, fewer than half were successful; by grade 12, the correct response rate was still less than two-thirds. Shultz and Austin (1983) have identified the increased difficulty of working with oblique reflection lines.

Performance on symmetry was quite mixed. More than 90 percent of students at grade 4 correctly selected the nonsymmetric figure in item 5. However, fewer than 50 percent of students at grade 8 successfully drew lines of symmetry for given figures in item 7; the success rate for grade 12 was just over 50 percent. The difficulty of this item for students in the upper two grades, as compared to that of item 5 for fourth-grade students, may be attributed at least in part to several factors. First, students were asked to produce the lines of symmetry in item 7, a short constructed-response item, rather than choose them as in item 5, a multiple-choice item. Second, item 7 uses the formal vocabulary of symmetry, rather than the informal vocabulary of "matching parts." Third, the lines of symmetry in item 7 were oblique, whereas those in item 5 were horizontal. Students in grade 12 did much better at choosing lines of symmetry in item 6; more than 80 percent were successful.

The correct response rate for twelfth-grade students increased significantly from previous NAEP assessments on three of the four items (items 1, 6, and 7). Performance on item 7 showed particularly large increases from 1990 to 1996, up from 30 percent to 54 percent. Correct-response rates increased significantly on both items 1 and 6, with the increase for item 1 up from 86 percent in 1990 to 91 percent in 1996 and that for item 6 up from 71 percent in 1992 to 84 percent in 1996. The increases in performance at grade 12 may suggest increased emphasis on transformational geometry in the school mathematics curriculum, as suggested in the NCTM *Standards* (1989).

Pythagorean Theorem

Five NAEP items assessed students' knowledge of the Pythagorean theorem. Descriptions and performance results for these items appear in table 8.6. Because all but one of the Pythagorean theorem items were administered at grade 12, the focus of this section is on twelfth-grade students' performance on this topic.

Table 8.6
Performance on Items Involving the Pythagorean Theorem

Item Description	Percent Correct	
	Grade 8	Grade 12
1. Given the length of two legs of a triangle, determine the length of the hypotenuse (integral lengths).	31	—
2. Given the length of two legs in a triangle, determine the length of the hypotenuse to the nearest tenth (decimal lengths).	—	57
3. Identify a special right triangle.	—	24
4. Sketch a right triangle based on given information about the lengths of the legs and hypotenuse.	—	16
5. Apply the Pythagorean theorem in a problem situation.	—	8

Note: Items 4 and 5 were short constructed-response items, and the rest were multiple-choice items.

Of the four items relating to the Pythagorean theorem, none were answered correctly by more than 60 percent of the students. The twelfth-grade students performed best on item 2, which asked them to calculate the length of the hypotenuse given the lengths of the legs, rounding the answer to the nearest tenth; 57 percent answered correctly. Students had access to a scientific calculator while working on this item. Perhaps the response rate for item 2 might have been somewhat higher had the given lengths been integral. Moreover, the performance on this item could have been affected by the rather high nonresponse rate (16 percent). That is, because the item was the seventeenth item in a twenty-item block, some students were not able to work on this and all subsequent items due to time constraints. The NAEP results, then, are based on only that portion of the students who responded to this item, thus limiting the interpretation of the performance.

The performance of the students in grade 12 dropped as the demands of the tasks increased. In item 4, students were asked to sketch a right triangle given certain information about the lengths of its sides; only 16 percent answered correctly. When asked to use the Pythagorean theorem to solve a word problem in item 5, only 8 percent of twelfth-graders answered correctly. Another 9 percent were able to apply the Pythagorean theorem but did not correctly interpret the results in answering the question. Finally, fewer than one-fourth were able to identify a special right triangle in item 3. One positive sign is a significant increase in the rate of correct responses from 1990 on item 2, from 47 to 57 percent.

The Pythagorean theorem is one of the most famous and useful theorems of geometry. Yet according to NAEP results, students possess limited knowledge about it. Performance levels for even direct application of the theorem are quite low, and the rates are even lower in situations that ask students to use the theorem in a problem-solving situation. Although one can imagine that the vast majority of students may be able to parrot "a squared plus b squared equals c squared," they are not developing usable knowledge about the theorem.

Similarity and Trigonometry

Five NAEP items assessed students' knowledge of similarity and trigonometry. One of these items was administered at both grades 8 and 12, one item was administered only at grade 8, and the remaining three items were administered only at grade 12. Descriptions and performance results for these items appear in table 8.7.

Students' performance on items involving the concept of similarity was rather low, except for item 1. This item presented a straightforward situation involving similarity, one that appears in most textbooks—given two similar triangles, find the length of an unknown side of one triangle. The side lengths were related by an integral scale factor, and the figures were oriented on the page in exactly the same way so that the corresponding sides could be easily identified. Performance on this item varied by grade level; about three-quarters of twelfth-grade and one-half of eighth-grade students obtained the correct answer. Although these performance levels can be viewed as rather low for such a straightforward item, they are greater than the average performance on short constructed-response items for grade 12 (34 percent) and grade 8 (49 percent), with the grade 12 comparison the more striking of the two. Another positive sign concerning this item is that performance increased significantly from 1990 to 1996 at both grades. In particular, performance rose from 70 to 77 percent at grade 12 and from 46 to 53 percent at grade 8.

The other item involving similar figures given to eighth-grade students proved more difficult; only 5 percent answered item 2 correctly. This item

Table 8.7
Performance on Items Involving Similarity and Trigonometry

Item Description	Percent Correct	
	Grade 8	Grade 12
1. Find the length of a side of one of two similar triangles using given lengths.	53	77
2. Draw a figure similar to a given figure on the basis of a ratio involving the areas of the figures.	5	—
3. Given two similar triangles, one within the other, find the length of one side of the smaller triangle.	—	37
4. Find the length of a side of a right triangle, given the degree measure of an angle and the length of the other side.	—	17
5. Find the measure of an angle in a right triangle, given the lengths of two of its sides.	—	8

Note: Item 3 was a multiple-choice item, and the rest were short constructed-response items. Item 3, a released item, is shown in table 8.8.

includes two features that may have added to its difficulty—students were asked to produce a triangle similar to a given triangle, not just find the length of a missing side, and the scale was given in terms of area instead of length. Even so, the correct-response rate is far below the grade 8 average of 49 percent for short constructed-response items.

Performance by twelfth-grade students was also much lower on another item involving similarity. In item 3, a released item shown in table 8.8, students were presented with two similar triangles, one within the other, and were asked to determine the length of one of the sides of the smaller triangle. To solve the problem correctly, students needed to recognize that the lengths of the sides of the smaller triangle were one-third the lengths of the corresponding sides of the larger triangle. Fewer than 40 percent of the students were successful. Few students selected the answer of 10, an answer suggesting an additive strategy (i.e., 4 − 2 = 2, so 12 − 2 = 10), a common misconception for students working with proportional situations (Tourniare and Pulos 1985). However, 47 percent of students selected 6 as the correct answer, which is based on an incorrect reading of the scale factor relating the two figures—that is, comparing segment *AD* to segment *DB* instead of to segment *AB*, which is the side of triangle *ABC* corresponding to segment *AD*. This particular mistake may reflect a lack of informal experiences with similarity—particularly dilations—which would allow students to see the relationships of the shapes in the figure instead of comparing the numbers without regard to the correspondence of the sides (Okolica and Macrina 1992).

Table 8.8
Similar: Sample Item

Item	Percent Responding Grade 12

[Figure: Triangle with C at top-left, B at top-right, A at bottom. Segment CB = 12. Point E on CA and point D on BA, with segment ED = x. EA = 3, AD = 2, DB = 4.]

If traingles ADE and ABC shown in the figure are similar, what is the value of x?

A. 4*	37
B. 5	9
C. 6	47
D. 8	3
E. 10	1

*Indicates correct response.
Note: Percents may not add to 100 because of rounding or omissions. This item is from an item block for which students were provided with, and permitted to use, scientific calculators.

Items 4 and 5 involved simple applications of trigonometry to solving a right triangle. Very few twelfth-grade students were successful on either of these items. Both items had omission rates exceeding 20 percent, which may suggest that students had not been exposed to trigonometry or did not recognize that trigonometry would be useful in the situations depicted in the items. This may reflect a lack of attention to trigonometry in the U.S. school mathematics curriculum, where its study is often delayed until at least the second or third year of high school mathematics. There is evidence from the student questionnaire data that in addition to a delay in taking a trigonometry course, many students report not taking trigonometry. In particular, in 1996 only 20 percent of the twelfth-grade students reported studying trigonometry for one year and 55 percent reported not studying trigonometry (Mitchell et al. 1999).

Properties of Angles in Geometric Figures

Whereas students' knowledge about angle measure is addressed in a later section of this chapter, a set of NAEP items assessed knowledge of angle properties in the context of geometric figures such as triangles and circles.

We examined performance on seven NAEP items, which are described in table 8.9. All items were administered only at grades 8 and 12.

Table 8.9
Performance on Items Involving Properties Related to Angles and Circles

Item Description	Percent Correct	
	Grade 8	Grade 12
Angles in triangles and other geometric figures		
1. Determine the measure of one angle in a triangle, given the measures of the other two.	45	81
2. Determine the measure of one angle in a triangle, given the measures of the other two.	—	78
3. Determine the measure of a remote interior angle of a triangle.	32	—
4. Using a protractor, draw a perpendicular line on a given line segment and measure the angle formed by two lines.	—	23
Circles		
5. Find the measure of a central angle in a circle.	34	—
6. Using a protractor, draw an arc with a measure of 235°.	—	26
7. Determine a method for finding the center of a circular paper disk and explain in geometric terms why the method is correct.	—	14[a]

[a]Percent of students scoring at the satisfactory or extended level.
Note: Items 4 and 6 were short constructed-response items, item 7 was an extended constructed-response item, and the rest were multiple-choice items. Item 6, a released item, is shown in figure 8.3; item 7, called Center of Disk, appears in chapter 11 by Silver, Alacaci, and Stylianou.

Eighth-grade students did not do well on two items involving finding a missing angle in a triangle. Fewer than half of the students in grade 8 were able to solve a problem like that in item 1. In item 3, in which students were asked to find a missing exterior angle given two remote interior angles, the percent of correct responses dropped to below one-third. Performance on item 1 increased slightly (but not significantly) from 1990, up from 43 to 45 percent, but performance on item 3 remained unchanged from 1992.

Twelfth-grade students did much better on the tasks relating to angles of a triangle, with the performance levels for both item 1 and item 2 at around 80 percent. Moreover, significantly more twelfth-grade students got these items correct in 1996 than in 1990 and 1992. They did not do as well on item 4, in which they had to find an angle in a complex figure; fewer than one-quarter were successful.

Three items related to the properties of circles are described as items 5–7 in table 8.9. Only about one-third of students in grade 8 were able find the measure of a central angle for an arc in item 5, perhaps reflecting a lack of experience with circles and arcs. Students in grade 12 were given two items related to circles. Item 6, a short constructed-response item, asked students to use a protractor to draw an arc with measure 235° on a given circle. To respond correctly, students not only needed to know the relationship between arcs and central angles but also needed to be familiar with the use of protractors. Since most protractors only provide measurements up to 180°, students needed to draw an arc 55° beyond 180°. Only about one-fourth of the students were able to draw an arc of 235° in item 6.

Examples of different types of responses from students to item 6 appear in figure 8.3. Responses 1 and 2 in the figure illustrate two types of correct responses from students, according to the scoring guide for this item. Response 1 shows an obtuse angle ABC measuring within ±5° of 235°. Response 2 was also counted correct, even though a sector is marked instead of an arc. Incorrect responses included not only those in which point B was placed incorrectly on the circle (the arc was not within ±5° of 235°) but also those where the student failed to mark the arc, as response 3 illustrates, or failed to mark point B as in response 4. Taking the broadest interpretation, where responses such as 3 and 4 are counted as partially correct, the percent-correct value rises to only 30 percent. This suggests that many twelfth-grade students may not be familiar with arcs of circles.

The second twelfth-grade item related to circles (item 7, table 8.9) asked twelfth-grade students to describe a mathematically correct procedure for finding the center of a circular disk. The results show that students had difficulties in finding the center of the given disk and also in describing the procedures they used, as fewer than 1 percent of the students responded at the extended level. An extended analysis of students' responses to this item, which can be found in chapter 11 by Silver, Alacaci, and Stylianou, indicates that most students used informal methods for finding the center of the disk. For example, many students used folding methods instead of formal compass-and-straightedge constructions. However, even when less-formal methods were used, the students failed to explain their reasoning and why their informal method may actually be mathematically correct.

Summary of Performance on Geometry Topics

In 1996 students, especially those at the twelfth-grade level, showed improvements from previous NAEP assessments (1990 and 1992) on a number of topics. However, their performance continued to be quite low when they were asked to apply their knowledge in less-familiar contexts or in more-complex situations. For example, although students were very successful on straightforward items involving similarity, their perform-

GEOMETRY AND MEASUREMENT 213

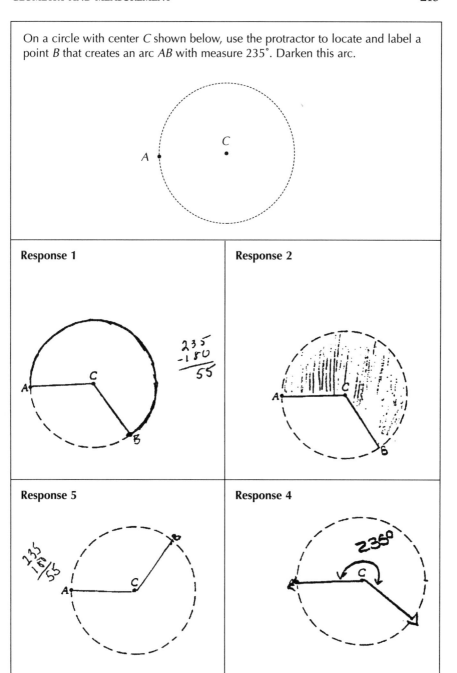

Fig. 8.3. Item involving drawing an arc and sample responses

ance dropped dramatically as the figures became more complicated. Similar observations can be made about identifying and naming shapes, transformations, and the Pythagorean theorem. Additional attention to tasks that across the grades increase in cognitive demands placed on students may be necessary to address this situation (Clements and Battista 1992).

MEASUREMENT OF GEOMETRIC FIGURES

According to the NCTM *Curriculum and Evaluation Standards* (1989, p. 51), "Measurement is of central importance to the curriculum because of its power to help children see that mathematics is useful in everyday life and to help them develop many mathematical concepts and skills." The NCTM *Standards* also point out that students need to pay attention to the attributes of different geometric shapes as well as to what it means to measure. In this section we examine NAEP items that assessed students' understanding of measurement as it relates to geometric figures. Thus, our focus is on angle measure, length, perimeter, area, and surface area and volume. Throughout this section, we attempt to understand how students' concepts of measurement interact with their understanding of geometry.

Angle Measure

Five NAEP items, described in table 8.10, assessed students' knowledge of angle measure. All five items were given to eighth-grade students, and two of the items were also given to fourth-grade students.

Table 8.10
Performance on Items Involving Angle Measure

Item Description	Percent Correct	
	Grade 4	Grade 8
1. Identify an instrument for measuring angles.	—	84
2. Identify angles that are smaller than right angles.	34	62
3. Draw an angle larger than a given degree measure.	12	70
4. Choose the correct ordering for a set of angles according to angle measures.	—	67
5. Draw an arrow in the direction given in degrees.	—	23

Note: Items 3 and 5 were short constructed-response items, and the rest were multiple-choice items. Items 4 and 5 were from item blocks for which students were provided with, and permitted to use, ruler-protractors.

Neither of the items given to fourth-grade students required the use of a protractor but, instead, focused on comparing angles to a right angle.

Nonetheless, results for these items suggest that fourth-grade students have limited experiences with angle measure; the percent-correct values for both items was below 35 percent.

As might be expected, the eighth-grade students scored higher than the fourth-grade students on the common items (items 2 and 3). The percent-correct values on four of the five items exceeded 60 percent, with 84 percent of the eighth-grade students correctly identifying the instrument used for measuring angles. They had far more difficulty with item 5, in which they were asked to produce an arrow in a direction given in degrees. Although fewer than one-quarter produced a correct response, nearly 70 percent were able to produce the correct angle but in the wrong direction.

Although these results may suggest that students in grade 8 are doing quite well with angle measure—especially when a protractor is available—some reason for caution might be found in the results for item 4, a multiple-choice item in which students were asked to order a set of angles according to their measure. This item could be easily answered by visual inspection, using a right angle as a standard, although students also had access to a protractor. Wilson and Adams (1992) suggest that students need more informal experiences with angle measure, such as iterating wedges to measure angles, before they begin using protractors. Such experiences may help students create more-meaningful conceptions of angle measure instead of focusing primarily on the use of a measuring instrument.

Length

Nine NAEP items assessed students' ability to identify measurement tools for length and to exhibit their knowledge about measuring length. Six items were administered to only fourth-grade students, two were administered to both the eighth- and twelfth-grade students, and one was administered to all three grade levels. Table 8.11 contains the performance data for these items.

On item 3, only 72 percent of the fourth-grade students selected the correct measuring instrument for length when given the choice of a thermometer, a clock, a ruler, and scales. Fourth-grade students appeared to be more familiar with the English system of measurement, with 86 percent and 77 percent of the students selecting appropriate units of measurement in items 1 and 2, compared to 34 percent selecting the appropriate unit of measurement for item 5, involving the metric system. Eighth-grade and twelfth-grade students did well on a similar item (item 4) that asked them to select the best unit to use when measuring the growth of a plant every other day, with 78 percent of the eighth-grade students and 87 percent of the twelfth-grade students answering correctly.

Fourth-grade students had difficulty completing item 8, which required them to use a ruler to draw a geometric shape with two given lengths; only

Table 8.11
Performance on Items Involving Length

Item Description	Percent Correct		
	Grade 4	Grade 8	Grade 12
1. Identify an appropriate unit of length (English units).	86	—	—
2. Identify an appropriate unit of length (English units).	77	—	—
3. Identify an instrument to measure length.	72	—	—
4. Identify an appropriate unit of length for measuring the growth of a plant (mixed English and metric units).	—	78	87
5. Identify an appropriate unit of length (metric units).	34	—	—
6. Determine the length of an object pictured above a ruler; end of the object and ruler not aligned.	22	63	83
7. Use a ruler to find the differences in distances between given locations.	28	—	—
8. Use a ruler to draw a figure with two side lengths given.	18	—	—
9. Determine possible length for a given object based on accuracy of measurement.	—	24	39

Note: : Items 6, 7, and 8 were short constructed-response items, item 3 was part of a multiple-response item, and the rest were multiple-choice items. Items 6 and 7 were from an item block for which students were provided with, and permitted to use, rulers.

18 percent completed the task successfully. Similarly, in item 7, fourth-grade students had difficulty using a ruler to determine distance; only 28 percent answered correctly. There is evidence that reading and using a ruler are difficult tasks for young children. In many instances children count the number of numbers on the ruler instead of counting the units of measurement (Wilson and Rowland 1993). For example, Lindquist and Kouba (1989a) reported that third-grade students had a difficult time reading a ruler when the object being measured was not aligned at the beginning of the ruler.

Perimeter

Perimeter is a special application of length: the measure of the distance around a region (Reys et al. 1998). Thus, there are many real-world applications related to this measure. Later in their careers, students may be asked to design structures with particular perimeters to maximize or minimize area. The 1996 NAEP mathematics assessment contained five items that assessed students' knowledge of perimeter, and these items are described in table 8.12. Two items were given at grade 4 and one item at grade 12, with an overlap item at grades 8 and 12 and an overlap item across all three grades. The results in the table reveal that although performance was rather high on some items (for example, item 1 for grades 8 and 12), performance on other items suggests that students do not have a good understanding of perimeter.

Table 8.12
Performance on Items Involving Perimeter Items

Item Description	Percent Correct		
	Grade 4	Grade 8	Grade 12
1. Given the perimeter of a geometric figure, find the length of a side.	45	64	76
2. Given the side lengths of a triangle, find the lengths of the sides in a square with the same perimeter.	26	—	—
3. Use a ruler to draw a figure with a given perimeter.	19	—	—
4. Use a ruler to find the circumference of a circle (value for π given to two decimal places).	—	—	29
5. Determine which of three geometric shapes has the longest perimeter; explain.	—	6	12

Note: Items 1 and 2 were multiple-choice items, and the rest were short constructed-response items. Items 2 and 5, both released items, are shown in table 8.13.

Only 19 percent of the fourth-grade students responded correctly to item 3, which asked them to draw a geometric figure with a specific perimeter. This rate of correct response is similar to the rate for another item (18 percent correct for item 8, table 8.11) in which students were asked to draw two sides of a triangle based on given measurements. Students' scores on these two items may indicate that they were having problems using a ruler to draw the sides of the shapes or to measure the length of an object. Item 1 was given across all three grades and asked students to determine the length of a side of a shape given its perimeter. Not surprisingly, as the grade level increased, the rate of correct responses increased, with 45 percent of the fourth-grade students responding correctly, about 65 percent of the eighth-grade students, and 76 percent of the twelfth-grade students.

Twenty-nine percent of the twelfth-grade students responding to item 4 were able to find the circumference of a circle given a centimeter ruler and an approximation of π (3.14). However, the formula for the circumference of a circle was not provided, nor were calculators. It is NAEP policy not to provide formulas that students are expected to know, thus if students did not know the formula for circumference ($C = 2\pi r$), they were unlikely to find a correct solution to this item. On the basis of a small sample of students' responses, we found that some students may have confused perimeter and area because in this item they applied the formula for finding the area of a circle ($A = \pi r^2$), whereas others had computational errors involving decimals.

Table 8.13 contains two released items that provide additional insight about students' understanding of perimeter. For item 1, a little more than 25 percent of the fourth-grade students were able to determine the side of a square that had the same perimeter as a triangle with sides of length 4, 7, and 9. A higher percentage of students (36 percent) chose 4, the length of one of the sides of the triangle, rather than the correct response. This situation may have occurred because the side of the triangle with a measure of 4 appears to be the same length as the sides of the square drawn in the item. Thus, students giving this response may have been reasoning from the drawing instead of using the definition of perimeter.

In item 2, eighth-grade and twelfth-grade students were given shapes labeled *N*, *P*, and *Q* (shown in table 8.13) and asked to determine the shape with the longest perimeter. For this item the students were told that perimeter is the "distance around," but they did not have access to a ruler. However, students still did not do well on this item. Only 6 percent of the eighth-grade students and 12 percent of the twelfth-grade students answered correctly. A correct response had to state that *P* has the longest perimeter, and the explanation had to indicate how the student decided that *P* was the correct choice.

We examined students' work from a small set of nonblank responses to this item to find examples of different types of strategies used to explain why *P* had the longest perimeter. Sample responses are shown in figure 8.4. Responses 1 and 2, both produced by twelfth-grade students, illustrate two types of responses that would be scored as correct on the basis of the official NAEP scoring guide. Response 1 represents a fairly typical empirical approach to comparing the perimeters of the three shapes. Here the student drew three line segments, one for each shape, by tracing the length of each side of each figure. A comparison of the length of each segment resulted in the conclusion that *P* had the longest perimeter.

Response 2, also produced by a twelfth-grade student, used a more sophisticated approach based on the Pythagorean theorem. Notable in this response is the establishment of relationships between the lengths of the sides among the figures: the side of the square *N* is the same length as one leg of the triangle *P*; the length of one leg of *P* is twice that of its other leg; and the length of a leg of *Q* is twice the length of the side of *N*, and its height

GEOMETRY AND MEASUREMENT

Table 8.13
Perimeter Items

Item	Percent Responding		
	Grade 4	Grade 8	Grade 12

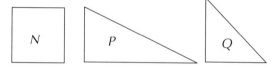

1. If both the square and the triangle above have the same perimeter, what is the length of each side of the square?

A. 4	36	—	—
B. 5*	26	—	—
C. 6	10	—	—
D. 7	25	—	—

[These shapes were available for use as manipulatives.]

2. Which of the shapes N, P, and Q has the longest perimeter (distance around)?

Shape with the longest perimeter: _____
Use words or pictures (or both) to explain why.

Correct response of P with an adequate explanation (for example, tracing the sides of each piece in aline and showing that P has the longest perimeter)	—	6	12
Incorrect response stating that P has the longest perimeter, but explanation inadequate or missing	—	54	49
Any other incorrect response	—	39	32
Omitted	—	1	7

*Indicates correct response.
Note: Percents may not add to 100 because of rounding or omissions. Item 2 was from an item block for which students were provided with manipulatives in the form of geometric shapes.

Response 1

Q |—————|————|————|————|

P |————|————————|————————|

N |————|————|—————|————|

I traced each side of each shape into a line. P is longest. therefore, P has longest perimeter.

Response 2

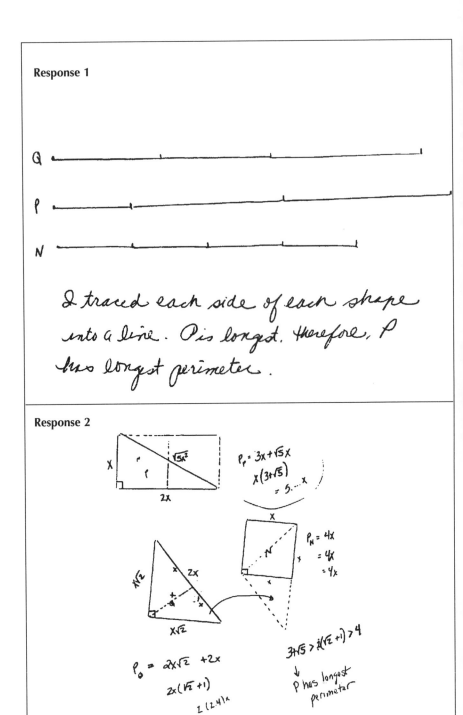

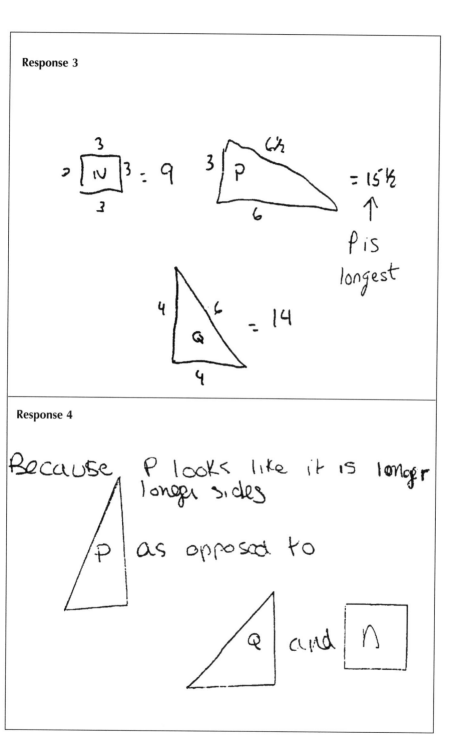

Fig. 8.4. Sample responses to the perimeter comparison item

is half that of its base. Once the student established these relationships, then showing that P had the longest perimeter was a matter of a relatively straightforward use of the Pythagorean theorem (not directly shown in the work, however), and the numerical comparison of the coefficients of the expressions for the perimeters of P, Q, and N, respectively:

$$3 + \sqrt{5} > 2(\sqrt{2} + 1) > 4.$$

This response stands in contrast to the low performance reported previously about students' ability to use the Pythagorean theorem in problem-solving situations (for example, fewer than 10 percent of twelfth-grade students produced a correct response to item 8, table 8.6), although it is likely that such sophisticated responses were rare in the NAEP sample.

The response from an eighth-grade student, labeled response 3 in figure 8.4, illustrates a more informal approach to finding the lengths of the sides of each figure so that their perimeters could be compared. Here the student, who may or may not have studied the Pythagorean theorem in middle school mathematics, used a self-imposed measurement system probably based on a side length of 3 units for the square N. The lengths of the sides of shapes P and Q appear to be based on how they compare to 3 units. For example, one of P's legs is the same length as a side of N (3 units), the other leg is twice as long (6 units), and the hypotenuse is longer than the 6-unit leg, about 6 1/2 units. The two legs of Q are a little more than 3 units (approximately 4 units), and the hypotenuse of Q is the same length as the longer leg of P, 6 units. Compared to the values derived from using the Pythagorean theorem (that is, approximately 6.71 and 5.66 for the hypotenuses of P and Q, respectively, using the student's values for the other two legs), the estimates of 6 1/2 and 6, respectively, are quite reasonable and support the correct conclusion that P has the longest perimeter. This response shows that reasoning about this problem does not depend exclusively on knowing and being able to use the Pythagorean theorem.

As noted in table 8.13, in the national sample, 54 percent and 49 percent of eighth- and twelfth-grade students provided incorrect responses that identified P as having the longest perimeter but that provided inadequate explanations. For example, as illustrated in response 4, some students simply wrote that P "looked like" it had longer sides but provided no evidence of comparing the *sum* of the lengths of the sides in all three figures.

NAEP results from the perimeter items reveal that fourth-grade students have difficulty with perimeter concepts, and eighth- and twelfth-grade students show a lack of understanding in more-complex contexts such as circumference and in situations that require a conceptual understanding of perimeter. For students to understand perimeter, they must be placed in situations where they can develop a meaningful understanding of the concept of perimeter. For example, when students are introduced to perimeter, they should be allowed to use string and measuring tape to measure the perimeters of objects. This activity will help the students to make the connection between measuring the length or width of an object and measuring the object's perimeter. Students should experience more difficult tasks after they understand the concept (Reys et al. 1998).

Area

The concept of area is very important because it is one of the most commonly used measurements and it is also the basis for many models used by teachers and textbook authors to explain computational procedures (Woodward and Byrd 1983; Hirstein, Lamb, and Osborn 1978). However, the concept of area is often difficult for students to understand, perhaps due to their initial experiences in which it is tied to a formula (such as area = length × width) rather than more conceptual activities such as counting the number of square units it would take to cover a surface (see, for example, Van de Walle 1997).

The 1996 NAEP contained nine items related to area, and these items are described in table 8.14. Only one item was administered exclusively at the fourth-grade level. Four items were administered only to eighth-grade students, three items were administered across grade levels, and two items were administered to only one grade level. Because most of the area items administered in the 1996 NAEP were secure, performance on those items cannot be discussed in detail.

On some area items, especially on those items that assessed area in simple ways, performance was rather high. About 60 percent of the eighth-grade students correctly answered item 3, which asked them to choose the best estimate and unit of measurement for area. As shown in the percent-correct values for item 1, some fourth-grade students and a majority of eighth-grade students were also able to choose the area of a figure embedded in a centimeter grid; 41 percent and 78 percent, respectively, gave correct responses. This item required students to have a conceptual understanding of area. They could not rely on a formula; instead, they needed to count square units.

Three items required students to know formulas for area. In item 4, only 44 percent of the eighth-grade students and 60 percent of the twelfth-grade students were able to choose the correct numerical expression for area for a given situation. Furthermore, in item 2, only 40 percent of the fourth-grade students correctly found the area of a figure, given the area of an inscribed figure. In item 5, about 50 percent of the twelfth-grade students chose the correct length for the side of a square, given the dimensions of a rectangle with the same area.

Students were much less successful on the remaining area items. In item 7, only 12 percent of the eighth-grade level students could determine the

Table 8.14
Performance on Items Involving Area

Item Description	Percent Correct		
	Grade 4	Grade 8	Grade 12
1. Choose the correct value for the area of a polygon depicted on a centimeter grid.	41	78	—
2. Determine the area of a geometric figure, given the area of an inscribed figure.	40	—	—
3. Choose the best estimate for an area of a region.	—	59	—
4. Choose the correct numerical expression for the area of a given geometric figure.	—	44	60
5. Given the dimensions of a geometric figure, determine the length for another geometric figure with the same area.	—	—	52
6. Compare the areas of two shapes.	6	27	35
7. Determine the number of square tiles it would take to cover a region of given dimensions.	—	12	—
8. Determine the number of boxes of square tiles it would take to cover a region of given dimensions.	—	9[a]	—
9. Show a number of different ways a region can be divided to find the area.	—	8[a]	—

[a]Percent of students scoring at the satisfactory or extended level.
Note: Items 6 and 7 were short constructed-response items, items 8 and 9 were extended constructed-response items, and the rest were multiple-choice items. Item 6, a released item, is shown in table 8.15.

number of square tiles of a given size it would take to cover a region with given dimensions. Similarly, very few eighth-grade students were successful on two extended constructed-response items, described as items 8 and 9 in the table. On item 8, they had to determine how many boxes of tiles it would take to cover a floor, given a scale and a ruler, and on item 9, they had to determine how an irregular shape could be divided to find its area. Fewer than 10 percent scored at the satisfactory or extended levels on these items.

Table 8.15 shows item 6 from table 8.14, a released item that was administered to all three grade levels, that asked students to evaluate statements comparing the areas of shapes N (a square) and P (a right triangle). Students each had two of each shape to use as manipulatives. Fourth-grade students were given a slightly different version of this item, which used cartoons of the children who were comparing the areas of N and P. Students did not do well on this item at any of the grade levels on the 1996 or 1992 assessments. However, as one might expect, twelfth-grade students performed better

Table 8.15
Area Item

Item	Percent Responding		
	Grade 4	Grade 8	Grade 12
[These shapes were available for use as manipulatives.]  Bob, Carmen, and Tyler were comparing areas of N and P. Bob said that N and P have the same area. Carmen said that the area of N is larger. Tyler said that the area of P is larger. Who was correct? _____ Use words or pictures (or both) to explain why. [Note: The statements of Bob, Carmen, and Tyler were presented as pictures in the grade-4 version of this item.]			
Correct response of Bob with an appropriate explanation (for example, showing by superimposition of shapes that two of shape P match two of shape N, and therefore they have the same area; thus, one P has the same area as one N)	6	27	35
Response of Bob, but explanation inadequate or missing	21	16	14
Any other incorrect response	74	54	46
Omitted	<1	2	5

Note: Percents may not add to 100 because of rounding. Item is from an item block for which students were provided with manipulatives in the form of geometric shapes.

Response 1

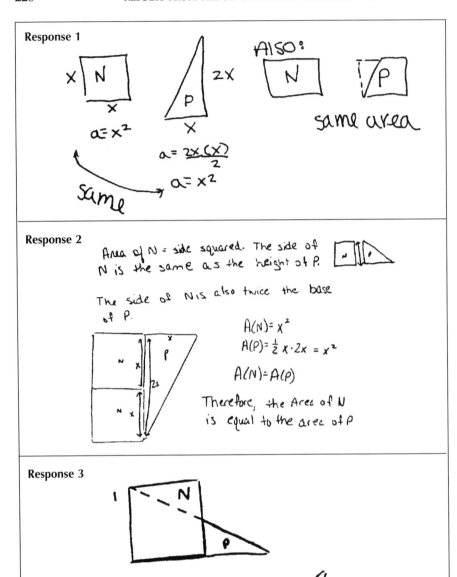

Response 2

Area of N = side squared. The side of N is the same as the height of P.

The side of N is also twice the base of P.

$A(N) = x^2$

$A(P) = \frac{1}{2} x \cdot 2x = x^2$

$A(N) = A(P)$

Therefore, the Area of N is equal to the area of P

Response 3

—continued on next page

GEOMETRY AND MEASUREMENT 227

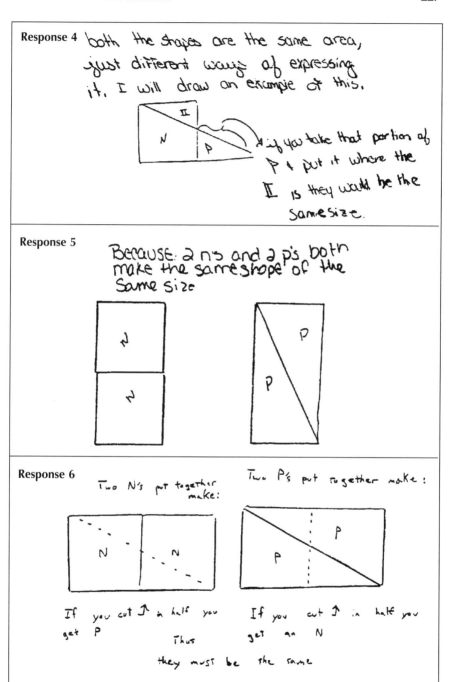

Fig. 8.5. Sample responses to the area comparison item

than students in the other two grade levels. A comparison of performance between assessments in 1996 and previous years shows that eighth-grade students did significantly better in 1996 than in 1992, but the percent correct rose only from 32 percent to 35 percent.

Figure 8.5 contains sample students' responses across grade levels to illustrate the ways in which students tried to solve this area comparison problem. All responses in the figure noted that Bob was correct. Responses 1 and 2, produced by an eighth-grade student and a twelfth-grade student, respectively, are examples of solving the problem through a comparison of the dimensions of the figures and the use of formulas. Because students did not have access to a ruler, it was important for them to discover by comparing the shapes that the side of the square N is the same length as one leg of the triangle P and that the length of one of the legs of P was twice that of its other leg. Once the students established these relationships, then a solution to the task using a formula became possible—as long as students knew the formulas for the areas of a square and a triangle.

Another way to solve this problem is by manipulating the shapes to determine whether the areas are the same or not. Responses 3 and 4 are examples of this method, with the latter produced by a twelfth-grade student and the former by a fourth-grade student. The eighth-grade student who produced response 1 also included a partially developed example of how shape N could be used to form shape P. The fourth-grade students' response, which is particularly intriguing, uses an animated style drawing to demonstrate that the two shapes have the same area. Here, the two figures are superimposed, and part of shape P is moved so that it fits over the part of shape N not already covered by shape P. Response 4, done by the twelfth-grade student, uses a similar technique of "flipping" part of shape N onto shape P, but includes a verbal explanation that leaves little doubt that this student understands that the two shapes have the same area.

Responses 5 and 6, both of which were produced by eighth-grade students, illustrate another way to manipulate the shapes to show that they have the same area. In this case, students used two of each shape to form two rectangles and showed that two squares and two triangles cover the same area. The eighth-grade student who produced response 5 depended entirely on the picture produced, whereas the other student used both words and pictures to justify the choice of Bob's claim as being correct.

One may infer that the results in this section are directly related to students' experiences with the concept of area. That is, most students primarily experience the concept of area through the memorization of formulas; they may never really understand what the concept means in terms of determining the size of a region. Woodward and Byrd (1983) and Van de Walle (1997) suggest ways of making the concept of area more meaningful to students that include introducing the concept of area by covering a region with subregions and determining the area by counting; next, students should begin to use things like graph paper and geoboards to determine the area of regions by counting units embedded in the figures; finally, formulas should be developed inductively using the counting process.

Surface Area and Volume

Surface area and volume are both attributes of three-dimensional figures. In the 1996 NAEP assessment, there were only a few items that assessed these topics. In the case of surface area, two items were given at the twelfth-grade level, and one was given at both grades 8 and 12; all items are secure. The percent-correct values on the surface area items were quite low, ranging from 3 percent on an eighth-grade item to 36 percent for twelfth-grade students on an item given to both eighth- and twelfth-grade students. Fourth-grade students were not given an item related to surface area.

One volume item was given at the fourth-grade level, and two volume items were given to both eighth- and twelfth-grade students. As was the case with surface area, students had difficulty answering volume items correctly, with performance levels at 30 percent correct or lower.

An understanding of surface area is directly related to an understanding of area. In order to understand surface area, students need to understand how to count area units, and they also need to understand that a polyhedron has faces that can be looked on as separate two-dimensional surfaces. On the basis of this understanding, surface area can then be computed by finding the areas of each of the faces and then summing those together for the total surface area of the solid. Since it is difficult for students to section an irregular plane figure into smaller parts to determine the area of the whole figure (Van de Walle 1997), it appears that the notion of deconstructing a geometric solid into its net to find the surface area would be more challenging. Moreover, many students confuse procedures for finding the volume of a three-dimensional figure with finding the surface area of the figure and vice versa. And, as with area, students' experiences with volume are often limited to the use of formulas rather than to activities related to filling the interior of a solid (Van de Walle 1997). Thus, it may not be surprising that students did not do well on items asking them to find the volume or surface area of a geometric solid.

Table 8.16 shows two released volume items that give us some insight into students' understanding of volume. Item 1 presented students with a diagram of a three-dimensional array of cubes and asked them to determine its volume by determining the number of cubes in it. To respond correctly, students needed to be able to "read" the two-dimensional representation of the solid and to visualize the cubes that do not appear in the figure. Fourth-grade students had difficulty with this item; in fact, fewer fourth-grade

Table 8.16
Volume Items

Item	Percent Responding		
	Grade 4	Grade 8	Grade 12

1. In this figure, how many small cubes were put together to form the large cube?
   A. 7
   B. 8*
   C. 12
   D. 24

A	4	—	—
B	33	—	—
C	32	—	—
D	29	—	—

2. A cereal company packs its oatmeal into cylindrical containers. The height of each container is 10 inches and the radius of the bottom is 3 inches. What is the volume of the box to the nearest cubic inch? (The formula for the volume of a cylinder is $V = \pi r2h$.)

   Answer: _____ cubic inches

Correct response of 283 with or without accompanying work	—	13	29
Partially correct response that shows substitutions into the formula but does not round or an unrounded answer with no work	—	17	25
Any other incorrect response	—	57	36
Omitted	—	12	8

* Indicates correct response
Note: Percents may not add to 100 because of rounding. Item 2 was from an item block for which students were provided with, and permitted to use, scientific calculators.

students were able to tell how many cubes were put together to form a larger cube in 1996 than in 1992, 33 and 37 percent respectively.

Previous studies have documented students' difficulties in problems of this type (that is, problems that present a diagram of a three-dimensional array of cubes and ask them to determine the number of cubes in it). For example, Ben-Haim, Lappan, and Houang (1985) showed that fewer than 25 percent of fifth-grade students could solve such a problem. Common errors of students who participated in this study included counting the cube faces shown in the picture and finding the surface area of the solid. Ben-Haim and his colleagues attributed students' errors to a difficulty in making inferences about a three-dimensional solid from a two-dimensional figure and to confusion of the concepts of volume and surface area.

In item 2 in table 8.16, 13 percent of the eighth-grade students and 29 percent of the twelfth-grade students were able to find the volume of a cylindrical container given the height and the radius of the container. For this item, students were given the formula for finding the volume of a cylinder and asked to round the answer to the nearest cubic inch. This item was also one in which students were allowed to use their calculators. It is interesting to note that about 17 percent of the eighth-grade students and 26 percent of the twelfth-grade students were able to make the correct substitutions into the formula, but did not correctly round their answers to the nearest cubic inch.

A Summary of Performance on Measurement Topics

Students' performance on measurement was quite low, with little improvement from the NAEP assessments of 1990 and 1992. Many fourth-grade students had trouble using rulers, and a substantial number of eighth-grade students were not able to order according to their measures angles that were visibly different in size. At all levels, students' concepts of area, surface area, and volume tended to be tied to the use of formulas. Additional attention to the conceptual foundations of measurement might be useful in helping students become more successful with measurement.

CONCLUSION

As reported by Dossey in chapter 2, the average scale scores for the content strand Geometry and Spatial Sense have increased significantly from 1990 and 1992 to 1996. The largest gain was at grade 4, where scale scores rose from 213 in 1990 to 225 in 1996. Gains for grades 8 and 12 are more modest, with a difference of below 10 scale-score points between years. In the content strand Measurement for grades 4 and 8, there was a significant increase in performance from 1990 to 1996 but not between 1992 and 1996; at grade 12, the 1996 scores were significantly higher than those in both 1990 and 1992.

As shown throughout this chapter, students' performance in geometry and measurement for the 1996 NAEP assessment was mixed, with gains in item-level performance in some areas from previous NAEP assessments but with performance in other areas showing little change. Positive changes occurred on geometry items related to nets of geometric shapes, on items related to symmetry, and on an item related to similarity. Other areas of geometry are of continued concern. For example, while students at all three grade levels seemed to be quite successful in identifying and naming familiar two- and three-dimensional figures and their properties, their performance dropped dramatically when asked to identify figures in less familiar contexts or to recognize more-complex properties. Performance on items related to the Pythagorean theorem was also disappointing; performance levels for even direct application of the theorem are quite low, and the rates are even lower in situations that ask students to use the theorem in a problem-solving situation. These mixed results in geometry mirror findings from the Third International Mathematics and Science Study [TIMSS], in which signs of progress were seen in the ranking of U.S. fourth-grade students above the international average in geometry (National Center for Education Statistics [NCES] 1997a), although eighth-grade students performed below the international average (NCES 1997b), and advanced twelfth-grade students who were taking precalculus or calculus were last in the international comparison (NCES 1998).

Students' performance continues to lag in measurement. Fourth-grade students had difficulty with items that asked them to use a ruler to measure an object or to draw a shape with particular dimensions. Students across all three grade levels had difficulty with perimeter and area concepts, especially situations in which they had to explain or justify answers. Twelfth- and eighth-grade students did not do well on items related to surface area. Similar results were seen in TIMSS, in which students at both fourth and eighth grades performed below the international level in measurement (NCES 1997a, 1997b).

These findings suggest that additional attention needs to be given to geometry and measurement across the curriculum. Although students generally performed well on the easiest tasks in which a property could be applied to straightforward situations where the figures were not complex, performance dropped dramatically in situations involving more than one step or figures that were more complex, in situations requiring the production of, or the comparison, of figures, and in situations requiring explanations. In order for this trend to be reversed, growth in the cognitive demands placed on students needs to occur across the grades (Clements and Battista

1992). This low performance is particularly disappointing because these are areas in which students have consistently scored poorly in the past (Strutchens and Blume 1997; Kenney and Kouba 1997; Lindquist and Kouba 1989a, 1989b) and limited evidence of progress can be seen.

REFERENCES

Ben-Haim, David, Glenda Lappan, and Richard Houang. "Visualizing Rectangular Solids Made of Small Cubes: Analyzing and Effecting Students' Performance." *Educational Studies in Mathematics* 16 (November 1985): 389–409.

Clements, Douglas H., and Michael T. Battista. "Geometry and Spatial Reasoning." In *Handbook of Research on Mathematics Teaching and Learning*, edited by Douglas A. Grouws, pp. 420–64. New York: MacMillan Co., 1992.

Hirstein, James J., Charles E. Lamb, and Allan Osborne. "Student Misconceptions about Area Measure." *Arithmetic Teacher* 25 (March 1978): 6, 10–16.

Kenney, Patricia Ann, and Vicky L. Kouba. "What Do Students Know about Measurement?" In *Results from the Sixth Mathematics Assessment of the National Assessment of Educational Progress*, edited by Patricia Ann Kenney and Edward A. Silver, pp. 141–64. Reston, Va.: National Council of Teachers of Mathematics, 1997.

Lindquist, Mary M., and Vicky L. Kouba. "Geometry." In *Results from the Fourth Mathematics Assessment of the National Assessment of Educational Progress*, edited by Mary M. Lindquist, pp. 44–54. Reston, Va.: National Council of Teachers of Mathematics, 1989a.

———. "Measurement." In *Results from the Fourth Mathematics Assessment of the National Assessment of Educational Progress*, edited by Mary M. Lindquist, pp. 35–43. Reston, Va.: National Council of Teachers of Mathematics, 1989b.

Mitchell, Julia H., Evelyn F. Hawkins, Pamela M. Jakwerth, Frances B. Stancavage, and John A. Dossey. *Student Work and Teacher Practices in Mathematics*. Washington, D.C.: National Center for Education Statistics, 1999.

National Center for Education Statistics. *Pursuing Excellence: A Study of U.S. Fourth-Grade Mathematics and Science Achievement in International Context*. Washington, D.C.: National Center for Education Statistics, 1997a.

———. *Pursuing Excellence: A Study of U.S. Eighth-Grade Mathematics and Science Achievement in International Context*. Washington, D.C.: National Center for Education Statistics, 1997b.

———. *Pursuing Excellence: A Study of U.S. Twelfth-Grade Mathematics and Science Achievement in International Context*. Washington, D.C.: National Center for Education Statistics, 1998.

National Council of Teachers of Mathematics. *Curriculum and Evaluation Standards for School Mathematics*. Reston, Va.: National Council of Teachers of Mathematics, 1989.

Okolica, Steve, and Georgette Macrina. "Integrating Transformation Geometry into Traditional High School Geometry." *Mathematics Teacher* 85 (December 1992): 716–19.

Reys, Robert E., Marilyn N. Suydam, Mary M. Lindquist, Nancy L. Smith. *Helping Children Learn Mathematics*. 5th ed. Boston, Mass.: Allyn and Bacon, 1998.

Sherard, Wade H. "Why is Geometry a Basic Skill?" *Mathematics Teacher* 74 (January 1981): 19–21.

Shultz, Karen A., and Joe Dan Austin. "Directional Effects in Transformation Tasks." *Journal for Research in Mathematics Education* 14 (March 1983): 95–101.

Strutchens, Marilyn E., and Glendon W. Blume. "What Do Students Know about Geometry?" In *Results from the Sixth Mathematics Assessment of the National Assessment of Educational Progress*, edited by Patricia Ann Kenney and Edward A. Silver, pp. 165–94. Reston, Va.: National Council of Teachers of Mathematics, 1997.

Tournaire, Francoise, and Steven Pulos. "Proportional Reasoning: A Review of the Literature." *Educational Studies in Mathematics* 16 (1985): 181–204.

Van de Walle, John A. *Elementary and Middle School Mathematics: Teaching Developmentally*. 3rd ed. New York: Longman, 1997.

van Hiele, Pierre. *Structure and Insight*. Orlando, Fla.: Academic Press, 1986.

Vinner, Shlomo and Rina Hershkowitz. "Concept Images and Common Cognitive Paths in the Development of Some Simple Geometric Concepts." In *Proceedings of the Fourth International Conference for the Psychology of Mathematics Education*, edited by Robert Karplus, pp. 177–84. Berkeley, Calif.: Lawrence Hall of Science, 1980.

Wheatley, Grayson. "Spatial Sense and Mathematical Learning." *Arithmetic Teacher*, 37 (February 1990): 10–11.

Wilson, Patricia S., and Verna M. Adams. "A Dynamic Way to Teach Angle and Angle Measure." *Arithmetic Teacher* 39 (January 1992): 6–13.

Wilson, Patricia S., and Ruth E. Rowland. "Teaching Measurement." In *Research Ideas for the Classroom: Early Childhood Mathematics*, edited by Robert J. Jensen, pp.171–94. New York: Macmillan, 1993.

Woodward, Ernest, and Frances Byrd. "Area: Included Topic, Neglected Concept." *School Science and Mathematics* 83 (April 1983): 343–47.

9

Data and Chance

Judith S. Zawojewski and J. Michael Shaughnessy

DATA analysis, statistics, and probability are mathematics topics that are gaining in importance in the United States as evidenced in the emphasis in the NCTM (1989) *Curriculum and Evaluation Standards for School Mathematics* and the emphasis in recent curricula from projects funded by National Science Foundation (for example, Connected Mathematics Program [Lappan et al. 1997]; Investigations into Number, Data, and Space [Technical Education Research Center 1997]). The shifting emphasis is timely as decisions and actions taken in business, industry, and government are increasingly based on statistical and probabilistic information. An increase in emphasis on statistics and probability is also evident in the National Assessment of Educational Progress (NAEP). In the past decade, the percent of items classified by NAEP as Data Analysis, Statistics, and Probability has more than tripled at grade 12 (from 6 percent in 1986 to 20 percent in 1996) and almost doubled in grade 8 (from 8 percent to 15 percent), although remaining almost the same in grade 4 (10 percent in 1986 to 11 percent in 1996) (National Assessment of Educational Progress 1986; National Assessment Governing Board 1994).

With respect to topics in data and chance in the 1996 NAEP mathematics assessment, Reese et al. (1997, p. 77) report that the NAEP content strand called Data Analysis, Statistics, and Probability was designed to emphasize "the appropriate methods for gathering data, the visual exploration of data, various ways of representing data, and the development and evaluation of arguments based on data analysis." There were fifty-six items on the 1996 NAEP mathematics assessment classified according to topics within this strand, and a subset of those items were administered to multiple grade levels (that is, to students in grades 4 and 8, grades 8 and 12, and all three grade levels). Items administered at grade 4 required that students use and construct graphs, make predictions from data, deal informally with measures of central tendency, and use basic concepts of chance. The assessment at grade 8 required students to analyze statistical claims, design experiments, use simulations, and make predictions on the basis of experiments. It also addressed the topics of sampling and formal terminology associated with probability and statistics. The grade 12 items focused on students' ability to apply the concepts of probability, to use formulas and more-formal terminology to describe situations, and to use mathematical equations and graphs to interpret data.

This chapter reports on students' performance with data and chance by examining individual NAEP items or clusters of related items and, where available, samples of students' responses to constructed-response questions. The focus of the chapter is on four categories that are often interrelated: central tendency, reasoning with data, graphical data displays, and probability and chance. In addition to reporting and interpreting performance on the basis of NAEP results, for some short and extended constructed-response items we also examined a set of sample student responses. This sample was not a representative sample of all responses; rather, it was a convenience sample of nonblank responses. Some of the items described in this chapter were extended constructed-response tasks, a type of NAEP item that is discussed extensively in chapter 11 by Silver, Alacaci, and Stylianou.

CENTRAL TENDENCY

There were seven items in descriptive statistics administered to students in grades 8 and 12, and all but one dealt with measures of central tendency, primarily emphasizing finding and using *mean* and *median*. Only a few of the items included *mode* as a choice, but since it was never the correct answer, the concept of mode was never assessed directly. Only one item at grade 12 dealt with a measure of dispersion. Thus, the focus in this section is on students' performance on five NAEP items assessing mean and median as measures of central tendency.

Past NAEP results indicated low performance on items involving mean and median (Zawojewski and Heckman 1997), and this trend continued in 1996. Performance results from 1996 for the mean and median items appear in table 9.1. There is evidence that performance improved on the first two items, which were administered in all of the last three NAEP mathematics assessments (1990, 1992, and 1996). Item 1 involved finding the mean of a grouped frequency distribution, which may have added to the item's difficulty. While working on this item, students were permitted to use a scientific calculator, which may have led to many unrounded responses despite the requirement that answers be rounded to the nearest whole number to be

DATA AND CHANCE

Highlights

- There was significant growth from 1992 to 1996 in eighth- and twelfth-grade students' performance on NAEP items that required them to find the mean and median of particular data sets. However, when given a choice about which statistic to use, students tended to select the mean over the median, regardless of the distribution of the data.

- Over half of the students in grade 8 appropriately considered the potential for bias and the number of data points when asked questions about survey samples.

- Although about three-fourths of students in grade 12 were able to read line graphs, only about one-third could read a box-and-whisker plot. Further, many eighth-grade students had difficulty using information from tables, charts, and graphs to make decisions and interpretations.

- More than half of the fourth-grade students were able to work with probability concepts involving one favorable outcome out of the total number of outcomes as compared to only about one-fourth who dealt successfully with more than one favorable outcome.

- Although able to list a sample, eighth- and twelfth-grade students had difficulty reasoning from sample spaces.

- Proportional reasoning was a source of difficulty for students in reasoning with data, graphs, and chance.

- Students in grade 4 and 12 have made significant improvement in the Data Analysis, Statistics, and Probability content strand since 1992, as have students in grade 8 since 1990.

scored as correct. To focus on the students' correct procedures for finding the grouped mean, the NAEP scoring categories of "correct" and "unrounded but correct calculation" were collapsed for this analysis. While the performance level for 1996 was low on this item (22 and 46 percent in grades 8 and 12, respectively), it is significantly higher than performance on that same item in 1990, in which only 16 percent of students in grade 8 and 36 percent of students in grade 12 answered correctly or gave unrounded responses.

Similarly, although the performance on item 2, involving the median of a set of numbers, is also disappointing (31 and 33 percent correct in grades 8 and 12, respectively), the performance levels are significantly higher than they were on the same item in 1990, when only 20 percent of eighth-grade and 22 percent of twelfth-grade students selected the correct response. It is interesting that in 1996 the performance levels of twelfth-grade students are higher than that of eighth-grade students for the item involving finding the mean, whereas they are similar for finding the median. It may be that many secondary school students encounter problems involving the mean in their mathematics courses, whereas the median usually is not addressed during the commonly taken high school courses of algebra 1, algebra 2, and geometry.

Table 9.1
Performance on Items Involving Mean and Median

Item Description	Percent Correct	
	Grade 8	Grade 12
1. Determine the mean from a grouped frequency distribution; round answer to the nearest whole number.	22[a]	46[a]
2. Determine the median of a set of numbers given in nonsequential order.	31	33
3. Select the summary statistic that best describes a given situation (median and mean among the choices).	19	28
4. Select the statistic (median or mean) that best represents a given situation; explain.	2	—
5. Select the statistic (median or mean) that best represents the average daily attendance at each of two movie theaters, given attendance data in tables.	—	4[b]

[a] Includes percent of students who gave an unrounded answer, which was considered incorrect according to the NAEP scoring guide.
[b] Percent of students responding at the satisfactory or extended levels.
Note: Items 1 and 4 were short constructed-response items, and items 2 and 3 were multiple-choice items. Item 5, a released extended constructed-response item, is shown in table 9.2.

Items 3, 4, and 5 in table 9.1 all required students to select the appropriate summary statistic to represent data. This selection is usually complex and involves both an analysis of the distribution of data and an understanding of the problem context. However, it may be that when faced with a choice of mean or median, students tend to select the mean, apparently without regard for the distribution or the context. For those items, mean and median were among the choices available to the student, and in the instances where median was the correct choice, there was one outlier in the set of data (apparently using distribution to direct the choice of median). The preference for mean was evident. For example, item 3 required that students choose the best summary statistic to describe a situation. NAEP results show that more eighth-grade students chose the mean (33 percent) instead of the correct answer of the median (19 percent). The pattern of preferring the mean over the median was repeated by the twelfth-grade students, with 28 percent choosing the median and 37 percent the mean.

For item 4, a short constructed-response item, students were asked to select either the mean or the median to represent a situation and to explain their choice. Only 26 percent of eighth-grade students correctly chose the median, but of that percentage only 2 percent included a correct explanation for their choice. Sixty percent of the students gave incorrect responses, but unfortunately the scoring guide for this item was not structured to provide specific information on the percent of those students who chose the mean. It is not unreasonable to speculate, however, that many of the incorrect responses involved choosing the mean, given that the choice was between that summary statistic and the median.

Only one item (item 5 in table 9.1) was released, and it may provide some understanding about why students preferred the mean over the median. The item, shown in table 9.2 and hereafter referred as Mean or Median, required students to understand a situation, examine two different data sets, and then select and explain their choice of the mean or median to describe the situation. Performance on this extended constructed-response question was very low, with only 4 percent of students in grade 12 producing responses that were scored at the satisfactory or extended levels.

To try to understand why students may prefer the mean over the median, we examined a set of about 200 responses to this item. From the responses, it became evident that many students apparently believed the mean was the better choice regardless of data distribution or context, claiming that the mean *is* the typical value or the average. This type of response was illustrated by one student who wrote, "The mean. To make generalization of 'typical' attendance, averages are used, not middle points." Another common type of response implied that the mean was superior to the median because it is more precise or more accurate, as illustrated by a student who responded, "The mean. An average gives a more accurate # because it involves all the #'s." The prevalence of these types of explanations seem to indicate an absolute belief in the superiority of the mean over the median. Perhaps earlier experiences with and greater emphasis on the mean (or arithmetic average) in traditional curriculum is a contributing factor to the belief by some that mean is *always* the best measure of central tendency.

Another contributing factor in students' frequent selection of the mean over the median may be the design of the three tasks (that is, items 3 and 4 described in table 9.1 and the item shown in table 9.2), none of which provides a clear purpose or context for deciding between median or mean. That

Table 9.2
Mean or Median

Item	Percent Responding Grade 12
[General directions]	

This question requires you to show your work and explain your reasoning. You may use drawings, words, and numbers in your explanation. Your answer should be clear enough so that another person could read it and understand your thinking. It is important that you show *all* of your work.

The table below shows the daily attendance at two movie theaters for 5 days and the mean (average) and the median attendance.

	Theater A	Theater B
Day 1	100	72
Day 2	87	97
Day 3	90	70
Day 4	10	71
Day 5	91	100
Mean (average)	75.6	82
Median	90	72

(a) Which statistic, the mean or the median, would you use to describe the typical daily attendance for the 5 days at Theater A? Justify your answer.
[Correct choice: median]

(b) Which statistic, the mean or the median, would you use to describe the typical daily attendance for the 5 days at Theater B? Justify your answer.
[Correct choice: mean]

Extended response	1
Satisfactory response	3
Partial response	10
Minimal response	28
Incorrect	25
Omitted	31

Note: Percents may not add to 100 because of rounding or off-task responses. This item is from a block for which students were provided with, and permitted to use, scientific calculators.

is, the students were asked to respond without knowing why they needed to choose between two measures of central tendency. In order to be scored as "extended," a student's response had to include a statement that the extremely low attendance on Day 4 for Theater A was an outlier and an explanation about how it could affect the mean. This line of reasoning implies that whenever there is an outlier in a set of data, the median is the best statistic to use regardless of purpose. For example, in the Mean or Median task in table 9.2, since no reason for making the choice was given, the student might assume that the statistics are going to be used to compare the attendance figures directly between the two theaters. Here, it could be argued that the identical statistic should be reported for both theaters. In fact, reporting both the mean and the median for each theater would be very effective for making such a direct comparison. Or if the attendances were to be compared using a t-test, only the mean (not the median) would be required. Because the students were not given (or asked to create) a purpose for choosing the mean or median, and because the task may have failed to alert students to consider whether the median might at times be a better indicator of central tendency than the mean, many may have simply chosen the statistic with which they were most familiar.

REASONING WITH DATA

The 1996 NAEP assessment included twelve items that required students to reason with data. The topics included data sampling; reading data from tables and charts; making data-based inferences, predictions, and decisions; and conducting a data-producing simulation. Two clusters of items from this set of twelve items provide specific insights into students' reasoning with data. One cluster of items addressed ways to sample data, requiring students to consider how both context and quantity are related to potential inferences. Another cluster required students to use information from tables to make decisions or solve multistep problems.

Collecting and Analyzing Samples

Purely random sampling from a target population is the most desirable method for determining a survey sample, but it is often impossible or impractical to generate a random sample. When nonrandom samples are used, attention needs to be paid to how well the qualitative features of the sample selected match the target population and also to whether an adequate number of data points have been collected. Three items in the 1996 NAEP that address these two aspects of selecting nonrandom samples for surveys reveal strong performance by most eighth- and twelfth-grade students. The released item, called Identify a Representative Sample and shown in table 9.3, assessed the qualitative features of a sample by asking students to decide which school setting would provide the most representative sample for a survey. About two-thirds of eighth-grade students recognized that selecting a sample from the cafeteria would produce the least biased results, although it was surprising that so many students (13 percent) selected "the faculty room," which was the only choice that did not offer a *student* population.

Table 9.3
Mean or Median

Item	Percent Responding Grade 8
A poll is being taken at Baker Junior High School to determine whether to change the school mascot. Which of the following would be the best place to find a sample of students to interview that would be most representative of the entire student body?	
A. An algebra class	11
B. The cafeteria*	65
C. The guidance office	10
D. A French class	1
E. The faculty room	13

*Indicates correct response.

Similarly, on a secure, short constructed-response item that assessed the potential for sampling bias, more than half of the students in grade 8 and three-fourths of the students in grade 12 produced correct responses. The aspect of sample size was assessed on a secure short constructed-response item on which, again, more than half of the eighth-grade students responded correctly.

Reading Information and Reasoning from Data in Tables

On items that required reading and using information from data tables and charts, most twelfth-grade students were successful, with about two-thirds responding correctly to the two secure items that involved proportions and more than 80 percent responding correctly to a secure item that required that a data point be inserted into a table of existing data. Two items, both of which were administered to students in grade 12 (and one of which was administered to other grade levels as well) appear in table 9.4.

The performance levels of students on item 1 in the table demonstrate the expected improvement in performance across grade levels with regard to

DATA AND CHANCE

Table 9.4
Reading Information and Reasoning from Data in Tables: Sample Items

Items	Percent Responding		
	Grade 4	Grade 8	Grade 12

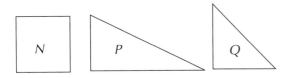

1. This question refers to pieces N, P, and Q [shown above].

   In Mr. Bell's classes, the students voted for their favorite shapes for a symbol. Here are the results.

	Class 1	Class 2	Class 3
Shape N	9	14	11
Shape P	1	9	17
Shape Q	22	7	2

   Using the information in the chart, Mr. Bell must select one of the shapes to be the symbol. Which one should he select and why?

   The shape Mr. Bell should select: _____

   Explain:

	Grade 4	Grade 8	Grade 12
Correct response of shape N with explanation	32	58	67
Incorrect response: Shape N chosen but with inadequate explanation or no explanation	12	15	17
Incorrect response: Shape Q chosen with an explanation that refers to the number of votes	32	16	9
Any other incorrect response	21	11	6
Omitted	3	<1	1

(continued)

Table 9.4
Reading Information and Reasoning from Data in Tables: Sample Items (continued)

Items		Percent Responding	
	Grade 4	Grade 8	Grade 12

TIME CARD Name: J. Jasmine	Number of Hours	Average Hourly Wage	Total Daily Earnings
Mon. 10:00 a.m. - 3:00 p.m.	5	5.50	27.50
Tues. 9:00 a.m. - 4:00 p.m.	7	5.50	38.50
Wed. 3:00 p.m. - 7:00 p.m.	4	5.75	23.00
Thurs. 2:00 p.m. - 8:00 p.m.	6		
Fri. 5:00 p.m. - 10:00 p.m.	5	6.00	30.00

1. According to the information above, what is the average hourly wage for Thursday's earnings if the total earnings for the five days was $153.50?

   Answer: _____

   The hourly wage rate changes at some hour during the day. At what time does the hourly wage rate change?

   Answer: _____

Two correct answers ($5.75 and 5:00 p.m.)	—	—	13
One correct answer	—	—	43
Incorrect response	—	—	40
Omitted	—	—	3

Note: Percents may not add to 100 because of rounding. Item 1 is from an item block for which students were provided with geometric shapes as manipulatives, and item 2 is from an item block for which students were provided with, and permitted to use, scientific calculators.

students' ability to use information from tables for making a decision. This item was administered at all three grades in both 1992 and 1996, and performance was stable over the years. Notable results reveal that the choice of shape Q was quite high for students in grade 4 (32 percent), but this choice was less appealing to students at grade 8 (16 percent) and least appealing to students in grade 12 (9 percent). It is likely that these students chose shape

Q because it is the shape with the greatest number of votes in one of the three classes. The decrease in attention to shape Q from grade 4 to grade 12 may signal students' increasing attention to various data points simultaneously in a decision-making situation. By grade 12 the overwhelming choice was the correct answer, shape N, accompanied by a complete and correct explanation. An examination of some actual twelfth-grade students' responses revealed a variety of justifications for choosing the correct response of shape N, ranging from emphasizing the sum of the cells across the table to distribution-based explanations, such as, "N. None of the classes like that the least."

Unlike the other items administered to students in grade 12, item 2 in the table was the only one for which their performance was not strong. More than one-half of the students were successful in using information from the table to obtain at least one correct answer—either the average hourly wage or the time at which the hourly wage changes—but only slightly more than 10 percent obtained two correct answers. This may be due to different levels of difficulty with respect to the two parts of the item; that is, perhaps finding the hourly wage for Thursday's earnings was more difficult than finding the time at which the hourly rate changed, or vice versa. Unfortunately, NAEP results do not separate the 43 percent of students who got one answer correct according to the percent who found only the correct hourly wage and the percent who found only the correct time.

Another source of difficulty may have been that students were correctly reading information from the table but then implementing an incomplete procedure for solving the problem. For example, an examination of a small set of responses to this item revealed that some students answered $34.50 to the first question, which is the result from first step of the multistep problem. A propensity to complete only one step in a multistep problem was noted in the previous analyses of students' performance on NAEP (for example, Kouba, Zawojewski, and Strutchens 1997), which may partially explain the lower performance level on this item than on the other three items discussed above.

UNDERSTANDING AND USING GRAPHICAL DATA DISPLAYS

The 1996 NAEP mathematics assessment included fourteen items that tested students' understanding of graphs as well as their ability to reason about graphs involving data or chance. These items reflect Curcio's (1987) notion of "graph sense," a theory which was developed by Friel, Curcio, and Bright (1988). Curcio speaks of three levels of graphical understanding: reading information from graphs, reading across graphs or across representations of data, and reading beyond graphs. The NAEP items involving graphs fall into three categories that somewhat resemble Curcio's levels of graphical understanding. One set of items simply asked students to read information from various types of graphs, such as line graphs, pictographs, box-and-whisker plots, and stem-and-leaf plots. Another cluster of items required students to complete bar graphs or to construct pictographs to represent information that had been presented in tabular form. There was also a set of items that required students to do some extended reasoning from graphs or to make some interpretation or decision on the basis of data; that is, going beyond the information presented.

Reading from Graphs

The cluster of NAEP items on reading information from graphs is described in table 9.5. The results reveal fairly strong performances on different forms of graphs—especially pictographs in grade 4, stem-and-leaf plots in grade 8, and line graphs in grade 12. Box-and-whisker plots and stem-and-leaf plots, both of which are graphical displays that feature the distribution of a data set, were included for the first time in the 1996 NAEP mathematics assessment. Twelfth-grade students demonstrated a poor understanding of box-and-whisker plots; almost two-thirds were unable to interpret the scale of the plot correctly to estimate a percentile. However, stem-and-leaf plots were easier for eighth-grade students than box-and-whisker plots were for twelfth-grade students, with about 60 percent of eighth-grade students responding correctly.

Table 9.5
Performance on Items Involving Reading from Graphs

Type of Graph Appearing in Item	Percent Correct		
	Grade 4	Grade 8	Grade 12
1. Pictograph	61	—	—
2. Stem-and-leaf plot	—	61	—
3. Line graph	—	—	77
4. Line graph	—	49	73
5. Box-and-whisker plot	—	—	38

Note: All were multiple-choice items. Items 1 and 3, both released items, are shown in table 9.6.

The relative ease with which students dealt with stem-and-leaf plots may be due to the innate simplicity of that kind of plot, which has a direct representation for each and every data point. The box-and-whisker plot, however, requires data reduction in which the data are transformed to quartiles prior to creating the graphic display. It may be that many twelfth-grade students have not been introduced to box-and-whisker plots in the mathematics or science curricula or that some students have been exposed to different types of box plots. There is an implicit assumption that the box-and-whisker plot, as used in the NAEP item, is based on quartiles, with the middle 50 percent of the distribution represented by the "box" and the upper and lower quartiles represented by the "whiskers." However, 90 percent box plots are also commonly used, and boxes can represent *different* types of distributions. For example, the box may represent data points that are within one standard deviation from the mean rather than some distance away from the median.

Two released items described in table 9.5 (items 1 and 3) involving graph reading appear in table 9.6. Performance on item 1 in table 9.6 shows that just over 60 percent of fourth-grade students were successful in reading information from a pictograph. A potential cause of the incorrect responses is of particular interest. Since the distracters for this item are all based on "13" (i.e., 13 ones, 13 tens, 13 hundreds, and 13 thousand), it is plausible that incorrect responses reveal a lack of understanding of proportions, specifically in place value and numeration. Because pictographs, by their very nature, represent proportional reasoning, much of students' difficulties may be attributed to lack of understanding of proportions. (For additional information about proportional reasoning, see chapter 7 by Wearne and Kouba.)

About three-fourths of students in grade 12 responded correctly to each of two line-graph items, but only 49 percent of students in grade 8 responded correctly to the line-graph item that was administered in both grades (see table 9.5). The strong performance of twelfth-grade students on line graphs continued a trend noted in the 1992 NAEP (Zawojewski and Heckman 1997). However, both line-graph items on the 1996 NAEP were relatively easy and representative of the types of graphs that appear in newspapers, magazines, and advertisements every day, so it is somewhat puzzling that performance was not higher. For example, for item 2 in table 9.6, 22 percent of students in grade 12 chose an incorrect answer even though the item required only that they read information in the graph and use subtraction to answer the question posed. Of those who responded incorrectly, 10 percent selected choice D (2.5 degrees). Misinterpreting the scale may have been the main source of difficulty for these students.

Reading across Graphs

The 1996 NAEP included additional clusters of items that are similar to tasks found in Curcio's (1987) category of "reading across graphs." In particular, the set of secure items described in table 9.7 asked students to construct or complete different types of graphs based on data presented in

Table 9.6
Reading from Graphs: Sample Items

Item	Percent Responding	
	Grade 4	Grade 12

CARTONS OF EGGS SOLD LAST MONTH

Farm A ○ ○ ○
Farm B ○ ○ ○ ○ ○ ○
Farm C ○ ○ ○

Each ○ = 100 cartons

1. According to the graph, how many cartons of eggs were sold altogether by farms A, B, and C last month?

A. 13	14	—
B. 130	10	—
C. 1,300*	61	—
D. 13,000	13	—

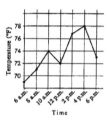

2. According to the graph above, the temperature at 10 a.m. is approximately how many degrees greater than the temperature at 8 a.m.?

A. 1	—	2
B. 1.5	—	5
C. 2	—	5
D. 2.5	—	10
E. 3*	—	77

*Indicates correct response.
Note: Percents may not add to 100 because of rounding.

tables; all items described in the table were also administered as part of the two previous NAEP assessments, in 1990 and 1992. Although these items cannot be discussed in detail, it is notable that eighth- and twelfth-grade students showed strong performance on them—77 percent to 90 percent responding correctly. More than half of the fourth-grade students were successful on items 1 and 3, which required them to complete a pictograph and bar graph, respectively. However, item 4, which also involved completing a

Table 9.7
Performance on Items Involving Reading across Graphs

Item Description	Percent Correct		
	Grade 4	Grade 8	Grade 12
Pictograph			
1. Complete a pictograph from data presented in a table.	63	89	—
2. Determine the number of symbols to be used in a pictograph.	—	80	—
Bar Graph			
3. Complete a bar graph based on data presented in a table.	54	85	90
4. Complete a bar graph based on given data.	36	77	—

Note: Item 2 was a multiple-choice item, and the rest were short constructed-response items.

bar graph, was particularly difficult for these students because it required some proportional reasoning.

Reading beyond Graphs

NAEP also included a cluster of items that matched Curcio's (1987) category of "reading beyond graphs." These are items which require students to interpret graphs and make decisions using information from graphs. As suggested by the performance data in table 9.7, most eighth- and twelfth-grade and many fourth-grade students can construct and complete graphs and therefore do quite well on items involving reading across graphs. However, results from the items described in table 9.8 reveal that fewer than half of students can interpret graphs, make decisions from information presented in graphical form, or provide extended reasoning for their interpretation or decision. For example, in item 1, fourth-grade students had to read and interpret information in tabular form, then perform some arithmetic operations, and finally construct a pictograph. This item was answered correctly by only 44 percent of the fourth-grade students. Such multistep items require a higher level of thinking than simply constructing a pictograph. An additional 15 percent of the fourth-grade students provided a correct response for only part of the problem. Although these students may have understood how to construct pictographs, the additional computational steps may have been a source of their difficulties.

In a circle-graph item, item 2 in table 9.8, only 27 percent of eighth-grade and just under half of twelfth-grade students were completely successful in correctly evaluating all four statements about the information in the

Table 9.8
Performance on Items Involving Reading beyond Graphs

Item Description	Percent Correct		
	Grade 4	Grade 8	Grade 12
1. Interpret data and create a pictograph.	44	—	—
2. Interpret information from a circle graph and choose the correct answer.	—	27[a]	49[a]
3. Interpret information from a scatterplot and make a decision; explain answer.	—	—	22
4. Interpret information from a histogram.	—	—	11
5. Decide which of two graphs presenting the same information is misleading.	—	2	—

[a]Percent of students answering all four statements correctly.

Note: Item 2 was a multiple-response item, and items 1, 3, and 4 were short constructed-response items. Item 5, a released extended constructed-response task, is shown in table 9.9.

circle graph. Difficulties on this item may include students' inability to compare proportions represented as percents or difficulty with reading and understanding the item. Performance on items 1 and 2 in the table indicate that using information from graphs to answer multistep problems is difficult, but interpreting graphical information to make decisions is even more difficult.

The three items that required students to make a decision or prediction on the basis of information from a graphic display were very difficult for students, as illustrated by the poor performance on items 3, 4, and 5 described in table 9.8. In item 3, students were shown a scatterplot and asked to judge the accuracy of two possible predictions. Only 22 percent of the twelfth-grade students chose the correct prediction and gave an acceptable justification. An additional 33 percent selected the correct prediction but included no justification for their decision. It is clear from examining a small sample of students' responses that some students gave justifications that either went well beyond the data in the graph or gave reasons that had nothing to do with the graph at all but were simply arguments associated with the context of the task.

The performance by twelfth-grade students on item 4, involving a decision based on information given in a histogram, was also extremely low. Only 7 percent answered correctly in 1992, and 11 percent in 1996. However, the increase in performance is statistically significant, offering some hope that students' performance on interpreting histograms may be improving. Students may just lack experience with histograms, or it may be the complexity of this particular task that accounts for students' lack of success on this short constructed-response item. For example, the item required students to read and interpret a histogram and then to make predictions beyond the given graph to another population. The histogram itself presented information in two continuous variables and required the use of proportional reasoning to interpret it. Reasoning about histograms appears to be a high cognitive demand activity.

Among their components of graph sense, Friel, Curcio, and Bright (1988) include the ability to recognize when one graph is more useful than another or which type of graph may be useful in a particular circumstance. Graph sense also includes the ability to use and to recognize misuses of variations of the *same* type of graph in situations where the shape and scale of the graphs can highlight different aspects of the data. The item called Metro Rail Graph and shown in table 9.9 was a released graph-comparison task. This extended constructed-response question asks students to choose between two different versions of the same line graph and to defend their choice. In general, eighth-grade students were unable to supply coherent reasons why someone might consider graph B to be misleading. Only 2 percent of the students provided a correct response accompanied by reasoning that was judged to be complete, indicating that critical analysis of graphs is not a strength among eighth-grade students. This item also produced a large percent of omitted (blank) responses, 31 percent. Students may not have responded to the item because it involved several parts that each required extended explanations, and as with all extended constructed-response items, it was at the end of the block of items. The extremely low level of performance may in part be due to students' lack of communication skills, since an additional 19 percent provided evidence of a partially correct explanation but could not adequately explain their reasoning.

To gain some insight into how students approached this problem, we examined a set of about ninety nonblank students' responses. Different types of reasoning were revealed by the analysis of the students' responses, some of which are illustrated in figure 9.1. Detailed explanations such as those provided in responses 1 and 2 were rare in our sample set. A number of students justified a choice of Graph B because it "went up so much higher" or "showed a higher increase" than Graph A, as illustrated by response 3. Some students indicated that Graph B was "easier to read," "more detailed," or "just showed things better"; response 4 is an example of this kind of explanation. There were other students in the sample, as illustrated in response 5, who apparently did not understand that both graphs depicted the same information, even though they were told this. There were also students who were unable to extend their reasoning to the second part of the item or who simply repeated responses to the first

Table 9.9
Metro Rail Graph

Item	Percent Responding Grade 8

[General directions]

This question requires you to show your work and explain your reasoning. You may use drawings, words, and numbers in your explanation. Your answer should be clear enough so that another person could read it and understand your thinking. It is important that you show *all* of your work.

METRO RAIL COMPANY

Month	Daily Ridership
October	14,000
November	14,100
December	14,100
January	14,200
February	14,300
March	14,600

The data in the table above has been correctly represented by both graphs shown below.

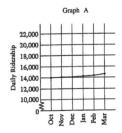
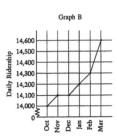

Which graph would be best to help convince others that the Metro Rail Company made a lot more money from ticket sales in March than in October? Explain your reasoning for making this selection.

Why might people who thought there was little difference between October and March ticket sales consider the graph you chose to be misleading?

Correct response	2
Partial response	19
Minimal response	34
Incorrect response	14
Omitted	31

Note: This item is from an item block for which students were provided with, and permitted to use, scientific calculators.

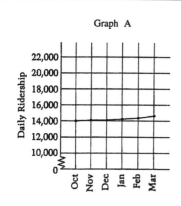
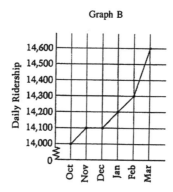

Part 1: Which graph would be best to help convince others that the Metro Rail Company made a lot more money from ticket sales in March than in October? Explain your reason for making this selection?

Part 2: Why might people who thought that there was little difference between October and March ticket sales consider the graph you chose to be misleading?

Response 1
Part 1: B, The less numrical space between the bracets will more acuratly show the increase

Part 2: Because it shows a small amount of increase in a big way

Response 2
Part 1: Graph B because the line goes almost straight up sometimes, and to me that would be more convincing.

Part 2: because it goes all the way up

—continued on next page

Response 3
Part 1:

Graph B because it shows how the graph goes up so much.

Part 2:

because it shows a big jump because all they did was make each square worth more ridership

Response 4
Part 1:

Graph B because its the best and it shows more detial in it

Part 2:

it shows more detial and it shows more in it.

Response 5
Part 1:

graph B because graph B has a higher Daily Ridership

Part 2:

I DON't think so.

Fig. 9.1. Selected responses to Metro Rail Graph

part; for example, "because it goes up higher" or "it showed it better." Results from this item suggest that more attention needs to be paid to critiquing graphs, discussing multiple graphical representations of the same data, and communicating reasoning in writing.

PROBABILITY AND CHANCE

The thirteen NAEP items that dealt with probability and chance fell into two categories: items dealing with simple probability and items requiring reasoning about probability. The eight simple probability tasks included calculating probabilities from the ratio of a selected number of outcomes to the total number of outcomes, choosing outcomes from given sample spaces, generating small sample spaces, or combining several of these skills within the same task. The five items involving reasoning about chance required students to solve a problem, show their work, and justify their answer. In these items students were asked to make predictions, inferences, or decisions on the basis of some information from a probability experiment or chance situation.

Simple Probability

On examination of the items that addressed simple probability, several issues emerged. Students' relative strength in "one out of N" situations as opposed to "several out of N" situations was apparent, and we refer to this as the "one out of N" phenomenon. There were also indications of a potentially confounding influence associated with students' weak proportional reasoning skills. Finally, there was evidence of students' inability to use information from sample spaces, even from those that students correctly generated.

The One-out-of-N Phenomenon

The items described in table 9.10 address two types of simple probability situations: those in which exactly one outcome is compared to the total number of possible equally likely outcomes (one out of N); and those in which more than one outcome is compared to the total number of possible equally likely outcomes (several out of N). A 1992 NAEP finding (Zawojewski and Heckman 1997) is also evident in the 1996 data: fourth-grade students are better at correctly answering one-out-of-N probability tasks than several-out-of-N tasks. Furthermore, on item 4 (a several-out-of-N task), more than 20 percent of eighth-grade students and 27 percent of fourth-grade students incorrectly chose the one-out-of-N foil from among the distracters. Thus, even though more than half the eighth-grade students

Table 9.10
Performance on Items Involving "One out of N" and "Several out of N"

Item Description	Percent Correct	
	Grade 4	Grade 8
1. Given a sample space of N equally likely outcomes, determine the probability of "one out of N" occurring.	59	—
2. Given another sample space of N equally likely outcomes, determine the probability of "one out of N" occurring.	51	—
3. Choose the correct probability in a "several out of N" situation.	31	—
4. Choose the correct probability in another "several out of N" situation.	24	59

Note: All were multiple-choice items. Item 3, a released item, is shown in table 9.11.

correctly responded to the several-out-of-N item, a sizable percent remained entrenched in one-out-of-N reasoning. Why does this happen?

Perhaps the problem is due to a lack of proportional reasoning skill, the inability to form and use ratios appropriately. Perhaps it is a lack of experience with probability as an application of proportion. Zawojewski and Heckman (1997) suggest that the work of Piaget and Inhelder (1975) on the development of probability concepts might explain students' difficulties. Certainly proportional reasoning is one of the indicators of the formal operational stage. However, perhaps students' difficulties are an early manifestation of what Konold has often referred to as "the outcome approach," in which any event can happen on a given trial (for example, Konold 1989; Konold et al. 1993). Konold's research, which has been done primarily with secondary school and college students, suggests that people make evaluations of the probability of outcomes based on a single, solitary trial and that anything *could* happen on any one trial. This type of thinking does not consider probability as a ratio, nor does it recognize probability as the relative frequency of an outcome in a repeatable probability experiment. Outcome approach thinkers believe that on any one trial any one result could happen from among the total number of possibilities, perhaps accounting for some of the tendency for students to select the one-out-of-N foil. This is a potentially interesting research opportunity, since the outcome approach has not been investigated with elementary school students.

Proportional Reasoning

The released item in table 9.11 provides some clues about fourth-grade students' use—or nonuse—of proportional reasoning on probability tasks. In particular, the one-out-of-N phenomenon might have motivated the 22 percent who chose 1 out of 5 (choice A) as their answer. However, the 28 percent who selected 2 out of 3 (choice D) may have done so because it uses the two numbers that appear in the question. Interestingly, this response would be correct if the item had asked for "odds," but it is unlikely that most fourth-grade students have been exposed to "odds" as a topic in mathematics. Although performance on this item in 1996 (31 percent correct) was significantly higher than it was in 1992 (26 percent correct), there is still reason to be concerned about fourth-grade students' performance on simple probability problems involving proportional reasoning in the several-out-of-N situation.

Table 9.11
A "Several out of N" Item

Item	Percent Responding Grade 4
There are 3 fifth graders and 2 sixth graders on the swim team. Everyone's name is put in a hat and the captain is chosen by picking one name. What are the chances that the captain will be a fifth grader?	
A. 1 out of 5	22
B. 1 out of 3	16
C. 3 out of 5*	31
D. 2 out of 3	28

*Indicates correct response.
Note: Percents may not add to 100 because of rounding or off-task responses.

Sample Spaces

Items on sample space were administered only to eighth- and twelfth-grade students. These items involve using information from a sample space to find a probability or using the reverse process of generating the sample space when given the probability. One secure item (not described in the tables) administered at both grade levels gave students the sample space and asked them to find the probability of a selected event. Further complicating this item was the need to deal with a ratio involving the several-out-of-N situation. Only 22 percent of eighth-grade and 39 percent of twelfth-grade students answered correctly; however, the performance for each grade was significantly higher than it was in 1992 when 18 percent of eighth-grade and 29 percent of twelfth-grade students answered correctly.

It is unlikely that the several-out-of-N phenomenon was the sole source of the difficulty that students had with this sample space item. Fifty-nine percent of the eighth-grade students chose the correct probability on item 4 in table 9.10, a several-out-of-N problem, compared to only 22 percent correct on the secure sample-space item. Students in grade 8 appear to have difficulty gleaning information from sample spaces and then correctly using it. Results from another secure multiple-choice item involving sample spaces provides corroborating evidence. On that item, only about one-third of eighth-grade students were successful in choosing the correct sample space when given the probability. An examination of a small set of responses to a secure extended constructed-response item involving reasoning about chance and given to eighth-grade students revealed that even in cases where the students listed the complete and correct sample space, some students were still not able to use it to complete the problem successfully. It may be that these students are somewhat practiced in producing sample spaces in school mathematics but spend less time interpreting or reasoning from sample spaces.

Reasoning about Chance

Among the NAEP probability items were some short and extended constructed-response tasks that required reasoning about chance. These items provided an opportunity to gain some insights into students' reasoning processes. Two of these items were released, one for grade 4 and one for grade 12, and are discussed next. Performance was poor on these items, and there was evidence from samples of students' work indicating difficulties in reasoning about chance as well as students' inability to communicate their thinking. The difficulties are representative of those that were also found on a secure extended constructed-response item given to eighth-grade students and requiring reasoning about chance.

Grade 4 Task: Gum Ball Machine

The item, called Gum Ball Machine and shown in table 9.12, was an extended-constructed response task involving reasoning about chance, for which 21 percent of fourth-grade students produced a satisfactory or extended response. An additional 29 percent of the responses indicated partial explanations for the number of red gum balls, with the remaining responses suggesting that students had difficulty in reasoning about chance and communicating their reasoning in writing.

A close examination of about 200 fourth-grade students' responses provided some insights into students' thinking on this task. As shown in the sample responses that appear in figure 9.2, there were students who used proportional reasoning or less sophisticated explanations that revealed that some attention was being paid to the relative number of different colors in the bag. For example, response 1 used a proportional approach, predicting that exactly 5 red gum balls would be pulled because half the gum balls were red. Other responses indicated 5 gum balls but simply noted "because there were more red" as the explanation. Some students who predicted 6 or

Table 9.12
Gum Ball Machine

Item	Percent Responding Grade 4
[General directions]	

Think carefully about the following question. Write a complete answer. You may use drawings, words, and numbers to explain your answer. Be sure to show all of your work.

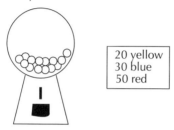

20 yellow
30 blue
50 red

The gum ball machine has 100 gum balls: 20 are yellow, 30 are blue, and 50 are red. The gum balls are well mixed inside the machine.

Jenny gets 10 gum balls from this machine.

What is your best prediction of the number that will be red?

Answer: _____ gum balls

Explain why you chose this number.

Extended response	7
Satisfactory response	14
Partial response	29
Minimal response	4
Incorrect response	33
Omitted	12

Note: Percents may not add to 100 because of rounding. This item is from an item block for which students were provided with, and permitted to use, four-function calculators.

---

more balls also used the argument that "there are more red balls," as illustrated in response 2.

Other types of reasoning were also apparent. For example, an explanation for the answer of "10 balls" in response 3 is consistent with Konold's "outcome approach" described earlier in this chapter. This student may be thinking, "The outcome of 10 reds *could* happen, so I'll predict it." There were also

Response 1

What is your best prediction of the number that will be red?

Answer: ____5____ gum balls

Explain why you chose this number.

It is half of ten

Response 2

What is your best prediction of the number that will be red?

Answer: ____6____ gum balls

Explain why you chose this number.

I got that answer because there are more reds so more changes of getting it.

Response 3

What is your best prediction of the number that will be red?

Answer: ____10____ gum balls

Explain why you chose this number.

I choose that number because there are 50 red gum balls and I can pick 10 red gum balls.

—continued on next page

Response 4

What is your best prediction of the number that will be red?

Answer: __4__ gum balls

Explain why you chose this number.

because if you got some yellow and blue too you wouldn't get very many reds. The colors where all mixed together they said.

Response 5

What is your best prediction of the number that will be red?

Answer: __6 red__ gum balls

Explain why you chose this number.

I divided and got ten and added 12 and times it by two and 44 plus six equals fifty.

```
    8 R 2          12
  6)50           + 10
    48            ──
    ──            22
     2           x  2
                  ──
                  44
                +  6
                  ──
                  50
```

Fig. 9.2. Selected responses to Gum Ball Machine

students who focused on the well-mixed nature of the gum balls in the machine and concluded that the colors should tend to balance out, or be fair, as illustrated by response 4. This kind of reasoning suggests that students believe that well-mixed populations should produce uniformly distributed samples of colors, regardless of the percents of colored balls in the machine. This is a manifestation of a belief that all outcomes in a probability experiment are equally likely to occur. It was not surprising that there were also a number of students who seemed to perform random arithmetic operations with the available numbers, as illustrated by response 5.

Low performance levels may also be due to causes other than students' actual reasoning within the task. For example, it may be that fourth-grade students are still not used to writing in mathematics, or perhaps they just cannot explain their reasoning in detail at this stage of their development (for example, Green 1983, as cited in Zawojewski and Heckman 1997). The design of the item may also have contributed to the low performance, since the wording encouraged students to predict a specific number, rather than a "likely range" of red gum balls. An emphasis on "likely intervals" is a more appropriate task for a sampling problem than requesting an exact numerical prediction (Shaughnessy et al. 1996; Shaughnessy 1997).

Grade 12 Task: Two Spinners

The released short constructed-response task, called Two Spinners and shown in table 9.13, deals with determining the probability of a compound event. Performance results showed that only 8 percent of the students disagreed with James and completely and correctly explained why they disagreed with him (for example, "the chances of winning are 1 in 4"). An additional 20 percent of the students also disagreed and gave a partial explanation that indicated some understanding of the sample space. This result supports an earlier conjecture that there are students who are able to generate sample spaces but are not necessarily able to use the sample space to reason about subsequent mathematical activity.

To gain some understanding of the ways in which students reasoned about this task, we used a set of about 300 nonblank student responses and classified the responses into categories representing different types of student responses. Some responses based on our classification appear in figure 9.3. Students who provided correct reasoning gave a variety of explanations. Some provided a representation of the sample space of four outcomes such as in response 1, and others provided a theoretical argument based on the product $1/2 \times 1/2$ (or $50\% \times 50\%$) for the two spinners such as in response 2. Another strategy was to reason that the chances *must* be less than 50-50, because one spinner alone was 50-50, and the second spinner was bound to reduce the chances, as illustrated by response 3.

DATA AND CHANCE

Table 9.13
Two Spinners

Item	Percent Responding Grade 12

The two fair spinners shown above are part of a carnival game. A player wins a prize only when *both* arrows land on black after each spinner has been spun once.

James thinks that he has a 50-50 chance of winning. Do you agree?

◯ Yes ◯ No

Justify your answer.

Correct response of No and correct explanation	8
Correct response of No with a partial explanation	20
Incorrect response—correct response of No with an incorrect explanation or no explanation	24
Incorrect response—incorrect answer of Yes with or without an explanation	44
Omitted	4

Note: This item is from an item block for which students were provided with, and permitted to use, scientific calculators.
Source for performance data: Mitchell et al. 1999, p. 183.

Incorrect reasoning was also evident in a variety of forms. Many responses indicated that since each spinner was half black, the chance of both spinners ending up black was 50-50, such as "half the area is black," and "there are two blacks out of four regions." Response 4 is typical of this kind of incorrect reasoning. This type of response is consistent with the "anchoring" strategy that has been described by Tversky and Kahneman (1974) in which students focus on the probability of the first event, which acts as an "anchor," and then generalize to predict the probability of a compound event. Other responses, such as in response 5, showed little evidence of

Response 1

James thinks he has a 50-50 chance of winning. Do you agree?

○ Yes ● No

Justify your answer.

He only has a 25% chance of winning

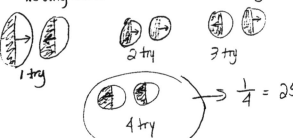

Response 2

James thinks he has a 50-50 chance of winning. Do you agree?

○ Yes ● No

only 50% **one** will, then
only 50% **of** the 50% (original)
that the other will.
25%

Response 3

James thinks he has a 50-50 chance of winning. Do you agree?

○ Yes ● No

Justify your answer.

No, because James must land on the black area with **both** spinners. His chances are reduced.

Response 4

James thinks he has a 50-50 chance of winning. Do you agree?

● Yes   ⊙ No

Justify your answer.

Because each circle is half Black. So he has a 50-50 chance of at least 1 arrow landing on the Black shaded area.

Response 5

James thinks he has a 50-50 chance of winning. Do you agree?

⊙ Yes   ● No

Justify your answer.

There are too many variables to consider like wind factor, smoothness of the spinners, and looseness of the arrows.

Fig. 9.3. Selected responses to Two Spinners

thinking mathematically and instead indicated that some sort of physical property of the spinners would keep the odds different from 50-50, such as the rate the spinners were spun or the initial position of the arrows. Jones et al. (1997) have documented similar statements among younger students in probability tasks in which students attribute outcomes to some sort of physical property of the apparatus. However, none of the twelfth-grade students' responses from the convenience sample attributed outcomes to "magic" as do many of younger students in the Jones et al. (1997) studies.

CONCLUSION

As Dossey reports in chapter 2, overall performance on the NAEP content strand Data Analysis, Statistics, and Probability is improving. In particular, average performance in grade 4 and 12 has improved significantly from 1992 to 1996, and in grade 8 significant gains have been made since 1990. The examination of clusters of related items in this chapter reveal that students are at least maintaining their performance in many areas and are in fact making small, but significant, gains in some areas. For example, there was significant improvement from 1990 to 1996 for eighth-grade and twelfth-grade students in finding the mean and median of a data set.

Although the performance levels are generally either improving or remaining stable, there are also areas of concern. For example, when proportional reasoning is required, students often experience difficulty on items involving reasoning about data, graphs, and chance. Further, even when students seem able to accomplish a particular technique or skill, such as reading a graph or listing a sample space, there is apparent difficulty when they are asked to reason beyond the immediate answer or to use the answer to make decisions. Zawojewski and Heckman (1997) reported that in the 1992 NAEP, fourth-grade students had more difficulty dealing with probability ratios that are not unit fractions, and that trend continues.

The items used in the 1996 NAEP have a number of strengths and show areas of improvement, but they also leave some areas of concern. One of the major strengths is the continued inclusion of short and extended constructed-response tasks for data analysis, probability, and statistics that provide the opportunity not only to evaluate student's ability to explain their reasoning but also to examine the actual ways in which students are reasoning about statistics and probability. However, the assessment could be improved by purposefully including clusters of items that address areas of difficulty that have surfaced from previous analyses of NAEP assessments, as well as those identified in the statistics education literature. For example, the role of proportional reasoning in students' performance on data analysis, statistics, and probability items would be greatly enhanced by designing sets of items that would reveal specific performance patterns and contribute to our understanding of students' use of proportional reasoning in concepts of statistics and probability. In the case of items that assess measures of central tendency (for example, mean and median), a cluster of items could be developed that requires students to consider the purpose for collecting the data to support the selection of an appropriate measure of central tendency.

It would also be helpful if more of the NAEP items on data and chance were administered to all three grade-level groups in order to provide more information on patterns of conceptual development in data and chance across the levels. For example, on the 1996 NAEP it would have been helpful to include responses from students in grades 8 and 12 on the Gum Ball Machine task and from fourth- and eighth-grade students on the Two Spinners task. Perhaps these suggestions can be incorporated into the NAEP mathematics assessment scheduled for 2004.

REFERENCES

Curcio, Frances R. "Comprehension of Mathematical Relationships Expressed in Graphs." *Journal for Research in Mathematics Education* 18 (November 1987): 382–93.

Friel, Susan N., Frances R. Curcio, and George W. Bright. "Making Sense of Graphs: A Theoretical Framework." Chapel Hill, N.C.: University of North Carolina at Chapel Hill, 1988.

Green, David R. A. "Survey of Probability Concepts in 3000 Pupils Aged 11–16 Years." In *Proceedings of the First International Conference on Teaching Statistics*, edited by D. R. Grey, P. Holmes, V. Barnett, and G. M. Constable, pp. 766–83. Sheffield, England: Organizing Committee of the First Conference on Teaching Statistics, 1983.

Jones, Graham A., Cynthia W. Langrall, Carol A. Thorton, and A. Timothy Mogill. "A Framework for Assessing and Nurturing Young Children's Thinking in Probability." *Educational Studies in Mathematics* 32 (March 1997): 101–25.

Konold, Cliff. "Informal Conceptions of Probability." *Cognition and Instruction* 6 (1989): 59–98.

Konold, Cliff, Alexander Pollatsek, Arnold Well, Jill Lohmeier, and Abigail Lipson. "Inconsistencies in Students' Reasoning about Probability." *Journal for Research in Mathematics Education* 24 (November 1993): 392–414.

Kouba, Vicky L., Judith S. Zawojewski, and Marilyn E. Strutchens. "What Do Students Know about Numbers and Operations?" In *Results from the Sixth Mathematics Assessment of the National Assessment of Educational Progress*, edited by Patricia Ann Kenney and Edward A. Silver, pp. 87–140. Reston, Va.: National Council of Teachers of Mathematics, 1997.

Lappan, Glenda, William Fitzgerald, James Fey, Susan Friel, and Elizabeth Phillips. *Connected Mathematics Program*. Palo Alto, Calif.: Dale Seymour Publications, 1997.

Mitchell, Julia H., Evelyn F. Hawkins, Pamela M. Jakwerth, Frances B. Stancavage, and John A. Dossey. *Student Work and Teacher Practices in Mathematics*. Washington, D.C.: National Center for Education Statistics, 1999.

National Assessment of Educational Progress. *Math Objectives: 1985–86 Assessment*. 17-MD-10. Princeton, N.J.: Educational Testing Service, 1986.

National Assessment Governing Board. *Mathematics Framework for the 1996 National Assessment of Educational Progress*. Washington, D.C.: National Assessment Governing Board, 1994.

National Council of Teachers of Mathematics. *Curriculum and Evaluation Standards for School Mathematics*. Reston, Va.: National Council of Teachers of Mathematics, 1989.

Piaget, Jean, and Barbel Inhelder. *The Origin of the Idea of Chance in Children*. Translated by Lowell Leake, Jr., Paul Burrell, and Harold D. Fishbein. New York: W. W. Norton, 1975.

Reese, Clyde M., Karen E. Miller, John Mazzeo, and John A. Dossey. *NAEP 1996 Mathematics Report Card for the Nation and the States: Findings from the National Assessment of Educational Progress*. Washington, D.C.: National Center for Education Statistics, 1997.

Shaughnessy, J. Michael. "Missed Opportunities in Research on the Teaching and Learning of Data and Chance." In *People in Mathematics Education: The Proceedings of the Twentieth Annual Meeting of the Mathematics Education Research Group of Australia*, edited by Fred Biddulph and Ken Carr, pp. 6–22. Waikato, New Zealand: The University of Waikato Printery, 1997.

Shaughnessy, J. Michael, Joan Garfield, and Brian Greer. "Data Handling." In *International Handbook of Mathematics Education*, edited by Alan J. Bishop, Ken Clements, Christine Keitel, Jeremy Kilpatrick, and Colette Laborde, pp. 205–37. Dordrecht, Netherlands: Kluwer Academic Publishers, 1996.

Technical Education Research Center (TERC). *Investigations into Data, Space and Number*. Palo Alto, Calif.: Dale Seymour Publications, 1997.

Tversky, Amos, and Daniel Kahneman. "Judgment under Uncertainty: Heuristics and Biases." *Science*, (27 September 1974), pp. 1124–31.

Zawojewski, Judith S., and David Heckman. "What Do Students Know about Data Analysis, Statistics, and Probability?" In *Results from the Sixth Mathematics Assessment of the National Assessment of Educational Progress*, edited by Patricia Ann Kenney and Edward A. Silver, pp. 195–223. Reston, Va.: National Council of Teachers of Mathematics, 1997.

10

Algebra and Functions

Glendon W. Blume and David S. Heckman

ALGEBRA receives substantial emphasis in the school mathematics curriculum. Elementary school mathematics curricula often develop algebraic ideas informally by engaging students in recognizing, describing, extending, and creating patterns; in representing and describing mathematical relationships; and in using variables and number sentences to express relationships. At the middle school level, many students complete an algebra course prior to taking high school mathematics courses. In 1996 about 25 percent of eighth-grade students reported being enrolled in a first course in algebra (Mitchell et al. 1999). It is also common for school districts to attempt to provide access to algebra for all students, and some states require algebra for high school graduation. Sixty-two percent of high school graduates in 1996 also took a second course in algebra, for example, Algebra II (National Center for Education Statistics 1997). Thus, it is important for educators to learn what students at all levels—elementary, middle school, and high school—know and understand about algebra.

This chapter summarizes results from the 1996 main NAEP mathematics assessment to develop a portrait of what students know about algebra and functions, including informal knowledge of algebraic ideas. We also examine students' knowledge of more advanced topics, for example, trigonometric functions, absolute value inequalities, and piecewise-defined functions. This chapter is intended to provide insights into students' performance that can inform teachers and researchers about a variety of issues that arise in the

---

We are grateful to Walter Deckert, Karen Flanagan, Linda Iseri, Margaret Kinzel, Cynthia Piez, Brad Slonaker, and Ismail Zembat (students in the first author's Projects in Mathematics Education Research, Curriculum Development, and Evaluation class) for providing comments on an earlier draft of this chapter.

teaching and learning of algebra. Among these issues are decisions about the nature of the early informal introduction of algebraic ideas, how students interpret and use algebraic symbolism, and how students use graphic, numeric, and symbolic representations of functions to model applied problems. Because many of the NAEP items classified in the Algebra and Functions content strand were administered across assessments, it was possible to examine clusters of items for growth in performance levels in topics such as working with number patterns, graphing in the coordinate plane, and solving equations.

Some of the NAEP items described in this chapter were extended constructed-response tasks, a type of NAEP item that is discussed extensively in chapter 11 by Silver, Alacaci, and Stylianou. In addition to reporting and interpreting performance based on NAEP results, for some short and extended constructed-response items we also examined a set of sample responses. This sample was not a representative sample of all the responses; rather, it was a convenience sample of nonblank responses.

In addition to assessing students' understanding of algebra topics in the main NAEP assessment, a number of algebra and functions topics were also assessed in a special study of students in grades 8 and 12 who were enrolled in mathematics courses considered to be advanced for a particular grade level (for example, algebra in grade 8 and precalculus in grade 12). The results from the Advanced Study are reported by Kilpatrick and Gieger in chapter 13.

ALGEBRA AND FUNCTIONS IN THE NAEP FRAMEWORK AND ITEMS

The 1996 NAEP mathematics framework (National Assessment Governing Board 1994) describes the content strand of Algebra and Functions as extending from "work with simple patterns at grade 4, to basic algebra at grade 8, to sophisticated analysis at grade 12, and involves not only algebra but also precalculus and some topics from discrete mathematics" (p. 33). The framework document emphasizes that students would be expected to "use algebraic notation and thinking in meaningful contexts to solve mathematical and real-world problems, specifically addressing an increasing understanding of the use of functions (algebraic and geometric) as a representational tool" (p. 33). The grade 4 assessment was to involve an informal demonstration of students' abilities to generalize, translate between representations, use simple equations, and do basic graphing. At grade 8 the assessment was to include more algebraic notation and require students to begin to use basic concepts of functions to describe relationships. By grade 12, students were expected to have an understanding of basic algebraic notation and terminology, to choose and apply various representations, and to use functions to represent and describe relationships.

The 1996 NAEP mathematics assessment contained seventy-eight algebra and functions items, with thirty-six items administered only to twelfth-grade students and forty-two items administered at more than one grade level to facilitate comparisons of students' performance between or among grade levels. All but one of the items discussed in this chapter were classified by NAEP in the content strand Algebra and Functions. One item that was classified by NAEP in the Data Analysis, Statistics, and Probability strand was included in this chapter because it illustrated an important aspect of graphing, namely, fitting a line to a scatterplot. Forty-eight of the items discussed in this chapter were in multiple-choice format, and the remaining thirty were constructed-response questions that required students to produce their own answers or explanations. Four of those thirty questions were extended constructed-response questions that required students to explain their answers and reasoning processes in writing. While working on certain algebra and functions items, fourth-grade students were permitted to use four-function calculators, and students in grades 8 and 12 were permitted to use scientific calculators, but not graphing calculators.

Although a large number of algebra and functions items were included in the 1996 NAEP, the paucity of released algebra and functions items makes it difficult to illustrate specific aspects of students' work on many of the items. The sections that follow discuss students' performance on individual NAEP algebra and functions items and clusters of related items in two categories: (1) informal algebra, including patterns and relationships, expressions, equations and inequalities, and graphing on the number line and in the coordinate plane; and (2) symbolic algebra, including expressions, equations, inequalities and systems, functions, and graphs.

INFORMAL ALGEBRA

Informal algebra involves work with algebraic entities, for example, variables, functional relationships, expressions, equations, and graphs. in relatively concrete ways. It includes working with functional relationships that are presented in the form of numeric patterns, use of simple symbolic notation (e.g., □ used to represent an unknown number) to represent arithmetic computations and relationships, use of numeric rather than symbolic approaches to solve equations written in number sentence form, and basic graphing that involves specific points on the number line or in the coordinate plane rather than graphs of symbolically defined functions or relations. Informal algebra serves as a precursor to the more general and abstract treatment of these algebraic entities in middle school and high school

Highlights

- Students at all grade levels generally could recognize patterns, and performance on many pattern items was higher in 1996 than it was in 1992 or 1990. However, few students performed well on questions that involved problem solving with patterns and required explanation or generalization.

- By grade 8 many students are successful on items for which informal algebraic approaches (for example, numeric rather than symbolic solutions of equations) are appropriate. Students had difficulty, however, with items involving expressions, equations and inequalities, and graphs that incorporated integers or rational numbers rather than just whole numbers.

- On items involving equations, inequalities, and systems, performance in 1996 exceeded, or was quite similar to, performance in 1992.

- Twelfth-grade students' performance on items involving functions and graphs was similar to their performance on those items in 1992 and 1990.

- Only a very small percentage of students were successful on extended constructed-response algebra and functions items that required communication in writing and justification of results. However, an examination of students' strategies suggests that it is likely that a much greater percentage of students employ strategies that could lead to correct solutions.

- For each of grades 4, 8, and 12, the 1996 performance levels in algebra and functions were significantly higher than corresponding levels in 1990 and 1992, suggesting sustained improvement over six years at all grade levels.

algebra courses. The following sections examine NAEP results related to patterns and relationships, informal interpretation and use of expressions, equations and inequalities, and graphs.

Patterns and Relationships

Patterns are central to mathematics at all levels. Students are taught to recognize and describe as well as extend sequences of numbers or objects. They also learn to create their own patterns, communicate those patterns to others, and represent patterns in, and relationships between, quantities

ALGEBRA AND FUNCTIONS

using tables, graphs, and rules that are stated verbally or symbolically. As students develop their capabilities to generalize about patterns and relationships between quantities, they develop an informal understanding of function that prepares them for subsequent, more formal work with numeric, graphic, and symbolic representations of functions in algebra. Patterns provide a convenient context in which students can move from primarily numeric work at the elementary school level to more general, symbolic algebra at the middle school and senior high levels.

Nine secure NAEP items and two released items assessed students' understanding of patterns. These items included three involving *repeating* patterns, in which elements in some way remained the same, and eight involving *growing* patterns, in which elements changed in some predictable way (Blume and Heckman 1997). Table 10.1 describes these items and gives the percents of students who responded correctly to them.

Repeating Patterns

Of the three repeating pattern items, described as items 1–3 in the table, one was administered at grade 4, one at grades 4 and 8, and one at all three grade levels. In each of these items the elements in the patterns were geometric figures or alphabetic characters rather than numeric elements. Students were asked to find a missing term, or terms, in the sequence rather than simply finding the next term(s). On both of the repeating pattern items administered at more than one grade level, performance improved substantially from one grade level to the next. Items that were answered correctly by only one-half to three-fifths of the fourth-grade students were answered correctly by approximately two-thirds to four-fifths of students in grade 8 or grade 12.

On item 1, 60 percent of the fourth-grade students and 85 percent of the eighth-grade students chose the correct orientation of the figure. Item 2, however, was more difficult for students in grade 4, with just over half of the students in grade 4 correctly choosing the missing figure. Failure to recognize that a consistent pattern was present in each of the collections of figures that comprised the terms was likely to have contributed to lower performance on this item. On item 3, administered at all three grades, performance at grade 4 also was lower than it was on item 1, perhaps because a correct response required the insertion of elements at various positions in the pattern and because there were only a few given elements from which students could discern the pattern. However, at grades 8 and 12, approximately two-thirds of the students correctly completed the pattern.

Growing Patterns

Eight of the 1996 NAEP items, items 4–11 in table 10.1, involved growing patterns. Although students' level of performance on some items (for

Table 10.1
Performance on Pattern Items

Item Description	Percent Correct		
	Grade 4	Grade 8	Grade 12
Repeating patterns			
1. Choose the *n*th term in a pattern consisting of orientations of a geometric figure.	60	85	—
2. Choose the figure that correctly completes a term in a pattern consisting of multiple geometric figures.	51	—	—
3. Complete a pattern of letters.	48	60	67
Growing patterns			
4. Choose the correct sum involving several terms in a pattern.	61	—	—
5. Choose a number that would be a term in a given number pattern.	35	—	—
6. Find the next several numbers in a given pattern and write the rule used.	19	51	—
7. Extrapolate a value from a table giving a relationship between two quantities and explain answer.	5[a]	—	—
8. Find the number of diagonals that can be drawn from any vertex of a 20-sided polygon.	—	54	—
9. Extend a pattern of sums and choose the correct answer.	—	32	—
10. Find the term number, *n*, for a subsequent term in a given pattern involving fractions and decimals.	—	23	43
11. Describe the 20th figure in a given pattern and provide a general description for any figure in the pattern.	—	—	4[a]

[a]Percent of students scoring at the satisfactory or extended level.
Note: Items 3, 6, and 8 were short constructed-response items, items 7 and 11 were extended constructed-response items, and the rest were multiple-choice items. Items 8 and 11, both released items, are shown in tables 10.2 and 10.3, respectively.

example, item 4) was comparable to that on some repeating pattern items, in general, students performed less well on growing pattern items than on repeating pattern items. In most instances, approximately one-fourth to one-half of the students at the various grade levels correctly answered the growing pattern items.

Performance was higher when students in grade 4 were asked to find missing terms in a pattern (item 4) than when they were to determine which of several given terms would appear subsequently in a pattern (item 5). However, when students were asked to supply the next terms in a pattern *and* write the rule that they used to find those terms, performance dropped substantially. On item 6 only 19 percent of the fourth-grade students and 51 percent of the eighth-grade students responded correctly by providing both the subsequent terms and the rule. However, an additional 19 percent at grade 4 and 22 percent at grade 8 provided the terms, but not the rule, or provided the rule but incorrectly supplied one or more of the terms. Despite being an item that required students to supply the immediately subsequent term(s) in a sequence instead of finding omitted terms or terms appearing somewhat later in the sequence—presumably more difficult tasks—this item may have been difficult because it involved a sequence that simultaneously embodied two different patterns. In such a sequence (for example, 1, 4, 3, 8, 5, 12, 7, 16, ...) students need to perceive the alternation—either the differences alternate between increasing and decreasing or the terms alternate between two sequences (in the preceding example, the sequence of consecutive odd numbers and a sequence of multiples of 4)—and then determine what the subsequent terms would be.

Item 10, a multiple-choice growing pattern item administered at grades 8 and 12, asked students to determine which term of a given sequence would be equal to a particular value (for example, determining that the term whose value is 32 is the fifth term of a sequence or, more generally, that the term of the sequence that has the value x is the nth term). This item also required students to convert between fractions and decimals. Only 23 percent of eighth-grade students and 43 percent of twelfth-grade students responded correctly. The most common distracter at both grade levels (chosen by 33 percent and 18 percent of students at grades 8 and 12, respectively) was a term number that corresponded closely to the particular value that was given. The students who chose this distracter may have done so because they confused the value of a term in a sequence with the number of that term. The conversion between decimals and fractions required in this item also may have contributed to its difficulty.

Three of the eight growing pattern items, items 9-11 in the table, involved nonconstant rates of change (that is, nonconstant differences between successive terms). Blume and Heckman (1997) reported that patterns based on nonconstant rates of change were more difficult for fourth-grade students than patterns based on constant rates of change. However, since none of the 1996 NAEP growing pattern items administered in grade 4 involved nonconstant rates of change, no corroboration of that result is possible. On the four growing pattern items administered in grade 8, nearly twice as many students responded correctly to items 6 and 8, those involving a constant rate of change (that is, constant differences between successive terms), than to items 9 and 10, those involving a nonconstant rate of change. However, it is somewhat difficult to argue that these pairs of items were comparable, since they required students to do somewhat different things, for example, finding the value of the nth term (item 8) as opposed to finding the term number, n, for a subsequent term in the pattern (item 10). Nevertheless, the suggestions made by Blume and Heckman remain valid: teachers need to provide all students with experiences in which they complete or extend, reason about, and identify the underlying rules for a variety of patterns that embody both constant and nonconstant rates of change.

At each grade level, one constructed-response growing pattern question required students to make use of patterns in a problem-solving context. Performance data for these questions, items 7, 8, and 11 in the table, suggest that students are not particularly successful in applying patterns in more complex problem-solving tasks. Only about half of the eighth-grade students were successful in solving the short constructed-response question, and less than five percent of the fourth-grade and twelfth-grade students were successful on their respective extended constructed-response questions.

The grade 4 question (item 7) involved extrapolation from a two-variable data table based on a realistic setting. It required students to find the value of one of the variables that corresponded to one of the subsequent values of the other variable. Students also had to explain how they found that value. Only 4 percent of the fourth-grade students produced an extended response in which they determined the correct value and provided an appropriate explanation based on ratio (for example, a change of c in one variable always results in a change of d in the other variable). Even if satisfactory responses that included a correct value without an explanation or an appropriate explanation with a minor error in producing the value are included, the percent correct rises only slightly to 5 percent. Although the setting for this question was a real (but perhaps somewhat unfamiliar) context, the amount of reading required in the question, the need for a written explanation, increments other than 1 in the independent variable, and the presentation of the data in tabular form in the order $(f(x), x)$ likely contributed substantially to its difficulty.

The released short constructed-response question involving problem solving with patterns (item 8) is shown in table 10.2. Fifty-four percent of eighth-grade students correctly answered this question. A small sample of written responses to this question provides some insight into students' approaches to this problem. Within this sample the majority of the students used an exhaustive or "brute force" approach, correctly generating the number of diagonals of each n-gon for values of n from 8 through 20, and a few students attempted such an approach but made an error in continuing the pattern. Some other students correctly used an arithmetic approach, subtracting 3 from 20. These students apparently noted a pattern and used the relationship that the number of diagonals was three less than the number of

ALGEBRA AND FUNCTIONS

sides. Among the kinds of errors evident in the responses were those that gave the number of *vertices* that yielded 20 diagonals or that inappropriately used multiplicative reasoning to conclude that there would be 8 diagonals for a 20-sided polygon because there are 2 diagonals for a 5-sided polygon. From the performance results question, one can conclude that about half of eighth-grade students are successful in extrapolating values (often one term at a time) from a simple pattern. Unfortunately, no information is available from this question about *how* eighth-grade students generalize patterns and how they represent and communicate about their generalizations.

Table 10.2
Growing Patterns: Sample Item

Item	Percent Responding Grade 8
From any vertex of a 4-sided polygon, 1 diagonal can be drawn.	
From any vertex of a 5-sided polygon, 2 diagonals can be drawn.	
From any vertex of a 6-sided polygon, 3 diagonals can be drawn.	
From any vertex of a 7-sided polygon, 4 diagonals can be drawn.	
How many diagonals can be drawn from any vertex of a 20-sided polygon?	
Answer: _____	
Correct response (17 diagonals)	54
Any incorrect response	38
Omitted	7

Note: Percents may not add to 100 because of rounding.

The released extended constructed-response growing pattern question administered to students in grade 12 (item 11 in table 10.1) and performance data for that question appear in table 10.3. Only 4 percent of the students gave an extended or satisfactory response. An extended response described the 20th figure correctly, noted that there would be 442 tiles in that figure, and provided a clear explanation with evidence of an accurate generalization. A satisfactory response described the 20th figure, gave the number of tiles, and provided some evidence of sound reasoning but perhaps included a computation error or lacked clarity. Descriptions of the criteria for the various performance categories, illustrative examples of students' responses, and additional discussion of students' performance on this question, called Extend Pattern of Tiles, can be found in chapter 11.

An analysis of a convenience sample of about 200 nonblank student responses revealed that although many students recognized the pattern and employed strategies that could lead to a correct solution, often those

Table 10.3
Extend Pattern of Tiles

Item	Percent Responding Grade 12
[General directions]	

This question requires you to show your work and explain your reasoning. You may use drawings, words, and numbers in your explanation. Your answer should be clear enough so that another person could read it and understand your thinking. It is important that you show <u>all</u> your work.

The first 3 figures in a pattern of tiles are shown below. The pattern of tiles contains 50 figures.

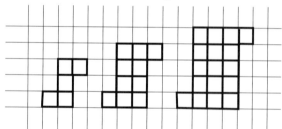

Describe the 20th figure in this pattern, including the total number of tiles it contains and how they are arranged. Then explain the reasoning that you used to determine this information. Write a description that could be used to define any figure in the pattern.

Extended response	2
Satisfactory response	2
Partial response	18
Minimal response	29
Incorrect response	25
Omitted	20

Note: Percents may not add to 100 because of rounding or off-task responses.

strategies were not implemented correctly or students were unable to generalize their results. For example, only a few students in this sample successfully determined a generalization that could be used to define any figure in the pattern. One possible contribution to the difficulty of this question might have been the wording of the problem. Since it stated that "the pattern of tiles contains 50 figures," it is possible that some students confused "50 figures" (50 terms in the sequence) with the idea that a particular figure contained 50 total tiles (50 tiles in one term of the sequence).

ALGEBRA AND FUNCTIONS

Students' differing perceptions of the pattern in this question led to several different uses of it in their problem solving. One visual-geometric strategy (for additional discussion of these strategies, see chapter 11) involved the decomposition of the figure into a large rectangle with two unit squares affixed at opposite corners. A second visual-geometric strategy involved viewing the pattern of squares as being composed of an $n \times n$ square with $1 \times (n + 1)$ rectangles affixed to its top and bottom sides. Other strategies included arithmetic-algebraic ones in which students attempted to generalize from the numbers of squares in each figure and ones in which they attempted to symbolize the pattern directly or recursively using a function or sequence. A relatively large number of students in this sample gave responses that were off-task or responses that indicated that the students misunderstood the problem.

Performance on this item provides some cause for concern as well as some evidence of students' capabilities with growing patterns. NAEP results show that on the Extend Patterns of Tiles task, only about 4 percent of the students gave extended or satisfactory responses, and a considerable number of students were unable to understand the problem, despite it entailing a setting—patterns of squares much like those students might have encountered when drawing polygonal regions on graph paper—that should not have been unfamiliar to students. Mathematics teachers should be concerned that although twelfth-grade students can recognize a relatively complex growing pattern, they have difficulties generalizing patterns, representing their generalizations, and communicating about those generalizations. Nevertheless, as discussed chapter 11, a number of students in this sample whose responses were incorrect demonstrated some appropriate reasoning on this problem. Those students appeared to recognize the pattern and used strategies that could have led to a correct response.

Performance on Pattern Items across Assessments

Some of the repeating pattern and growing pattern items were administered across NAEP assessments (1990, 1992, 1996), thus permitting performance comparisons. Table 10.4 presents the percent correct values for seven items from table 10.1 that were administered across assessments. In the descriptions that follow, all reported differences that are statistically significant use the .05 level.

On item 3, the repeating pattern item, performance improved significantly from 1990 to 1996 at all three grade levels, with increases in percent correct ranging from 7 percent at grade 12 to 14 percent at grade 4. Performance on that item also significantly improved from 1990 to 1992 at grades 4 and 12. On the six growing pattern items administered across assessments (items 4, 5, and 8–11), performance improved significantly for four of the seven items; however, increases in percent correct ranged from only about 4 to 7

Table 10.4
Comparison of Performance on Pattern Items Administered across Assessments

Item Description[a]	Grade	Percent Correct		
		1990	1992	1996
Repeating patterns				
3. Complete a pattern of letters.	4	34	40*	48*†
	8	50	53	60*†
	12	60	67*	67*
Growing patterns				
4. Choose the correct sum involving several terms in a pattern.	4	56	55	61*†
5. Choose a number that would be a term in a given number pattern.	4	—	29	35†
8. Find the number of diagonals that can be drawn from any vertex of a 20-sided polygon.	8	—	50	54†
9. Extend a pattern of sums and choose the correct answer.	8	—	33	32
10. Find the term number, n, for a subsequent term in a given pattern.	8	18	21	23*
	12	36	38	43*
11. Describe the 20th figure in a given pattern and provide a general description for any figure in the pattern.	12	—	5[b]	4[b]

[a]Item numbers correspond to those in table 10.1.
[b]Percent of students scoring at the satisfactory or extended level.
*Indicates a statistically significant difference from 1990.
†Indicates a statistically significant difference from 1992.

percent. Significant increases occurred for both growing pattern items at grade 4 (items 4 and 5), one of the three growing pattern items at grade 8 (item 10), and one of the two growing pattern items at grade 12 (item 10). In general, percent-correct values for these pattern items were slightly higher in 1996 than in 1990 or 1992. This suggests modest, but steady, progress in improving students' capabilities to recognize and work with patterns.

Data gathered on the 1990, 1992, and 1996 NAEP grade 4 and grade 8 teacher questionnaires regarding emphasis on algebra and functions topics potentially may provide some insight into performance differences on pattern items across assessments. Changes in teachers' emphasis on algebra and functions topics may help to explain the improvement in performance on the pattern items from 1990 to 1996. Particularly at grade 4 one might hypothesize that an increase in instructional emphasis on algebra and functions includes an increase in emphasis on patterns. Unfortunately, any attempt to use data from the NAEP teacher questionnaires as a way to interpret changes in performance results on pattern items must take into account the change in wording associated with the question that asked teachers about their classroom practices regarding informal introduction of algebra and functions concepts. On the 1996 questionnaire, the wording was different from that used on previous versions. In particular, on the 1990 and 1992 questionnaires, teachers were asked about the *emphasis* they placed on informal algebra and functions topics, with response categories of "heavy," "moderate," and "little or no" emphasis. On the 1996 questionnaires, the question was rephrased in terms of frequency—that is, *how often* teachers addressed such topics, with response categories of "a lot," "some," "a little," and "none." Despite these wording differences, it is possible to compare informally (that is, without referring to statistical significance) the results from 1990 and 1992 with those from 1996 under assumptions that the category of "little or no emphasis" and the combination of the frequency categories "a little" and "none" are similar enough to be roughly comparable with the same assumption made for the categories "heavy emphasis" and "a lot."

In 1996, 9 percent of students in grade 4 had teachers who reported that they addressed algebra and functions "a lot," as compared to only 3 percent and 2 percent of students in 1992 and 1990, respectively, whose teachers reported giving heavy emphasis to algebra and functions topics. Although this increase is modest, a more striking decrease occurred in the percent of students in grade 4 who had teachers who reported a lack of attention to or emphasis on such topics. In 1996, 60 percent of students in grade 4 had teachers who reported that they addressed algebra and functions "a little or none," as compared to 66 percent and 82 percent of students in 1992 and 1990, respectively, whose teachers reported giving "little or no emphasis" to algebra and functions. This suggests that, at grade 4, increased exposure to informal algebra and functions topics (presumably including work with patterns) might be responsible for improved performance on both repeating and growing pattern items.

The teacher questionnaire results for algebra and functions emphasis in grade 8 were similar, but less striking. In 1996, 57 percent of students in grade 8 had teachers who reported that they addressed algebra and functions "a lot," as compared to 48 percent in both 1992 and 1990 who reported giving "heavy emphasis" to algebra and functions topics. In 1996, only 9 percent of students in grade 8 had teachers who reported that they addressed algebra and functions "a little or none," as compared to 12 percent and 18 percent of students in 1992 and 1990, respectively, whose teachers reported giving "little emphasis" to algebra and functions. It is possible that eighth-grade teachers increased their emphasis on patterns as much as fourth-grade teachers did, but such an increase may not have been reflected in what was reported. At grade 8 many more topics not involving patterns are included in algebra and functions than at grade 4, so with teachers of eighth-grade students an increase in emphasis on patterns might not have been perceived as a very large increase in emphasis on the broader topic of algebra and functions.

From the data on pattern items it appears that many students can perceive and work with both repeating and growing patterns. Performance on pattern items improved from 1990 to 1996, and additional instructional emphasis on patterns may be one explanation for that improvement. Problem solving with patterns, however, remains difficult for students, particularly when students are asked to generalize or explain their answers.

Expressions, Equations, Inequalities, and Graphs

Fifteen NAEP items addressed informal algebra topics that involved representing written statements in numeric or symbolic form, solving equations or inequalities presented in the form of number sentences (for example, $3 + \square = 7$ or $3 + \_\_ = 7$) and graphing points on the number line and in the coordinate plane. Three items involved representing or evaluating expressions that were stated verbally, and four items involved solutions of equations or inequalities. The remaining eight items addressed locating points on the number line or points in the coordinate plane.

Representing and Evaluating Expressions

Table 10.5 describes the informal algebra items involving expressions (items 1-3) and summarizes the performance of students in grades 4, 8, and 12 on them. Item 1 assessed translation from a written statement to a symbolic expression. On this released item, which appears as the first item in table 10.6, two-thirds of the students in grade 4 chose the correct expression, with 15 and 12 percent choosing the most plausible distracters, $12 - N$ and $N + 12$, respectively. Items 2 and 3 in table 10.5 presented information from which students in grade 8 could evaluate a numeric expression involving integers or rational numbers to produce their answers. Approximately 25 percent of students in grade 8 chose the correct value for item 2, with a similar percent of students producing the correct value for item 3 (a released item shown as the second item in table 10.6) that involved the calculation of the difference between positive and negative integers.

From performance on items that involved writing and using expressions, it appears that although many fourth-grade students were successful at choosing a symbolic expression involving whole numbers that matched a written description, eighth-grade students were much less successful in using written information to produce and evaluate expressions involving integers or rational numbers. Examining sample responses to item 2 in table 10.6 can provide some insight into eighth-grade students' understandings and errors regarding the use of written information to produce algebraic

ALGEBRA AND FUNCTIONS

Table 10.5
Performance on Informal Algebra Items Involving Expressions, Equations, and Inequalities

Item Description	Percent Correct		
	Grade 4	Grade 8	Grade 12
Writing and using expressions			
1. Choose a symbolic expression that matches a given written description.	67	—	—
2. Given a written description of an expression involving rational numbers, evaluate the expression.	—	28	—
3. From a written description of a context based on feet below and above sea level, create and evaluate an expression involving integers.	—	22	—
Informally finding solutions of equations and inequalities			
4. Determine the solution of a simple linear equation presented in number sentence format	61	83	—
5. Select the correct solution to an equation presented pictorially by balancing sides of the equation.	35	72	85
6. Given a situation modeled by an equation in two variables, find possible solution pairs.	23	—	—
7. Choose the whole number solutions to a simple linear equality presented in number sentence format.	21	59	83

Note: Items 3, 4, and 6 were short constructed-response items, and the rest were multiple-choice items. Items 1 and 3, both released items, are shown in table 10.6.

expressions involving integers. Selected responses to this item appear in figure 10.1. As response 1 illustrates, some eighth-grade students wrote a correct expression, 1277 − (− 294)) and then rewrote that expression as 1277 + 294. Other students, as shown in response 2, created a correct expression and gave the correct answer but did not indicate how they produced 1571. Responses 3 and 4 are representative of those in which students did not write a correct expression but still managed to produce the correct answer or created a correct expression but did not compute the answer correctly. In some cases, students did not write an expression involving negative integers, perhaps because they misunderstood the concepts of "above and below sea level"; response 5 is an example of this situation.

Table 10.6
Representing and Evaluating Expressions: Sample Items

Items	Percent Correct	
	Grade 4	Grade 8
1. N stands for the number of stamps John had. He gave 12 stamps to his sister. Which expression tells how many stamps John has now?		
A. N + 12	12	—
B. N – 12*	67	—
C. 12 – N	15	—
D. 12 × N	3	—
2. The lowest point of the St. Lawrence River is 294 feet below sea level. The top of Mt. Jacques Cartier is 1,277 feet above sea level. How many feet higher is the top of Mt. Jacques Cartier than the lowest point of the St. Lawrence River? Show your work.		
Correct response of 1571 with appropriate work	—	22
Correct procedure (addition) but does not answer 1571	—	3
Incorrect response	—	70
Omitted	—	4

*Indicate correct response.
Note: Percents may not add to 100 because of rounding or omissions. Item 2 is from an item block for which students were provided with, and permitted to use, scientific calculators.

Solving Equations and Inequalities

Items 4–7 in table 10.5 addressed informal (presumably numeric rather than symbolic) solutions of equations and inequalities. Items 4 and 7, for example, involved number sentences such as $\Box - 9 = 17$ and $9 - \Box > 5$, respectively. Only 21 to 35 percent of fourth-grade students succeeded in finding solutions of equations and inequalities in items other than item 4, which 61 percent of those students answered correctly. On the cross-grade items (items 5 and 7), the performance of students in grades 8 and 12 improved substantially from that for students in grade 4. However, even the 85 percent correct in grade 12 is disappointing, given the simplicity of these items; they were not presented in terms of formal algebraic notation (for example, using variables). Also, given that the student questionnaire data show that 81 percent of students in grade 12 reported studying first-year algebra for one year and 66 percent reported studying second-year algebra for one year (Mitchell et al. 1999), better performance on these items might be expected.

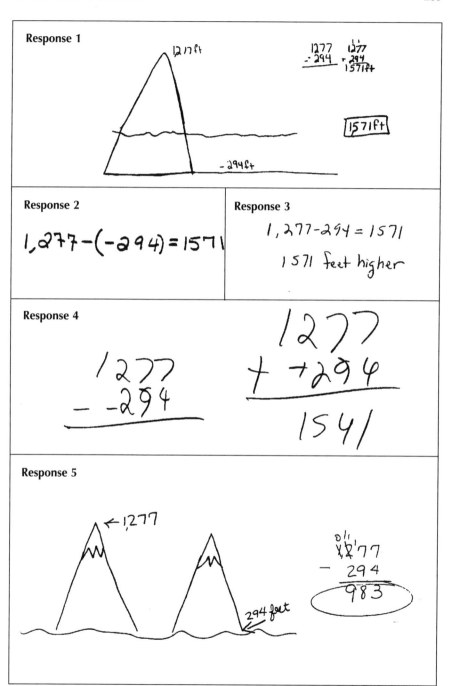

Fig. 10.1. Selected responses to a released item involving expressions with integers

Graphing on the Number Line and in the Coordinate Plane

Just as there are relationships between graphing points on the number line and representing solutions to equations and inequalities in one variable, there are relationships between graphing points in the coordinate plane and representing solutions to equations and inequalities in two variables. Hence, it is important to compare students' performance on items involving informal solution of equations and inequalities and students' early attempts to represent points on the number line and in the coordinate plane.

Eight secure NAEP items administered to students in grades 4, 8, and 12 assessed students' capabilities to identify points on the number line and locate points in the coordinate plane. Two of these items were in multiple-choice format, and six were short constructed-response items. Table 10.7 gives descriptions of these items and summarizes students' performance on them. On these eight items there were substantial improvements in performance across grade levels, most likely reflecting the increased attention given to graphing in the later grades. In general, the percent of correct responses to items addressing graphing was lower than it was for informal algebra items addressing equations and inequalities.

Performance on item 2 suggests that many students are not proficient at locating points on the number line when *mixed number* coordinates are required. In this item students were given coordinates of several points on a number line (for example, 4 2/3, 5 1/3, and 6) and were asked to name additional points that fell between the given points (for example, 5 and 5 2/3). The percent-correct value on this item for students in grade 4 (19 percent) is particularly low. However, if one includes the responses that had at most one error among the several coordinates produced by the student, the percent-correct values increase slightly for both grades, from 19 to 24 percent for grade 4, and for grade 8, from 64 to 71 percent. Performance on this item contrasts sharply to that on item 4 in table 10.1, an item involving a pattern presented in the context of a number line with *whole number* increments other than 1, on which more than 60 percent of students in grade 4 responded correctly.

Items 1 and 7 were an item pair; in item 1, students were asked to draw a coordinate-plane path that was generated by a given rule, and in item 7, students needed to determine whether a particular point that was far removed from the portion of the coordinate plane that appeared in the figure would lie on that path. Both items appeared on the grade 8 NAEP test. Although nearly 60 percent of students in grade 8 were able to draw a correct path in item 1 and 52 percent were able to determine whether the point would lie on the path in item 7, only 23 percent were able to provide an appropriate reason in item 7. Even if one includes responses that indicated the recognition of a pattern but with erroneous information, the percent-correct value rises only to 30 percent. The disappointing result on this item is that 44 percent of the eighth-grade students had a completely

Table 10.7
Performance on Informal Algebra Items Involving Graphing

Item Description	Percent Correct		
	Grade 4	Grade 8	Grade 12
1. Given a rule for creating the path of an object in a coordinate plane, draw the path.	18	59	—
2. Supply the missing coordinates on a number line.	19	64	—
3. Choose the correct coordinates of a vertex of a polygon, given the coordinates of the other vertices.	—	52	78
4. Indicate on a number line the set of numbers satisfying a given condition.	—	45	68
5. Choose the correct coordinates of a vertex of a polygon, given the coordinates of the other vertices.	—	43	—
6. Determine the location of a particular point in the coordinate plane.	—	41	—
7. Given a rule for creating a path of an object in a coordinate plane, determine whether a particular point would be on that path; explain.	—	23	—
8. Locate one pair of possible coordinates of a vertex of a polygon when the other vertices are given.	—	—	32

Note: Items 3 and 5 were multiple-choice items, and the rest were short constructed-response items. Items 1 and 7 were a pair of related items based on a common scenario.

incorrect response, 9 percent gave an off-task response (for example, a unrelated drawing or doodle), and 17 percent chose not to answer the item. Results such as these support the need for additional curricular emphasis on reasoning and communication.

Comparisons to performance on previous NAEP assessments were possible for three of the seven graphing items. Although there was no significant difference in performance from 1990 to 1996 on item 6, for items 3 and 4 performance was significantly higher in 1996 than in 1990 at both grade levels. Increases in percent correct ranged from 5 to 11 percent.

Performance on the informal algebra items addressing expressions, equations and inequalities, and graphs suggests that representing expressions and solving number sentences that involve whole numbers are much easier than working with rational numbers or integers. Also, difficulty increased when items included inequalities or simple equations in two variables. Less than half of the eighth-grade students could locate points on the number line or in the plane when simple conditions determined their locations. Even at grade 12, only 78 percent of the students could locate a missing coordinate for the vertex of a polygon when the conditions for its location could be found from the description of the polygon. It would seem wise for teachers at all levels to spend more time helping students to understand and use one- and two-dimensional coordinate systems, particularly in light of increasing access to graphing calculators at the middle school and senior high school levels.

SYMBOLIC ALGEBRA

A variety of NAEP algebra and functions items administered only to eighth- and twelfth-grade students addressed symbolic algebra topics. In contrast to the informal algebra items, these relied more on students' capabilities to work symbolically with expressions, equations, and inequalities and to use function notation. Included were simplifying and evaluating expressions, solving equations and inequalities, solving systems of equations, and evaluating functions. The symbolic algebra items also included items that assessed students' use of graphical representation, for example, constructing a graph from a function definition given in symbolic form. Thirty-six of the symbolic algebra items are discussed in the sections that follow.

Expressions

Five secure NAEP items assessed eighth- and twelfth-grade students' abilities to produce expressions, both numeric and algebraic, and to evaluate or simplify them or to determine their equivalence. Table 10.8 presents a summary of performance on these items, each of which was presented in multiple-choice format. Performance varied greatly depending on what the item required of students.

On item 1, as was the case when it was administered in 1990 and 1992, nearly all students in grades 8 and 12 were able to evaluate correctly a simple expression involving order of operations on whole numbers. However, the presentation of this item included parentheses, which most likely made it easier for students. Had parentheses not been included, students would have had to make a decision about which operation to perform first rather than recognizing that parentheses determined which operation was to be performed first.

Other items involving expressions were much more difficult for students. In item 4, a secure item, twelfth-grade students were required to expand an expression that involved exponents. The secure item is similar to this example: Determine which of $10x^6$, $10x^8$, $25x^2$, $25x^6$, or $25x^8$ is equal to $(5x^4)^2$. Fewer than half the students in grade 12 responded correctly to item 4, with 25 percent choosing a response that reflected finding the product of the base and

Table 10.8
Performance on Items Involving Expressions

Item Description	Percent Correct	
	Grade 8	Grade 12
1. Choose the correct value of a numeric expression involving whole numbers.	94	96
2. Choose a symbolic expression in three variables that represents a given quantity.	52	—
3. Determine whether each of three symbolic expressions is equivalent to a given symbolic expression.	28[a]	—
4. Choose the symbolic expression that is equivalent to a symbolic exponential expression.	—	46
5. Choose the correct numeric answer to a numeric logarithmic expression.	—	24

[a]Percent of students answering all three statements correctly
Note: Item 3 was a multiple-response item, and the rest were multiple-choice items.

the exponent (in the example, multiplying 5 and 2), and 21 percent choosing a response that reflected adding the two exponents (in the example, adding 4 + 2) instead of finding their product. Item 5 may have been difficult because it required students to know properties of logarithms, a topic some twelfth-grade students may not have had an opportunity to learn. Students continue to have difficulty simplifying and evaluating expressions that are more complex, for example, those involving exponents, despite the fact that many algebra courses place heavy emphasis on such skills. This may result from, as Demby (1997) points out, students often constructing their own rules for simplifying algebraic expressions and not necessarily implementing rules provided to them by their teachers.

Equations, Inequalities, and Systems

Fifteen NAEP items administered to students in grades 8 and 12 involved writing and solving equations, solving inequalities, and solving systems of equations. Nine items addressed writing and solving equations, two items involved solving inequalities, and four items involved systems of equations. Generally, performance on these items in 1996 exceeded or was quite similar to performance in 1992. The sections that follow discuss the items and results related to equations, inequalities, and systems of equations that involved symbolic, but not graphical, representations.

Equations and Inequalities

Descriptions and performance data for the nine items that addressed writing and solving equations are shown in table 10.9. On these items, only when the equations were very simple or when it was possible to substitute the choices into the equation given in a multiple-choice item did the majority of students in grades 8 and 12 find the correct solution. For example, on item 4 students could substitute values from the table into the equations that constituted the choices, and on item 5 they could substitute the ordered pairs in the choices into the given equation. When equations had more complicated coefficients or when students had to produce an equation, performance declined considerably. These difficulties are likely related to students' understandings of equations. MacGregor (1998) notes that students' difficulties in constructing equations stem in part from their inability to grasp the notion of the equivalence between the two expressions in the left and right sides of the equation.

Table 10.9
Performance on Items Involving Equations

Item Description	Percent Correct	
	Grade 8	Grade 12
1. Solve a simple radical equation.	50	—
2. Solve a linear equation in two variables for one variable in terms of the other.	50	—
3. Determine which ordered pair is a solution to a linear equation in two variables.	41	—
4. Determine which equation in two variables fits the values given in a table.	37	61
5. Determine which set of ordered pairs is a solution to a nonlinear equation.	—	53
6. Choose the correct solution for a specified variable in a literal equation in three variables.	—	38
7. Given an equation related to a real-world situation, substitute given values to solve the equation.	—	30
8. Solve a quadratic equation in which the coefficient of the quadratic term is not equal to 1.	—	16
9. Write an equation for a word problem.	—	9

Note: Items 8 and 9 were short constructed-response items, and the rest were multiple-choice items.

The only released item of these nine was item 3, a grade 8 item that stated, "Which of the following ordered pairs (x, y) is a solution to the equation $2x - 3y = 6$?" The correct response, $(3, 0)$, was chosen by 42 percent of the students and the distracters $(6, 3)$, $(3, 2)$, $(2, 3)$, and $(0, 3)$ were chosen by 16, 16, 21, and 4 percent of the students, respectively. The ordered pair $(2, 3)$ may have been the most popular incorrect choice because it contained the coefficients in the order they were given or perhaps simply because $2 \times 3 = 6$. It is not likely that students found the correct solution but interchanged the x- and y-coordinates, since few chose the distracter (0, 3). No short constructed-response items provided information concerning students' understanding of ordered pairs and the nature of solutions of equations. If such items were included in the NAEP assessment, they would provide teachers and researchers with valuable information, not just information about students' capabilities to solve equations or choose from among potential solutions but information about what students think constitutes a solution and their understanding of what constraints exist (for example, number of solutions) when one solves a particular equation.

Items 2 and 6 in table 10.9 addressed the solution of literal equations—equations that define the relationship between or among two or more variables, for example, the relationship among the area, altitude, and lengths of the bases of a trapezoid given by $A = (1/2)h(b_1 + b_2)$. Half of the students in grade 8 could solve an equation of the form $a + b = c$ for one variable when two of a, b, and c were variables and the other was a numeric constant. On the more complex three-variable literal equation in item 6, only 38 percent of the students in grade 12 correctly solved for one variable in terms of the other two. Although most students may be learning to solve fairly simple literal equations, many students may not be encountering more complex literal equations in which they need to solve an equation for one variable in terms of the others.

There were eight instances in table 10.9 for which comparisons of performance on items across assessments were possible. In four instances (items 1, 3, 4 [for grade 12], and 6) performance results remained stable between the NAEP administrations in 1992 and 1996. In three instances (items 2, 4 [for grade 8 in 1990], and 5) performance on items involving equations was significantly higher in 1996, with the percent of students answering correctly ranging from about 50 percent to about 35 percent in 1996 as compared to a range of about 45 percent to about 30 percent on prior assessments.

On item 9, one of only a few algebra and functions items for which performance was significantly lower in 1996 than in 1992, performance declined from 14 percent correct in 1992 to 9 percent correct in 1996. In this item twelfth-grade students had difficulty writing an equation that could be used to solve a traditional algebra word problem. This decline might be attributable to decreased emphases in algebra curricula on traditional algebra word problems involving ages, values of certain numbers of coins, and the like. On a companion item that asked students to express two quantities in a problem in terms of one variable, only 3 percent of twelfth-grade students were successful, a significant decline from 9 percent correct in the 1990 and 1992 assessments. It appears that it is even more difficult for students to express quantities in terms of a variable, perhaps a more conceptual task and one in which they need to distinguish between letters and their referents (MacGregor and Stacey 1997; Stacey and MacGregor 1997), than it is for them to "translate" from word problem form to an equation that can be solved, which may be a more algorithmic task.

Two multiple-choice items assessed eighth-grade students' ability to solve simple linear inequalities, for example, ones of the form $ax \pm b > c$, where a, b, and c are positive integers. Nearly three-fourths of the students chose the correct solution of a simple linear inequality presented in verbal form and just over half of the students correctly chose the numerical solution to a simple linear inequality. Although a majority of students successfully solved these items, more complex inequalities on prior NAEP assessments have been considerably more difficult (Blume and Heckman 1997).

Systems of Equations

Four NAEP items summarized in table 10.10 assessed twelfth-grade students' ability to solve systems of equations; all items were administered in 1992 and 1996. Performance levels ranged from 30 percent correct to 88 percent correct. On item 2, performance in 1996 was significantly higher than in 1992, with the increase being approximately 6 percentage points. On the other three items involving systems of equations, however, performance was comparable across the 1992 and 1996 assessments.

Table 10.10
Performance on Systems of Equations Items

Item Description	Percent Responding Grade 12
1. Select the solution to a pair of one-variable linear equations of the form $a \times q = q$.	88
2. Using substitution, select the solution for one variable in a system of several simple quadratic equations.	58
3. Solve a system of equations in two variables and select the solution for one of them.	48
4. Solve a system of linear equation in two variables.	30

Note: Item 4 was a short constructed-response item, and the other three were multiple-choice items.

Functions

In addition to the informal algebra items that addressed functions indirectly in the context of patterns and relationships, four NAEP items administered at grades 8 and 12 more directly addressed symbolic representations of functions. Table 10.11 describes these items and students' performance on them.

ALGEBRA AND FUNCTIONS

Table 10.11
Performance on Function Items

Item Description	Percent Correct	
	Grade 8	Grade 12
1. Select an expression for a linear function described in a real-world context.	58	—
2. Given a table of values for a linear function described in a real-world context, complete a table for values of x and $f(x)$ and produce a symbolic function rule.	20	—
3. Given a quadratic function rule and a linear function rule, write the expression for the composite function.	—	20
4. Given a description of two linear functions, solve a problem based on their comparison.	—	6[a]

[a] Percent of students scoring at the satisfactory or extended level.
Note: Item 1 was a multiple-choice item, items 2 and 3 were short constructed-response items, and item 4 was an extended constructed-response item.

A released item (item 1 in the table) involved this situation: "A plumber charges customers $48 for each hour worked plus an additional $9 for travel." The question posed to the students was, "If h represents the number of hours worked, which of the following expressions could be used to calculate the plumber's total charge in dollars?" with the five choices being:

$48 + 9 + h$
$48 \times 9 \times h$
$48 + (9 \times h)$
$(48 \times 9) + h$
$(48 \times h) + 9$

Fifty-eight percent of students in grade 8 were able to select the correct expression, $(48 \times h) + 9$. The percent of students selecting each of the incorrect choices ranged from 8 to 15 percent. Unfortunately, the correct choice was the only one of the responses that directly mirrored the verbal sequence "48," "per hour," and "plus nine dollars," as presented in the problem, so students who translated literally and sequentially would have responded correctly without needing to select the appropriate sequence of operations or grouping of terms.

Generating and using symbolic representations of functions was difficult for students. Despite some success on item 1, only 20 percent of eighth-grade students were able to find values for x and $f(x)$ and produce a symbolic function rule in item 2, another linear function item presented in a real-world context. However, an additional 68 percent had one or two correct entries in the table or a correct function rule, but not both. This suggests that almost 90 percent of the students determined the relationship between the dependent and independent variables but many of them may not have been able to apply it or symbolize it. On item 3 only 20 percent of twelfth-grade students correctly produced a symbolic expression for the composition of two functions, and on item 4 only 6 percent of the students provided a satisfactory or extended response in a problem setting involving two linear function rules.

Graphs

Just as one can relate performance on informal algebra items that involve solutions to number sentences to performance on items that involve graphing points on the number line and in the coordinate plane, one can relate performance on items requiring solutions of equations, inequalities, and systems to performance on items involving graphical representation of equations, inequalities, and systems and graphical representation of functions. Graphical solutions, in the absence of access to graphing calculators, do not appear to be easier than symbolic solutions.

On an item that required the identification of a graphically represented solution to a linear inequality of the form $ax \pm b \geq c$, 62 percent of twelfth-grade students chose the correct solution. On a linear absolute value inequality of the form $|ax \pm b| \geq c$ with the same coefficients and operations as the preceding linear inequality, only 28 percent of twelfth-grade students chose the correct graphically represented solution. Similar performance was exhibited when twelfth-grade students were asked to choose the correct symbolically represented solution set for a quadratic inequality of the form $ax^2 + bx + c \geq 0$; only 33 percent chose the correct set. Aspects of equations and inequalities that increase their complexity, for example, nonlinearity or the inclusion of absolute value, often make items substantially more difficult for students, whether solutions are represented graphically or symbolically. It would be interesting for teachers to compare their students' performance when graphically solving linear or quadratic inequalities of these forms with a graphing calculator or spreadsheet (Brunner, Coskey, and Sheehan 1998; Niess 1998) to the NAEP results for which graphing calculators were not available to students. For example, they might determine whether students use the graph of $y = |ax \pm b|$ to determine when $|ax \pm b| \geq c$, whether producing a solution graphically differs in difficulty from producing the solution symbolically, and whether students demonstrate different understandings and skills when using a two-dimensional graph of a linear absolute-value function to produce a one-dimensional solution set for a linear absolute-value inequality than they do when using symbolic manipulation to produce a solution for a linear absolute-value inequality.

Other graphing items administered to students in grade 12 assessed students' understanding of the formal language of functions, representation of functions in graphical form, and interpretation of graphically represented functions. Table 10.12 describes these items and gives twelfth-grade students' performance on them.

Table 10.12
Performance on Items Involving Graphical Representation of Functions

Item Description	Percent Responding Grade 12
1. Use the graph of a function to choose the value of x, given $f(x)$.	68
2. Given the graphs of f and g, choose the value for which $f(x) = g(x)$.	63
3. Interpret the graph of a step function.	57
4. Given the graph of a polynomial function, choose the correct number of real roots.	49
5. Given a scatterplot, choose an equation that best fits the data.	39
6. Use the graph of a function to choose the value of $f(x)$.	35
7. Choose which of five graphs could be the graph of a function.	20
8. Graph a system of two linear equations in two variables.	13
9. Sketch the graph of a piecewise-defined function.	8
10. Given the graph of two functions, f and g, determine the transformation(s) needed to obtain the graph of g from the graph of f.	7[a]

[a]Percent of students scoring at the satisfactory or extended level.
Note: Items 8 and 9 were short constructed-response items, item 10 was an extended constructed-response item, and the rest were multiple-choice items. Items 1, 2, and 6 were related items based on the same graph of two functions. Item 7, a released item, is shown in table 10.13.

In items 1, 2, and 6, a family of related items, students were given the graphs of two functions, say f and g, that intersected at a point. In item 1 they were required to use the graph of f to find x for which $f(x) = c$. Sixty-eight percent of the students selected the correct choice, and 63 percent appeared to be able to locate the point of intersection of the two functions in item 2. Although in item 6 only 35 percent correctly found $f(c)$ from the graph, an additional 28 percent chose the correct value for $g(c)$ rather than $f(c)$. This suggests that when given graphs of functions, over 60 percent of students in grade 12 may be able to interpret the graphs to answer questions presented symbolically using function notation. Students performed similarly when required to fit a line to a scatterplot. In item 5, a graphing item classified by NAEP as a data analysis item, when required to select an equation that represented the line that best fit the data points in a scatterplot, 39 percent chose the correct equation and another 21 percent chose an equation with the correct slope but an incorrect y-intercept.

Item 7, a released item shown in table 10.13, required students to choose which of five graphs could be the graph of a function. Only 20 percent of the students chose the correct graph, and even among students who reported having taken calculus, only 55 percent selected the correct answer (Mitchell et al. 1999).

In item 4 in table 10.12, nearly half of the students could determine from the graph of a polynomial function the number of its real roots. Since only relatively small numbers of intersection points were reasonable and the item was multiple choice, guessing might have inflated the percent correct somewhat. However, students seemed to relate real roots to intersections of the graph with the axes, since the most popular distracter, chosen by 20 percent of the students, was the total number of times the graph intersected the horizontal and vertical axes. As with item 4, no specific function rule was given to define the function in item 10. On this item students were to describe transformations (for example, shift 3 units to the left, shift 5 units up, flip over the line $y = 3$) that were necessary to obtain one graph from another. Only 7 percent of the students produced a satisfactory or extended response on this item, which, like item 4, addresses a capability that is likely to become more important as students' access to graphing calculators and computer graphing tools increases and they determine properties of functions from their graphs and manipulate graphs by translating and reflecting them.

Producing graphs to solve a system of two linear equations in two variables was quite difficult for students. In item 8 only 13 percent of students in grade 12 correctly graphed the system and found the coordinates of the solution, although another 11 percent provided the correct solution but did not correctly graph the lines. Fewer students could produce the correct solution to a system of linear equations when required to solve the system graphically than when allowed to solve a comparable system using any available method, as in item 4 of table 10.10.

The remaining items in table 10.12 involved less commonly encountered functions such as piecewise-defined functions (for example, $f(x) = 2 - x^2$ for $x \leq -1$, $f(x) = x + 2$ for $-1 < x < 1$, and $f(x) = 3$ for $x \geq 1$). Item 3 involved interpretation of a step function. Although 57 percent responded correctly, this item only required students to read values from the graph and choose the value that resulted from doing a simple computation with them. Item 9 involved sketching a piecewise-defined function and was much more difficult, with only 8 percent of students responding correctly. On a related item that involved choosing the range of the same piecewise-defined function when given its definition in symbolic form, only 16 percent answered correctly. On this item students could have reasoned from the symbolic

ALGEBRA AND FUNCTIONS

Table 10.13
Function Item

Item	Percent Responding Grade 12
1. Which of the following could be the graph of a function?	
A.	9
B.	14
C.	32
D.	23
E.*	20

*Indicates correct response.
Note: Percents may not add to 100 because of rounding.

definition of the function or they could have graphed the various portions of the function to determine the range. Since piecewise-defined functions require students to be familiar with several different functions—potentially a different one in each portion of the definition—items involving them are more complex than many other items addressing functions and their graphs. This increase in complexity appears to have an adverse effect on students' performance when working with items involving either graphic or symbolic representations of piecewise-defined functions.

Twelfth-grade students' performance in 1996 on items involving graphic representation was quite similar to that in 1990 and 1992. There were no significant differences for five of the six items for which comparisons to 1990 or 1992 were possible: the two items involving graphing linear inequalities described at the beginning of this section and items 3, 4, 5, and 8 in table 10.12. On item 3, the item involving interpretation of a step function, performance in 1996 was significantly higher than in 1992, but the increase in percent correct was only 5 percentage points.

CONCLUSION

As Dossey reports in chapter 2, performance on the NAEP content strand Algebra and Functions is improving at all three grade levels. In fact, when compared to the average scale scores for 1990 and 1992, the average scale scores for 1996 were significantly higher, with the greatest gains appearing at grades 4 and 8. Average scale scores for fourth-grade students have increased from 214 in 1990 to 227 in 1996, and for eighth-grade students, the corresponding increases are from 261 in 1990 to 273 in 1996. These scale score increases are consistent with fourth- and eighth-grade teachers' reported increases in emphasis on algebra and functions topics from 1990 to 1996.

Looking at performance on clusters of related items reveals that performance on algebra and functions items in the 1996 NAEP was, in general, comparable to, and in some cases significantly exceeded, performance in 1992 or 1990. In only a few instances did performance on algebra and functions items decline significantly from 1992 to 1996. Although there was some progress in 1996 in exceeding the 1990 and 1992 performance levels, many students still do not perform well on the topics that usually receive the most emphasis in algebra courses. However, some of the information from students' responses to extended constructed-response tasks suggests that despite poor performance a substantial number of students may have employed reasonable solution strategies, a finding that is consistent with that reported by Silver, Alacaci, and Stylianou in chapter 11. Shifts in curricular emphasis on algebra in elementary school are, however, reflected in somewhat improved performance on items involving patterns and informal algebra. Some 1996 NAEP items provide insight into students' problem solving, reasoning, and communication, although, as with previous NAEP assessments, solving nonroutine problems, generalizing, and communicating about their reasoning is quite difficult for most students.

Although the 1996 NAEP provides a substantial amount of information about students' performance on various algebra and functions items, some areas are insufficiently addressed. For example, insight into students' capabilities to make connections between representations and connections between concepts is not easily obtained from most of the 1996 items. To get such information teachers can assign their students items, or families of related items, specifically designed to provide information about those students' abilities to make connections, and such items should be more prominent in future NAEP assessments. Also, students' understandings of, facility with, and ability to connect multiple representations (graphic-numeric, symbolic-graphic, etc.) are not easily determined from the NAEP algebra and functions items in the main NAEP assessment. Given that access to graphing calculators is expanding, it is increasingly important to assess students' uses of multiple representations. In fact, as it currently is structured, NAEP may not give students a chance to demonstrate their full capabilities because graphing calculators are not provided (Bethel and Miller 1998). This is one area in which future assessments could make a substantial contribution to our knowledge of students' capabilities with, strategies for, and understandings of algebra and functions.

REFERENCES

Bethel, Sandra C., and Nicolas B. Miller. "From an E to an A in First-Year Algebra with the Help of a Graphing Calculator." *Mathematics Teacher* 91 (February 1998): 118–19.

Blume, Glendon W., and David S. Heckman. "What Do Students Know about Algebra and Functions?" In *Results from the Sixth Mathematics Assessment of the National Assessment of Educational Progress*, edited by Patricia Ann Kenney and Edward A. Silver, pp. 225–77. Reston, Va.: National Council of Teachers of Mathematics, 1997.

Brunner, Ann, Kathy Coskey, and Sharon K. Sheehan. "Algebra and Technology." In *The Teaching and Learning of Algorithms in School Mathematics*, 1998 Yearbook of the National Council of Teachers of Mathematics, edited by Lorna J. Morrow, pp. 230–38. Reston, Va: National Council of Teachers of Mathematics, 1998.

Demby, Agnieszka. "Algebraic Procedures Used by 13-to-15-Year-Olds." *Educational Studies in Mathematics* 33 (June 1997): 45–70.

MacGregor, Mollie. "How Students Interpret Equations." In *Language and Communication in the Mathematics Classroom*, edited by Heinz Steinbring, Maria G. Bartolini Bussi, and Anna Sierpinska, pp. 262–70. Reston, Va: National Council of Teachers of Mathematics, 1998.

MacGregor, Mollie, and Kaye Stacey. "Students' Understanding of Algebraic Notation: 11–15." *Educational Studies in Mathematics* 33 (June 1997): 1–19.

Mitchell, Julia H., Evelyn F. Hawkins, Pamela M. Jakwerth, Frances B. Stancavage, and John A. Dossey. *Student Work and Teacher Practices in Mathematics.* Washington, D.C.: National Center for Education Statistics, 1999.

National Assessment Governing Board. *Mathematics Framework for the 1996 National Assessment of Educational Progress.* Washington, D.C.: National Assessment Governing Board, 1994.

National Center for Education Statistics. *The Condition of Education 1997.* Washington, D.C.: U.S. Department of Education, 1997.

Niess, Margaret. "Using Computer Spreadsheets to Solve Equations." *Learning and Leading with Technology* 26 (November 1998): 22–27.

Stacey, Kaye, and Mollie MacGregor. "Ideas about Symbolism that Students Bring to Algebra." *Mathematics Teacher* 90 (February 1997): 110–13.

11
Students' Performance on Extended Constructed-Response Tasks

Edward A. Silver, Cengiz Alacaci, and Despina A. Stylianou

THE goal of building students' capacity for mathematical problem solving, reasoning, and communication is widely accepted among professional educators, parents, and many members of the educationally attentive public. Although this goal has strong support, it is difficult to measure the extent to which it is being achieved in schools. Most large-scale assessments of mathematics achievement provide little direct evidence regarding these aspects of students' performance, since these tests typically rely almost exclusively on multiple-choice items. Although these tests can, and sometimes do, include items that are quite challenging, the multiple-choice item format typically permits only indirect inferences about students' thinking.

Among large-scale assessments, the National Assessment of Educational Progress (NAEP) mathematics assessment is an exception in this regard. NAEP has always included some items intended to assess important facets of students' mathematical problem solving, reasoning, and communication, although it has done so in different ways over time. In the first NAEP assessment in 1973, some mathematics-related tasks that required complex problem-solving performance and communication were conducted through interviews. From that time forward until 1990, however, NAEP's attempts at assessing mathematical problem solving and communication were more modest, using a combination of multiple-choice items and short constructed-response tasks, in which students provide their own answer instead of selecting one from a set of provided options. In recent mathematics assessments, NAEP has expanded its repertoire of tasks intended to assess directly some important aspects of students' mathematical problem solving, reasoning, and communication.

Beginning in 1992 and continuing in 1996, extended constructed-response tasks have been included as part of the NAEP mathematics assessment. As the name implies, an extended constructed-response (ECR) task is similar to a short constructed-response task in that it requires students to provide rather than choose an answer, as is the case in multiple-choice items, but it calls for a more extended response on the part of students. Typically, these tasks require students not only to solve problems but also to explain their method or to justify a solution or a mathematical claim. In this way, responses to these tasks offer glimpses at students' mathematical problem solving and reasoning and at the adequacy of their communication about mathematical ideas.

Although rich information about student thinking is available in the extensive set of data collected as part of the NAEP mathematics assessment, it is not so easily extracted from this data. NAEP's official reports of student performance are usually based on statistical, quantitative analyses of data rather than on qualitative analyses, and they rarely reveal data on which their analyses are based. One exception is an interesting report by Dossey, Mullis, and Jones (1993) that provided a summary of student performance on released short and extended constructed-response tasks from the 1992 NAEP mathematics assessment. Yet even this report did not contain a qualitative analysis of the kinds of reasoning and problem solving used or of the errors made by students. The richness of students' written responses to extended constructed-response mathematics tasks has been established by researchers investigating assessments other than NAEP (for example, Magone et al. 1994), and this provides good reason to expect that students' responses to the NAEP ECR tasks will be a similarly important data source.

In this chapter, we consider the 1996 NAEP extended constructed-response tasks. We begin with an overview of the ECR tasks administered as part of the 1996 NAEP and a summary of students' performance on these tasks. Then we take a close look at students' performance on several released constructed-response items, offering a qualitative analysis of facets of students' responses that illuminate their mathematical thinking.

NAEP EXTENDED CONSTRUCTED-RESPONSE TASKS: AN OVERVIEW

In 1996, twenty-three extended constructed-response (ECR) tasks were included as part of the main NAEP assessment. Typically, these tasks required students to explain their solutions or to justify their answers and conclusions in writing, by providing examples, or by drawing figures and diagrams.

As was the case in 1992, it was assumed that students would have at least five minutes for the completion of each ECR task in the 1996 NAEP

Highlights

- Students generally omitted or performed poorly on most extended-constructed response (ECR) tasks in the 1996 NAEP mathematics assessment. On average, the percent of students' responses judged to be satisfactory or better on the ECR tasks was very low at all grade levels, with as little as 2 percent, and no more than about 30 percent, of the students responding satisfactorily across twenty-three tasks at three grade levels.

- Analyses of actual students' responses to selected NAEP ECR tasks revealed that many students, including some who appeared to use reasonable approaches to solve a problem or to engage in sound mathematical reasoning, had difficulty providing clear and complete written explanations to justify their approach or their answer in a manner that was mathematically correct.

- Analyses of students' responses further revealed that some features of NAEP task and scoring rubric design appeared to contribute to the poor performance of students in unexpected ways. The design of some tasks seemed to constrain students' responses in ways that were not in line with expectations evident in the scoring rubrics. Also, variations in the cognitive complexity or the mathematical sophistication of different students' responses was not always captured by the scoring rubric.

- There was little overall difference between males and females at any grade level in the mean percent of responses judged to be satisfactory or better.

- At all grade levels the percent of Black and Hispanic students providing a response judged to be satisfactory or better was significantly lower than the percent of White students for the ECR tasks, although performance on the ECR tasks was low for all race/ethnicity subgroups.

assessment. These tasks were distributed across item blocks, with no more than one ECR task appearing within a single item block. When an ECR task appeared in an item block in the main assessment, it was always the last task in the block. An additional thirteen ECR tasks were included in the special studies blocks that were used in 1996. In this chapter we consider only the ECR tasks that were part of the main NAEP assessment in 1996; in chapter 12, Kenney and Lindquist discuss some ECR tasks in the theme blocks, and in chapter 13, Kilpatrick and Gieger discuss some ECR tasks found in the advanced studies of mathematics at grades 8 and 12.

In the main assessment, the ECR tasks were distributed across grade levels, with seven tasks at grade 4 and eight tasks at each of grades 8 and 12.

At each grade level the ECR tasks were also distributed across content topics. Across the three grades, the topic area of data analysis, statistics, and probability was the topic area most frequently assessed using ECR tasks (eight ECR tasks were included), and measurement was the topic area least frequently assessed with ECR tasks (only two ECR tasks were included).

Figure 11.1 contains two examples of released ECR tasks that were administered as part of the 1996 NAEP mathematics assessment. The first task, called Gum Ball Machine, was administered to fourth-grade students, and the second task, called Center of Disk, was administered to twelfth-grade students. Both of these examples are typical of NAEP ECR tasks, in that both require students to provide an explanation of their reasoning or their solution to a mathematical problem. Seven of the ECR tasks from the 1996 NAEP mathematics assessment have been publicly released: Compare Geometric Figures and Gum Ball Machine at grade 4; Number Tiles and Metro Rail Graph at grade 8; and Extend Pattern of Tiles, Center of Disk, and Mean or Median at grade 12. Later in this chapter we discuss students' responses to four of these tasks (Compare Geometric Figures, Number Tiles, Center of Disk, and Extend Pattern of Tiles). Our analysis and discussion complement the discussion of several of these tasks that can be found elsewhere in this volume—the Compare Geometric Figures tasks is also discussed in chapter 8 on geometry and measurement by Martin and Strutchens and the Extend Pattern of Tiles task is also discussed in chapter 10 on algebra and functions by Blume and Heckman. A discussion of the other tasks—Gum Ball Machine, Metro Rail Graph, and Mean or Median—can be found in chapter 9 on data and chance by Zawojewski and Shaughnessy.

Evaluating Students' Responses to ECR Tasks

Students' responses to the ECR tasks were evaluated by NAEP using a five-level scoring scheme; incorrect, minimal, partial, satisfactory, and extended. To ensure consistency in criteria across ECR tasks, five generic levels of performance, shown in figure 11.2, were specified and used to guide the development of the initial scoring guide for each task.

Students' Performance on the ECR Tasks

The level of students' performance on the 1996 NAEP extended constructed-response tasks was quite low. Table 11.1 summarizes the performance data for all twenty-three ECR tasks included in the main NAEP assessment in 1996 and gives the national percentages of students whose responses were scored as being at the level of satisfactory or extended for each of the ECR tasks at grades 4, 8, and 12 in 1996 (and in 1992, if the task was also administered then). Table 11.1 also gives the national percentages

Grade 4 Task: Gum Ball Machine

[General directions]

Think carefully about the following question. Write a complete answer. You may use drawings, words, and numbers to explain your answer. Be sure to show <u>all</u> of your work.

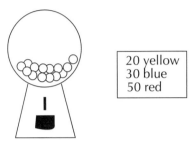

The gum ball machine has 100 gum balls: 20 are yellow, 30 are blue, and 50 are red. The gum balls are well mixed inside the machine.

Jenny gets 10 gum balls from this machine.

What is your best prediction of the number that will be red?
Answer: _____ gum balls.
Explain why you chose this number.

(Note: While working on this task, students were provided with, and permitted to use, a four-function calculator.)

Grade 12 Task: Center of Disk

[General directions]

This question requires you to show your work and explain your reasoning. You may use drawings, words, and numbers in your explanation. Your answer should be clear enough so that another person could read it and understand your thinking. It is important that you show <u>all</u> your work.

> Describe a procedure for locating the point that is the center of a circular paper disk. Use geometric definitions, properties, or principles to explain why your procedure is correct. Use the disk provided to help you formulate your procedure. You may write on it or fold it any way that you find helpful, but it will not be collected.

(Note: Students were provided with a paper disk with a diameter of 8.5 cm.)

Fig. 11.1. Sample extended constructed-response (ECR) tasks

No Response	There is no response
Incorrect Response	The work is completely incorrect or irrelevant. Or the response states, "I don't know."
Minimal	The response demonstrates a minimal understanding of the problem posed but does not suggest a reasonable approach. Although there may or may not be some correct mathematical work, the response is incomplete, contains major mathematical errors, or reveals serious flaws in reasoning. Examples are absent.
Partial	The response contains evidence of a conceptual understanding of the problem in that a reasonable approach is indicated. However, on the whole, the response is not well developed. Although there may be serious mathematical errors or flaws in the reasoning, the response does contain some correct mathematics. Examples provided are inappropriate.
Satisfactory	The response demonstrates a clear understanding of the problem and provides an acceptable approach. The response also is generally well developed and coherent, but contains minor weaknesses in the development. Examples provided are not fully developed.
Extended	The response demonstrates a complete understanding of the problem, is correct, and the methods of solution are appropriate and fully developed. Responses are logically sound, clearly written, and do not contain any significant mathematical errors. Examples are well chosen and fully developed.

Fig. 11.2. NAEP mathematics performance levels for extended constructed-response tasks. *Source:* Dossey, Mullis, and Jones 1993, p. 89

of omits (that is, blank responses) for each task. Of the twenty-three extended constructed-response tasks, the percent of satisfactory or better responses was less than 10 percent on about half of the tasks. On only two tasks was the percent of satisfactory or extended score levels' responses greater than 25 percent.

As the data in table 11.1 indicate, there was considerable variability in performance across tasks, but it is clear that students in all three grades generally had difficulty responding to the extended constructed-response tasks in

Table 11.1
National Percents of Satisfactory or Better Responses and Omits for the NAEP Extended Constructed-Response Tasks, 1992 and 1996

Task Description	Percent of Satisfactory or Better Responses		Percent of Omits	
	1992	1996	1992	1996
Grade 4				
Compare Geometric Figures	10	11	6	5
Gum Ball Machine	—	21	—	12
Data Analysis, Stat., Probability	—	31	—	11
Algebra and Functions 1	18	20	4	6
Algebra and Functions 2	—	5	—	27
No. Sense, Properties, Operations	—	15	—	7
Geometry and Spatial Sense	—	18	—	36
Grade 8				
Number Tiles	13	15	3	5†
Metro Rail Graph	—	2	—	31
Data Analysis, Stat., Probability 1	—	2	—	7
Data Analysis, Stat., Probability 2	13	17†	5	6
Measurement 1	—	9	—	25
Measurement 2	7	8	9	9
Geometry and Spatial Sense	—	8	—	29
No. Sense, Properties, Operations	—	8	—	31
Grade 12				
Extend Pattern of Tiles	5	4	23	20
Center of Disk	12	13	20	23
Mean or Median	—	4	—	31
No. Sense, Properties, Operations	—	11	—	26
Data Analysis, Stat., Probability 1	—	20	—	16
Data Analysis, Stat., Probability 2	28	31	4	4
Algebra and Functions 1	—	6	—	52
Algebra and Functions 2	—	6	—	19

— Indicates that the task was not administered in 1992.
† Indicates significant difference from 1992.

1996. In fact, on no ECR task at any grade level was the percent of satisfactory or better responses greater than the average percent correct for short constructed-response tasks or multiple-choice items at that grade. For fourth-grade students, the average percent of responses on the ECR tasks judged to be satisfactory or better was only about 17 percent; whereas, the mean percent correct was 54 percent for multiple-choice items and 38 percent for short constructed-response tasks. Similarly, for eighth-grade students, the mean percent of responses on the ECR tasks, judged to be satisfactory or better was only about 9 percent; the mean percent correct was

55 percent for multiple-choice items and 49 percent for short constructed-response tasks. Twelfth-grade students' mean percent of responses on the ECR tasks judged to be satisfactory or better was only about 12 percent; whereas, the mean percent correct was 60 percent for multiple-choice items and 34 percent for short constructed-response tasks.

Another indication of the apparent difficulty of the tasks for the students is the high rate of omission. The mean percent of omits for the ECR tasks at grade 4 was about 15 percent. At grades 8 and 12, the omission rate was even higher; the mean percent of omits for the ECR tasks at grade 8 was about 18 percent, and at grade 12, it was about 24 percent. Comparing these means to the average percentage of satisfactory or better responses at each grade level, we see that, on average, fourth-grade students were as likely to skip an ECR task as to give a satisfactory or better response, and eighth-grade and twelfth-grade students were *twice as likely* to skip an ECR task as they were to provide a satisfactory or better response.

As the data in table 11.1 also suggest, students' performance (both with respect to success and omission) changed very little between 1992 and 1996 on those tasks administered in both years. When there were changes, however, most of the changes were in a positive direction, although the magnitude of the changes was generally quite small, an average of less than two percentage points. A few of the changes were found to be statistically significant (there were more omissions in 1996 on one task and higher performance on one task), but these differences do not appear to signal matters of educational significance.

In addition to looking at the average performance for the entire national sample, it is also possible to examine performance by demographic subgroups. Table 11.2 summarizes the percent of responses judged to be satisfactory or better in 1996 for males and females and also for White, Black, and Hispanic students. One observation that can be made about these data is that the NAEP extended constructed-response tasks were difficult for everyone—for males and for females, for White students and for Black and Hispanic students.

There was little overall difference between males and females at any grade level in the mean percentage of responses judged to be satisfactory or better, but this overall average masks the variation across tasks—at each grade level, males outperformed females on some tasks, females outperformed males on some, and there was no gender difference on others. Of the twenty-three ECR tasks, a significant gender difference was found for seven tasks, with the difference in favor of males on five of these tasks and in favor of females on the other two tasks.

Interpretations for the variation in male and female students' performance on the NAEP ECR tasks must be made with care, due to the small number of such tasks administered and the even smaller number of tasks for

Table 11.2
National Percents of Satisfactory or Extended Responses to 1996 NAEP Extended Constructed-Response Tasks by Gender and Race/Ethnicity

Task Description	Nation	Male	Female	White	Black	Hispanic
Grade 4						
Compare Geometric Figures	11	10	13	13	5	6
Gum Ball Machine	21	21	20	25	7	11
Data Analysis, Stat., Probability	31	29	33	37	12	18
Algebra and Functions 1	20	18	23	26	7	8
Algebra and Functions 2	5	6	3	6	<1	2
No. Sense, Properties, Operations	15	17*	12	17	6	8
Geometry and Spatial Sense	18	21*	14	22	4	11
Grade 8						
Number Tiles	15	13	18*	18	7	4
Metro Rail Graph	2	1	2	2	2	<1
Data Analysis, Stat., Probability 1	2	2	2	3	<1	2
Data Analysis, Stat., Probability 2	17	17	17	21	4	13
Measurement 1	9	9	9	11	2	3
Measurement 2	8	8	8	10	1	3
Geometry and Spatial Sense	8	8	8	10	1	4
Number Sense, Properties, Operations	8	10*	6	10	<1	4
Grade 12						
Extend Pattern of Tiles	4	5	3	5	<1	1
Center of Disk	13	14	13	16	4	9
Mean or Median	4	5*	2	5	0	1
No. Sense, Properties, Operations	11	12	10	13	3	5
Data Analysis, Stat., Probability 1	20	18	22	24	8	14
Data Analysis, Stat., Probability 2	31	27	35*	35	16	21
Algebra and Functions 1	6	9*	4	7	3	2
Algebra and Functions 2	6	7	5	7	0	4

* Indicates significant performance difference compared to the other gender subgroup.

which significant gender differences were found. Nonetheless, a close look at the tasks on which gender differences were found suggests that performance on them resonates with the findings of several other studies of gender differences in mathematics test performance. For example, Magone (1996) found that the females in her sample of students in grades 6–8 were more likely than the males to be complete when displaying written work and providing verbal justifications on extended constructed-response mathematics tasks similar to those used in NAEP. As a consequence, females gained an advantage over males on these tasks when the scoring rubrics assigned value to the clarity of exposition or explanation provided. Moreover, in an analysis of performance on multiple-choice items involving routine mathematics, Marshall (1983, 1984) found that male students were more likely than female students to make errors due to carelessness, to not follow directions, to fail to execute completely a series of identical steps required by an algorithm, or to lack persistence with multistep procedural tasks.

From this prior research it could be predicted that females would outperform males on ECR tasks in which the mathematical content demands were relatively simple but in which repetitive, orderly execution of procedures was required or in which an elaborate verbal explanation was needed. In accord with this prediction, the 1996 NAEP mathematics ECR tasks on which females outperformed males differed from the tasks on which males outperformed females in that they usually required clear and complete verbal justifications of answers, and they required the careful, systematic, orderly execution of a series of steps involving routine procedures, such as combinatorial counting tasks. Although the data are not presented in the tables, it is also the case that females were more likely than males to respond to the ECR tasks; that is, the boys had a higher omit rate.

In addition to pointing toward gender differences, the data summarized in table 11.2 also indicate differences associated with race/ethnicity.[1] In particular, White students performed better than Hispanic and Black students on essentially all of the NAEP ECR tasks at each grade level. At grade 4, for example, an average of about 21 percent of the White students provided responses judged to be satisfactory or extended in contrast to about 9 percent of the Hispanic students and about 6 percent of the Black students. Similar differences were found at the other two grade levels, although the absolute levels of performance were lower for all groups at the other grades, and the differences in relative success were somewhat less pronounced. This

---

1. Unfortunately, it was not possible to determine the statistical significance of differences between groups for several of the tasks. For many tasks, the numbers of students, especially Black and Hispanic students, who provided satisfactory or extended responses were too small to allow a reliable determination of standard error bands around the point estimate of average performance. Therefore, it was not possible for us to evaluate the differences in a consistent manner across all the tasks, and we decided not to report the results of a partial analysis. Nevertheless, the generalization stated in the chapter appears to be correct. In all cases for which we were able to determine the error bands with reasonable certainty, there was a statistically significant difference between the performance of White students and that of both Black and Hispanic students. In each of those cases, no significant performance difference was detected between Black and Hispanic students.

performance pattern is also discussed by Strutchens and Silver in chapter 3, and they note that it is difficult to disentangle socioeconomic status from race/ethnicity in interpreting these results. As they show, quite similar findings are obtained if socioeconomic categories (for example, participation in Title I and the school lunch program) rather than race/ethnicity categories are used to analyze performance differences on these and other NAEP questions.

As we have seen, the data reported by NAEP about students' performance on the extended constructed-response tasks provides useful general information about the extent of student success, about relative success on the ECR tasks compared to the other types of NAEP items, and about performance differences between gender or race/ethnicity subgroups. But the percent of students at each of the score levels or the percent who did not respond tells us very little about students' mathematical problem solving, reasoning, or communication. In order to extract this kind of information, it is necessary to analyze in other ways the actual student responses, and this is the focus of the next section of this chapter.

NAEP EXTENDED CONSTRUCTED-RESPONSE TASKS: A CLOSER LOOK

In this section we consider in more detail samples of students' responses to four different ECR tasks from the 1996 NAEP main assessment. In particular, one task given at grade 4 (Compare Geometric Figures), one at grade 8 (Number Tiles), and two at grade 12 (Center of Disk and Extend Pattern of Tiles) are examined. For each of these tasks, a sample of actual students' responses was obtained and analyzed. These samples afford opportunities to glimpse the nature and quality of students' problem-solving strategies, the reasoning they displayed, and the errors they made. In this way, the qualitative analysis reported here opens a window into students' thinking and communicating about each task, and it provides a kind of information that is not readily available from the other summaries of student performance provided by NAEP. The sample of responses we analyzed was not a representative sample of all the responses; rather, it was a convenience sample of nonblank responses. To reduce the likelihood of improper generalizations based on the responses we examined, the percentages of types of student responses that exhibit particular characteristics are not reported here and no attempt is made to relate the sample of student work to the score levels in NAEP.

Grade 4 Task: Compare Geometric Figures

The task referred to as Compare Geometric Figures, shown in table 11.3, was administered to fourth-grade students. This task asked students to compare two shapes presented on grid paper. One of the figures was a rectangle and the other a nonrectangular parallelogram, though students were not given this information explicitly. Students were asked to list ways in which the two geometric figures were alike and ways in which they were different. Thus, the responses afford a glimpse at students' reasoning and communicating about geometric figures.

Table 11.3
Compare Geometric Figures

Task	Percent Responding Grade 4
[General directions] Think carefully about the following question. Write a complete answer. You may use drawings, words, and numbers to explain your answer. Be sure to show <u>all</u> of your work.	

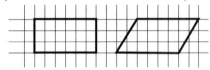

In what ways are the figures above alike? List as many ways as you can.

In what ways are the figures different? List as many ways as you can.

Extended response	<1
Satisfactory response	11
Partial response	29
Minimal response	31
Incorrect	23
Omitted	5

A description of, and an illustrative example for, each of the NAEP five score categories appear in figure 11.3. In the 1996 NAEP, fewer than 1 percent of the fourth-grade students gave a response that was scored as extended—providing two correct aspects of likeness and two correct aspects of difference. An additional 11 percent gave satisfactory responses—two correct aspects of either likeness or difference and one of the other.

To gain further information about students' reasoning and communication, we obtained and analyzed about 260 student responses and examined the aspects of likeness or difference found in these responses. A summary of these aspects is given in figure 11.4. Using a scheme suggested by Lehrer, Jenkins, and Osana (1998), the aspects of likeness or difference provided by

Incorrect—Incorrect response
In what ways are the figures above alike? List as many ways as you can.

In what ways are the figures above different? List as many ways as you can.

Minimal—Student gives a nonspecific response such as "the one on the right is skinnier," OR only one correct reason (alike or different)
In what ways are the figures above alike? List as many ways as you can.

① They can both be square
② They can both be slanted
③ They can both turn many ways

In what ways are the figures above different? List as many ways as you can.

① In the picture they are diferent.
② One is slanted.
③ One is firm and strate.

Partial—Student gives one correct reason why the figures are alike and a correct reason why they are different, OR two reasons alike, OR two reasons different.
In what ways are the figures above alike? List as many ways as you can.

- They both have 18 squares,
  they are both rectangles,
  they are both the same size.

In what ways are the figures above different? List as many ways as you can.

- one is slanted, one is not

—continued on next page

Satisfactory—Student gives two reasons why the figures are alike and one reason why they are different, OR one reason why they are alike and two reasons why they are different.

In what ways are the figures above alike? List as many ways as you can.

Their the same legth. They have the same amount of room

In what ways are the figures above different? List as many ways as you can.

They aren't the same shape. They don't have the same amount of full cubs.

Extended—Student gives at least two valid reasons why the figures are alike and at least two valid reasons why they are different.

In what ways are the figures above alike? List as many ways as you can.

They have 4 sides.
They have parll sides.

In what ways are the figures above different? List as many ways as you can.

One has square corners.
One is more slant.

Fig. 11.3. Compare Geometric Figures: Scoring categories and sample responses

the students were classified into three groups: (1) AF, appearance-based features (that is, informal, visual, or descriptive characteristics); (2) GF, conventional geometric features and figure component characteristics (for example, number of sides or corners, length of sides, perimeter, area), and (3) CM, class membership characteristics (that is, designation as belonging to a family of related geometric figures, such as rectangles or squares). In figure 11.4, the aspects of likeness and difference provided by students are listed in the order of their frequency of use in the sample of students' responses we analyzed, along with an indication of the correctness of the use of each aspect. In responding about likeness, students most frequently used reasoning based on conventional geometric features and figure component characteristics, but in responding about difference, they most frequently used appearance-based features.

Aspects of Likeness
(by frequency of occurrence)

	Category*
1. Four sides	GF
2. Same area	GF
3. Same length	GF
4. Both pictured on a grid	AF
5. Four corners	GF
6. Both rectangles**	CM
7. Straight sides	AF
8. Same perimeter**	GF
9. Both composed of little squares	AF
10. Both squares**	CM
11. Both "shapes"	AF

Aspects of Difference
(by frequency of occurrence)

	Category*
1. One straight, the other slanted (tilted, sideways, crooked)	AF
2. Different "shapes"	AF
3. Different areas**	GF
4. One has more "full cubes"	AF
5. One just looks different	AF
6. One is a square (or triangle), the other is not**	CM
7. One is a rectangle, the other is not	CM
8. Different perimeters**	GF

* AF: Appearance-based features; GF: Conventional geometric features; CM: Class membership
** Incorrect

Fig.11.4. Aspects of likeness and difference found in students' responses to Compare Geometric Figures

The distinctions made in Lehrer et al. (1998) are similar to van Hiele's (1986) model of the development of students' geometric reasoning. The van Hiele model suggests that a student progresses from reasoning about the appearance of figures to reasoning that views figures as having geometric properties and then to reasoning that considers figures as members of classes of related figures with specified properties. Relating the scheme of Lehrer et al. and the van Hiele model, it can be argued that appearance-based features (AF) represent a more elementary level of reasoning about geometric figures than do conventional geometric features (GF), since the latter are abstractions from the figures themselves. Class membership (CM) represents an even higher level of reasoning because classes of geometric figures are collections of abstract geometric features (for example, rectangles have two pairs of parallel sides). Accordingly, judgments related to conventional geometric features and class membership represent more-complex forms of reasoning.

Given these distinctions, we can conclude that the bases students provided for their judgments about difference between the two figures were generally somewhat less complex than were their bases for judgments about likeness. As shown in figure 11.4, likeness judgments were based largely on conventional geometric features and component characteristics (GF), in contrast to the difference judgments that were based largely on appearance-based features (AF).

Also, students' reasons varied within type by level of sophistication. That is, some judgments about likeness and difference were quite trivial (regardless of their place in the categorization scheme above) such as "four sides" and "four corners" for likeness and "slanty" versus "straight" and "different shapes" for difference. But other reasons, such as "they have the same area" for likeness and "one is a rectangle, the other is not" for difference, were more sophisticated.

Neither the geometric complexity of likeness and difference judgments nor the varying levels of sophistication in the kinds of reasons used by students is reflected in the NAEP scoring guide for this task. Instead, credit is assigned largely on the basis of the number of correct reasons given. For example, the sample response in the "partial" category in figure 11.3 contains two correct reasons: one for difference (one is slanted and the other is not) and one for likeness (they have the same area [expressed as "18 squares"]). The scoring scheme distinguishes between correct and incorrect reasons, since this response also contains an incorrect reason for likeness (they are both rectangles), and this was not given credit. But the scoring guide for Compare Geometric Figures was insensitive to variations in the sophistication of the reasons. For example, a response that had exactly two correct reasons, including the same basis for likeness in the response that received "partial" score discussed above and a more complex reason for difference (e.g., one is a rectangle and one is not; or the two figures have different perimeters), would have received the same score. Thus, variations in the relative sophistication of students' reasons did not enter into the scoring. Moreover, neither was the greater complexity (and hence difficulty) of difference judgments taken into account. A response that contained two correct reasons for difference and one for likeness was treated as being equivalent to a response that contained two correct reasons for likeness and one for difference. According to the scoring guide, both were judged to be satisfactory. Therefore, what was important in scoring a response using the NAEP guide for this task was the *number* of correct statements and not the relative sophistication or complexity of the reasoning used in the response.

Given that in this task students were required to list the similarities and differences, it is also interesting to investigate the vocabulary they used to express geometric ideas. For some of the aspects of likeness or difference that addressed conventional geometric features, it was not unusual for students to use standard geometric terminology (for example, sides and length). For other geometric features however, they used informal terms. For example, to express the idea that the figures have the same or different areas, students often used the terms *number of blocks* or *size*. Similarly, to express the idea that the two figures had four angles, they often used the term *corners*. This suggests that students' awareness of geometric features about area and angle was more extensive than was their formal vocabulary; that is, their knowledge of these geometric ideas was at an informal, intuitive level.

As reported in figure 11.4, the most frequently suggested aspect of difference between the figures was the fact that only one was "slanted." Students used a variety of words to express the "slanty" aspect of the parallelogram—words such as *leaning, tilted, sideways,* and *crooked* were common to many of the responses. This finding appears to confirm the suggestion of others (for example, Lehrer et al. 1998) that young children have a rich, informal, tacit knowledge base on which to build a more complete mathematical understanding of angles.

As an ECR task, then, Compare Geometric Figures appears to have been successful in eliciting students' reasoning about geometric figures and their facility with the language of geometry. The analysis of a set of responses revealed that students in grade 4 had the conceptual tools necessary to describe the rectangle and the parallelogram. In particular, students were able to attend to a variety of features such as sides, angles, area, length, and perimeter. And although the students often did not use standard geometric vocabulary in their descriptions, there is evidence that students possessed informal knowledge that can form the basis for building a more formal understanding of geometry in the later grades.

Grade 8 Task: Number Tiles

The task referred to as Number Tiles is shown in table 11.4. The task, which was administered to eighth-grade students, required them to use their understanding of whole-number place value in a problem-solving context. The task involved a game being played by two students (Maria and Carla) who were arranging number tiles within the frame of a problem involving subtraction of a two-digit number from a three-digit number. The NAEP task placed constraints on the digits to be used and asked students to determine which of the two girls would be able to create a subtraction problem with the larger result.

Table 11.4
Number Tiles

Task	Percent Responding Grade 8

[General directions]
This question requires you to show your work and explain your reasoning. You may use drawings, words, and numbers in your explanation. Your answer should be clear enough so that another person could read it and understand your thinking. It is important that you show <u>all</u> your work.

In a game, Carla and Maria are making subtraction problems using tiles numbered 1 to 5. The player whose subtraction problem gives the largest answer wins the game.

Look at where each girl placed two of her tiles.

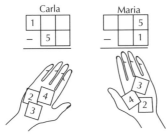

Who will win the game? _____
Explain how you know this person will win?

Extended response	1
Satisfactory response	14
Partial response	17
Minimal response	32
Incorrect	31
Omitted	5

According to the NAEP scoring guide, in order to solve the problem at the satisfactory or extended level, students needed to explain that Maria would win the game because the largest difference Carla could get had to be less than 100, whereas any difference obtained by Maria would be greater than 100. Although the mathematics involved in this task is quite simple for eighth-grade students, only 15 percent of the students were able to provide a satisfactory or extended answer, and 63 percent gave an incorrect or minimal answer. Figure 11.5 contains descriptions of the criteria for the five score levels and an illustrative example of each.

To investigate the modes of reasoning and strategies students used in this task, a set of about 670 students' responses were obtained and analyzed. A few students indicated that Carla would win the game and showed a misunderstanding of the conditions of the task (for example, adding instead of subtracting), and a few others picked Maria as the winner but either gave no written support for that conclusion or cited a reason that revealed a misunderstanding of the task. However, on the basis of a considerable number of responses, there was evidence that students in grade 8 understood the task; that is, they decided that Maria would win the game and provided a relevant reason for their conclusion. Nevertheless, there was ample variation in the nature and the adequacy of the basis for the conclusion that Maria would win the game. In our analysis, we focused on responses in which students indicated that Maria would win the game and supported that conclusion in some relevant way.

Samples of responses, representing different kinds of reasoning, appear in figure 11.6. Response 1 is representative of the responses that relied solely on place-value arguments to justify the conclusion that Maria would win the game. Most of these explanations referred to the fact that Maria could put larger digits in higher place-value positions, such as the hundreds digit of the minuend, whereas Carla was not able to do so. It was more common, however, for students to use specific examples in order to justify the conclusion that Maria would win. In fact, the vast majority of students did so, although some also provided a conceptual, place-value explanation.

Students' use of examples appeared to align with one of two general problem-solving strategies in this task. In a manner reminiscent of trial-and-error approaches to problem solving, about half of the students who used examples in their responses did so without any apparent attention to choosing a special example. These examples satisfied the problem constraints but involved a casual arrangement of number tiles without evidence of any special reason for the resulting arrangement. Response 2 in the figure represents one instance in which there appears to be no compelling reason for the choice of 142 as Carla's minuend and 425 as Maria's.

A second type of example given by students appeared to align with another problem-solving strategy; namely, considering a special case. These

Incorrect—Student answers Carla OR other incorrect response.

Who will win the game? _____tie_____

Explain how you know this person will win.

They both have the same numbers

Minimal—Student answers Maria and gives a simple arrangement of digits but with no explanation OR answers Maria with an incorrect explanation.

Who will win the game? _____MIRA_____

Explain how you know this person will win.

Becaus she has
$$\begin{array}{r} 435 \\ -\ 21 \\ \hline 414 \end{array}$$
hers is hier then Carla

Partial—Students answers Maria with partially correct or incomplete but relevant explanation.

Who will win the game? _____Maria_____

Explain how you know this person will win.

Well because

$$\begin{array}{r} 143 \\ -\ 52 \\ \hline 91 \end{array}$$ for Carla & $$\begin{array}{r} 345 \\ -\ 21 \\ \hline 324 \end{array}$$ or $$\begin{array}{r} 245 \\ 31 \\ \hline 214 \end{array}$$

Maria comes up with the highest score. even uf you change it different way.

—continued on next page

Satisfactory—Student answers Maria and gives an explanation such as "Carla has only 1 hundred but Maria can have 2, 3, or 4 hundreds," or "Maria can never take away as much as Carla."

Who will win the game? __Maria__

Explain how you know this person will win.

> Maria will win because no matter which one she plays as her first digit in the top number it will be more than 1 and on the bottom it will be less than 5 so Carla has no chance of winning.

Extended—Student answers Maria and gives an explanation such as "The largest possible difference for Carla is less than 100 and the smallest possible difference for Maria is 194," "Carla will only get a difference of 91 or less but Maria will get several larger differences," or other appropriate explanation.

Who will win the game? __Maria__

Explain how you jnow this person will win.

> It is definite that Maria will win because no other number can 5 be subtracted from without having to borrow from the one. This will cause all her answers to be less than 100. No matter what Maria does with her number tiles her answer will always be abov 100.

Fig. 11.5. Number Tiles: Scoring categories and sample responses

deliberate examples revealed a qualitatively different form of reasoning than the use of casual examples. Examples related to two kinds of special cases were evident in the set of student responses: the best possible arrangements for both Maria and Carla and the worst arrangement for Maria and the best arrangement for Carla.

The first special case involved the "best possible" arrangements of digits—that is, the arrangement that would maximize the result of the

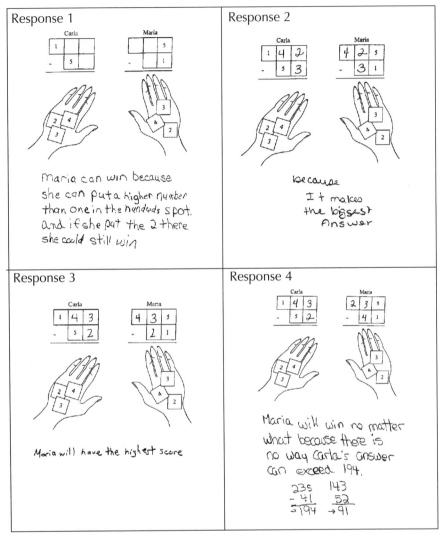

Fig. 11.6. Number Tiles: Additional student responses

subtraction for *both* Maria and Carla. Given the digits already placed, the best possible numbers were 435 for Maria's minuend and 143 for Carla's; and the result is a win for Maria. It is likely that students who used this strategy, as illustrated in response 3, assumed that Maria and Carla would deliberately try to place digits in such a manner as to maximize the difference, in accordance with the stated rules of the game. This strategy gives evidence of more thoughtfulness on the part of students than did the trial-and-error approach, and it demonstrates that Maria will win if she makes the correct moves, but it falls short of demonstrating that Maria will always win. Fewer students used this kind of example than used the casual examples described above.

The other special case involved the "worst possible" arrangement of digits for Maria and the "best possible" arrangement for Carla. Given the digits already placed, Maria's worst possible choice for the minuend was 235, and Carla's best possible choice was 143. Even in this case, Maria wins. This use of an example, as illustrated in response 4, demonstrated an efficient strategy for examining the extreme case and demonstrating that Maria would always win. Relatively few students provided this type of example.

It is interesting to note that the NAEP scoring guide differentiates among the five score levels largely on the basis of whether an explanation is given and on the thoroughness of that explanation. For example, if a response designated Maria as the winner and gave a specific example but no written explanation, then it was scored at the minimal level. In this view, all examples are treated equally. That is, despite differences in the level of sophistication associated with the different types of examples, a response of this form in which the best-and-worst case example was used would be scored exactly the same as a response in which a casual example was used. Both of these responses, containing an example without an explanation, would be scored as "minimal" using the scoring guide for the Number Tiles task.

Although most of the students were able to understand this task and determine that Maria would be the winner, few responses gave evidence of good mathematical reasoning either by constructing thoughtful examples or by providing more general explanations based solely on place-value principles. Example-based reasoning was the most prevalent approach among the eighth-grade students in our sample of responses. It is possible that the predominance of specific examples in students' explanations is a consequence of the way this task was presented; that is, there were number tiles in the figure with empty slots in which to place them, which might have suggested the production of specific examples. Nevertheless, the abundance of casual examples and dearth of deliberate examples suggests that many students did not understand how to use examples effectively to justify a conclusion. Perhaps students at this grade level have had too few opportunities to engage in mathematical argumentation and reasoning that make serious use of well-chosen examples.

It is also possible that the scarcity of good mathematical reasoning in the student responses to this task can also be attributed to the ease of concluding that Maria would win the game. If the task were designed such that Maria and Carla could each win under some conditions, it would have required more complex reasoning. Given the poor performance of students in the NAEP version of the Number Tiles task, it is reasonable to assume that performance would have been even lower if the task had called for higher-level reasoning. But it is also plausible that the ease of concluding that Maria would win the game actually suppressed the level of reasoning employed by the students. If so, then perhaps a more complex task would have evoked more complex reasoning.

Grade 12 Task: Center of Disk

The task referred to as Center of Disk, shown in table 11.5, was administered to twelfth-grade students. It asked them to describe a mathematically correct procedure for finding the center of a circular disk. According to the NAEP scoring guide, to solve the problem correctly students needed to describe a procedure for finding the center of a disk, using either a geometric-construction technique by locating the intersection of the perpendicular bisectors of two nonparallel chords, a folding technique by locating the intersection of two diameters, or a combination of the two methods (for example, folding the two chords and their perpendicular bisectors). In any case, students were expected to explain clearly and completely, using appropriate mathematical language, the geometric properties that formed the basis of the method they used. Students were provided with a paper disk with an 8.5 cm diameter that they could use as a manipulative but that would not be collected or used as part of the scoring scheme.

Table 11.5
Center of Disk

Task	Percent Responding Grade 12
[General directions] This question requires you to show your work and explain your reasoning. You may use drawings, words, and numbers in your explanation. Your answer should be clear enough so that another person could read it and understand your thinking. It is important that you show <u>all</u> your work.	
Describe a procedure for locating the point that is the center of a circular paper disk. Use geometric definitions, properties, or principles to explain why your procedure is correct. Use the disk provided to help you formulate your procedure. You may write on it or fold it any way that you find helpful, but it will not be collected.	
(Students were provided with a paper disk with a diameter of 8.5 cm.)	
Extended response	1
Satisfactory response	13
Partial response	9
Minimal response	25
Incorrect	28
Omitted	23

Note: Percents may not add to 100 because of rounding.

Figure 11.7 contains descriptions of the criteria for the extended, satisfactory, partial, minimal, and incorrect performance categories and an illustrative example of a student response for this task from each category. The percent of students who provided responses to this task at each of these score levels is shown in table 11.5, and the data clearly indicate that twelfth-grade students had difficulty with this task. Only 1 percent of the students gave an extended response, providing a complete, mathematically correct description of their procedure; and only 13 percent gave a satisfactory response, finding the center and describing the procedure used, although without including formal geometric principles or appropriate terminology. Almost one-fourth of the students provided no response at all.

To gain insights into students' reasoning about the Center of Disk task, we obtained and analyzed a sample of about 200 student responses. The intent of our analysis was to gain an understanding of the methods used by students to find the center, to examine the adequacy of their communication about their reasoning, and to identify their errors. The analysis identified a large number of responses in which students attempted to describe procedures that were mathematically correct but failed to explain their reasoning completely. Another equally large number of students described incorrect procedures that involved using various measurements rather than construction methods, and still other students gave responses that involved only rough sketches or responded incorrectly showing little understanding of either the problem situation or the geometric concepts involved. Examples of different types of solution methods and responses appear in figure 11.8.

Of all the responses in our sample, only three contained a description of a formal geometric-construction method to find the center. Students who produced this kind of response described the center of the disk as the intersection of the perpendicular bisectors of two nonparallel chords, although only one student gave a complete, mathematically correct description of, and justification for, the procedure. That student's description and justification appear as response 1 in figure 11.8.

Many more responses were based on an attempt to describe an empirical, paper-folding method in which the disk is twice folded in half, defining two diameters; the center is taken to be at their intersection. The satisfactory response in figure 11.7 is representative of this strategy. Of the responses of this type, very few included a mathematically correct description and explanation of the procedure. In fact, most students who based their responses on paper-folding were not successful in explaining the mathematical correctness of their method. For example, students often failed to mention that folding the disk in half identifies a line of symmetry for the disk (and the associated circle), thereby yielding a diameter. Many students also failed to mention explicitly that the point of intersection of two diameters is the

Incorrect—Incorrect response

> Measure around it and use that number for C in the equation.
> $C = 2\pi r$
> If the number is 6.5 for ex. it would be $6.5 = 2\pi r$
> $r = 1.04$ then you would measure to the middle w/a ruler

Minimal—Student gives a response that shows a line that appears to include the center of the circle; for example, a diameter or the perpendicular bisector of a chord; no explanation needed.

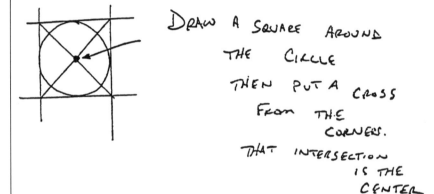

> DRAW A SQUARE AROUND THE CIRCLE THEN PUT A CROSS FROM THE CORNERS. THAT INTERSECTION IS THE CENTER

Partial—Student locates the center by folding or describes a compass and straightedge construction of the perpendicular bisectors of 2 nonparallel chords; explanation may be incomplete or incorrect—OR student explains a drawing of two diameters or the perpendicular bisector of 2 nonparallel chords; explanation must be complete and correct.

> You could take the ruler and draw a diameter across the disk. Draw another diameter perpendicular to the first diameter. Where the two diameters intersect is where the point is located.

—continued on next page

Satisfactory—Student locates the center by folding or by compass and straightedge construction and gives an explanation that describes the procedure but does not include formal geometric principles or does not use appropriate terminology (for example, bisector, midpoint, etc.)

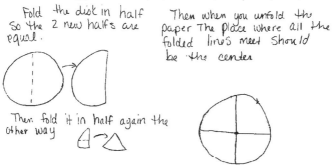

Extended—Student locates the center by folding or compass and straightedge construction and explains clearly and completely what geometric properties lead to the decisions involved in the method chosen:
—Two diameters intersect in the center of the circle;
—The intersection of 2 perpendicular bisectors of 2 nonparallel chords is the center of the circle

> Fold the disk exactly in half. Then fold it in half again. The fold lines are diameters. Two diameters intersect in the center of the circle.

Fig. 11.7. center of Disk: Scoring categories and sample responses

center of the disk. In general, these students did not communicate in writing why their paper-folding method was appropriate and mathematically correct.

Response 2 in figure 11.8 is another example of this type of response. The student who gave this response used the folding technique to find two diameters of the circle and considered the intersection of the two to be the center of the disk. However, the student failed to provide any mathematical justification for the folding procedure. The comment, "All I did was use

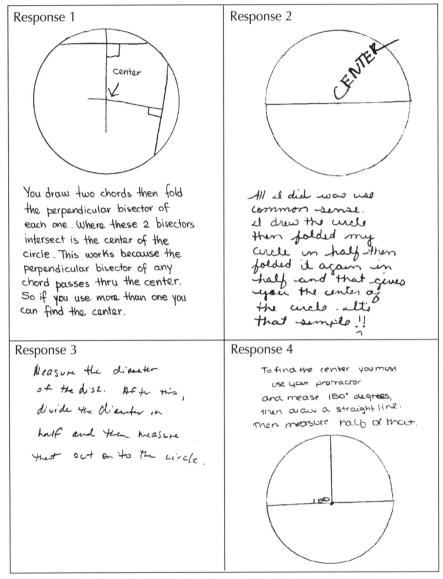

Fig.11.8. Center of Disk: Additional student responses

common sense ... It's that simple!!" suggests that this student did not sense a compelling need to connect this relatively easy procedure to the underlying mathematical theory that justified its correctness.

As noted previously in this section, a relatively large number of responses in the sample were based on the paper-folding method to find the center of the disk, in contrast to the small number based on a formal geometric-construction method. Students' preference for the paper-folding approach was likely due, at least in part, to the presentation of the task. Students were given a paper disk, and this made the paper-folding method feasible. Moreover, they were encouraged by the task directions to "fold it in any way that you find helpful." Compasses or rulers, which would have facilitated use of a geometric-construction method, were not provided to the students.

Another common method of finding the center involved measurement and approximation. Response 3 in figure 11.8 is representative of this measurement-approximation method. Students who gave this kind of response explained that first they drew one or two diameters (as the "longest chords" or "the line that divides the circle in two equal parts") by measuring or simply "finding" them. Then they proceeded to find the center of the disk by locating the intersection of the diameters they drew, by folding them or measuring them and dividing them in two equal segments, or by simply approximating the location of the center.

A small number of responses in the sample described procedures that involved various geometric facts that were not very useful for the purposes of this task. For example, a few students used the fact that the center of the disk is equidistant from all points on the circumference and drew several radii mentioning that "that point" is the center because "if you measure the distances are all equal." Response 4 in figure 11.8 represents this class of responses, in which irrelevant geometric facts were used. Further, some other responses consisted of a list of algebraic or numerical facts and expressions such as the area formula for a circle ($A = \pi r^2$), a parametric equation of a circle ($x^2 + y^2 = r^2$), and the value of π. Finally, a number of responses consisted of drawings of circles with either a dot or a cross indicating the center of the circle but without any written explanation or description of a procedure.

The results reported by NAEP for Center of Disk and our analysis of a convenience sample of students' responses to the task suggest that one of the major reasons for the generally poor performance on this task was the requirement that students explain their reasoning and their solution process. Responses often indicated that students had some knowledge and understanding of geometric concepts, but the responses also revealed inadequate written communication about the mathematics they knew. Requiring students to describe a procedure itself, as opposed to explaining the outcome of a procedure, further complicated students' communication difficulties.

However, difficulty with communication is not sufficient to explain entirely the students' low performance on this task. Our analysis of students' responses to the Center of Disk task indicates that many failed to connect the problem to mathematical knowledge and procedures that they knew. In fact, a relatively large number of students solved the problem using the empirical folding approach but failed to give a sufficient mathematical explanation concerning the correctness of their answer.

The analysis of students' responses to the Center of Disk task suggests that the task design added an unexpected complexity to the task. Students were probably led to a false impression that the task was an easy one which required little mathematics beyond the empirical paper-disk manipulation. On the one hand, the use of a paper disk may have afforded some students the opportunity to engage in a demanding task that they would have omitted had it been administered without a paper disk. On the other hand, the paper disk provided a tacit invitation to use an empirical rather than a rigorous and principled approach to the problem.

Additional performance data for the Center of Disk task (Mitchell et al. 1999) reveal how difficult the task was even for students in grade 12 who took mathematics classes beyond algebra 2, including those who took calculus. Only about 15 percent of students who reported taking a geometry course and 26 percent of students who reported studying calculus scored at the satisfactory or extended levels for this task. Because students who took geometry or calculus are likely to know the mathematics involved in the task, their low performance may suggest complexities related to the design of the task rather than poor problem-solving, reasoning, and communication skills. Overall, the Center of Disk was a task that elicited unsatisfactory results in performance scores, reasoning, and communication.

Grade 12 Task: Extend Pattern of Tiles

The task referred to as Extend Pattern of Tiles, shown in table 11.6, asked students in grade 12 to reason about a growing pattern, that is, a pattern in which a sequence of numbers or objects is extended in some predictable manner. For this particular pattern, students were given the first three figures drawn on a grid, and they were asked to find and describe the 20th figure and provide an explanation of their reasoning. No explicit instructions were given to the students as to whether the intervening figures should be included or omitted from their explanation. Further, students were not explicitly instructed to provide symbolic generalizations of the pattern.

Students' responses to the Extend Pattern of Tiles task were scored according to the five performance levels described earlier in this chapter. To solve the problem at the extended level, students needed to describe the 20th figure correctly, including the number of tiles in the figure and its shape, and to provide a clear explanation of their thinking with some evidence of accurate generalization based on inductive reasoning. Figure 11.9 contains descriptions of the criteria for the five score levels and an illustrative example of a student response for this task from each category.

Table 11.6
Extend Pattern of Tiles

Task	Percent Responding Grade 12
[General directions] This question requires you to show your work and explain your reasoning. You may use drawings, words, and numbers in your explanation. Your answer should be clear enough so that another person could read it and understand your thinking. It is important that you show <u>all</u> your work. The first 3 figures in a pattern of tiles are shown below. The pattern of tiles contains 50 figures. 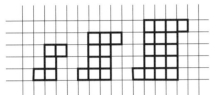 Describe the 20th figure in this pattern, including the total number of tiles it contains and how they are arranged. Then explain the reasoning that you used to determine this information. Write a description that could be used to define any figure in the pattern.	
Extended response	2
Satisfactory response	2
Partial response	18
Minimal response	29
Incorrect	25
Omitted	20

Note: Percents may not add to 100 because of rounding or off-task responses.

Results for the Extend Pattern of Tiles task indicate that students had considerable difficulty with this question. As table 11.6 indicates, only 2 percent of the students gave extended responses to this task. Also, only 2 percent of the students responded at a satisfactory level, which means that they provided responses similar to those at the extended level, but with slight computational errors, when finding the number of tiles of the 20th figure, or their explanations lacked some clarity. Another indication of the difficulty of this question is the fact that 20 percent of twelfth-grade students did not provide any response.

To gain further information about students' reasoning on the Extend Pattern of Tiles task, a sample of about 200 students' responses was obtained and analyzed. The analysis focused on students' reasoning and on the

Incorrect—Incorrect response

The pattern shape stayed the same. It just doubled in size. Not much else to say about it

Minimal—Student attempts to draw or describe the given pattern or an additional figure in the pattern; drawing or description goes beyond what is shown in the task.

[student work showing rows of numbers: top row 1–10, second row 2–11, mid 1–6; then 11–18, 12–19; 19, 20 / 20, 21; and a rectangular frame with numbers 1,2,3,4,5,6...20,21 along top and 1,2,3,4,5,6,7... along bottom]

Partial—Student illustrates or describes at least one additional figure in the pattern correctly or states there are 442 tiles in the 20th figure.

1 = 3
2 = 4
3 = 5

20 = 22

1 = 2
2 = 3

x = 21

It will be 22 tiles tall with 21 tiles on the top and bottom row with 20 tiles in the middle 20 rows.

—continued on next page

Satisfactory—Student describes the 20th figure and gives the number of tiles in that figure. Some evidence of sound reasoning must be present; there may be a computation error. Explanation may lack some clarity.

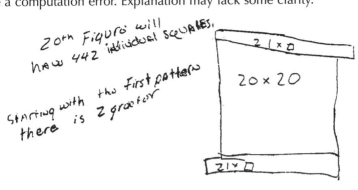

Extended—Student describes the 20th figure correctly, including the fact that there are 442 tiles. the explanation is clear, with evidence of accurate generalization (not necessarily symbolic) based on inductive reasoning.

> Each figure increases 1 layer in height and 1 middle layer in width for every succession relative to the first. For example, for the nth section, the figure will be $n+1$ units across the base, n units wide, $n+1$ units across the top, and $n+2$ units high. This is the pattern.
> The 20th figure will be 21 units across on the bottom length, 20 units wide in the middle, 22 units high, and 21 units at the top. The increase is linear.
> Total number of tiles it contains:
> $21 + (20 \times 20) + 21 = 442$
> The inner square is always $(n \times n)$ units

Fig. 11.9. Extend Pattern of Tiles: Scoring categories and sample responses

various approaches students used to solve the problem. Overall, two general approaches in students' responses were identified: a visual-geometric approach that focused on the shape of the given figure and an arithmetic-algebraic approach that used the dimensions of the first three figures either to form the general pattern of growth or to form a basis for a recursion algorithm. The students who were able to produce at least a partially correct answer were those who took into account both the geometric and the arithmetic pattern of growth. A few students focused on the overall arithmetic pattern of growth without implicit or explicit references to the visual pattern. Finally, a large number of students responded incorrectly, showing little understanding of the concepts involved and the problem situation. Examples of the different types of methods and responses appear in figure 11.10.

The first two examples in the figure, labeled response 1 and response 2, represent a visual-geometric approach to the problem. Students who used this kind of approach probably viewed the pattern as a 20 × 22 rectangle with two "flaps," one on the top right and one on the bottom left of the

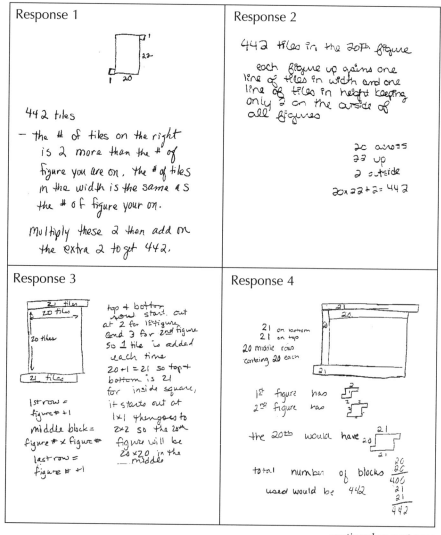

—continued on next page

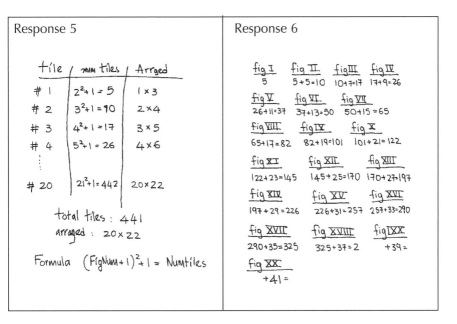

Fig. 11.10. Extend Pattern of Tiles: Additional student responses

rectangle. However, within this type of visual-geometric approach to the problem, two ways of reasoning about the 20th figure were identified. The first, demonstrated in response 1, focused on the pattern of growth of each dimension and indicates a tendency for generalization based on the shape and the number of tiles in the first three figures. The explanation indicates this pattern of growth and a tendency for this kind of generalization even though the response itself did not include a symbolic description of the pattern. Similarly, a number of other responses included tables to indicate this growth pattern on each of the dimensions of the rectangle.

Response 2 illustrates another type of reasoning, this one focusing on the recursive character of the pattern within the visual-geometric approach. According to this view, the 20th figure can be obtained from the 19th, which can be obtained from the 18th, and so on, in each case by adding a column and a row to the previous figure while keeping the two tiles on the top right and bottom left of the rectangle. The reasoning evident in both kinds of responses (responses 1 and 2) is based on the geometric shapes of the first three figures, but the approaches used to determine the area of the 20th figure differ. It is worth noting, however, that most student explanations were not as well developed as the ones shown here. In fact, most students limited their explanations to the dimensions of the rectangle or a rough sketch of the figure. Moreover, students often made computational errors in the generalization of the pattern; a number of students computed a 19 × 22 or a

20 × 23 rectangle, and as a result, they provided the wrong number of tiles for the 20th figure.

Responses 3 and 4 in figure 11.10 illustrate a second visual-geometric approach to the problem in which the pattern is viewed as a 20 × 20 square with an extra row on the top extending one tile to the right and another row on the bottom extending one tile to the left. These responses suggest that some students focused primarily on the pattern of growth of each dimension of the square and the bottom and top rows. There is evidence from the responses that a few students focused on the recursive pattern as is shown in the sample "extended" response in figure 11.9. As another student explained, "The tiles grow one in number each time in two different areas: number across on top and bottom and number of tiles in the middle lengthwise and in width. Using this, the 20th [figure] will be 21 tiles across on top and bottom and 20 tiles wide and long in the middle."

Another approach to finding the 20th figure involved a description of the pattern in an algebraic or arithmetic fashion. Response 5 in figure 11.10 represents this approach. Students who gave this kind of response appeared to focus on finding the total number of tiles of the 20th figure rather than emphasizing the shape of the figure. Some students constructed a table listing the number of tiles of the first three given figures, found an arithmetic pattern that described these numbers, and generalized it to the 20th figure. Other examples included attempts to describe the pattern in a more algebraic fashion in the form of a function, an arithmetic sequence, or a recursive algorithm. This class of responses, more than others, reflected students' obvious attempts to generate symbolic generalizations of the pattern such as algebraic descriptions. For example, one student's response gave an algebraic description of the nth figure as $a_n = n(n + 2) + 2 = n^2 + 2n = 2$—a description that corresponds to viewing the nth figure as a rectangle with two "flaps." A few of the responses that fall in this class, though, used some type of a counting or "brute force" approach, such as the example shown as response 6 in figure 11.10. Students who used this approach listed all twenty figures and the number of tiles they contained, or the dimensions of each figure, by simply increasing the number of tiles by a fixed amount until they reached the 20th figure. Some of these responses also provided a correct geometric description of the 20th figure, but there was less emphasis on the description of the shape than on the number of tiles it contained.

Further, a considerable number of the responses were limited to incomplete, and often incorrect, descriptions of the dimensions of the 20th figure without any clear suggestion of what shape the 20th figure would have or any indication of the bases for the conclusions. Students who gave this kind of response typically focused on the arithmetic-algebraic aspect of the pattern and overlooked its geometric component. However, a number of the responses that focused on the overall geometric features of the pattern failed to take into account the arithmetic-algebraic aspect of the pattern. Moreover, a number of students who apparently focused on the geometric aspect of the problem avoided drawing any diagrams that would explain their reasoning and relied completely on verbal descriptions of the shape of the 20th figure. Finally, a number of responses in the sample consisted of incomplete solutions (for example, lists of various numbers without justifications and unlabeled sketches without any written explanation or description of how the drawings pertain to the problem) that made it difficult to interpret the solution strategies these students used.

Viewed through the lens of the NAEP scoring guide (see figure 11.9), students' overall performance on the Extend Pattern of Tiles task was poor. Only 4 percent of the students who responded produced a solution that was scored either at the extended or a satisfactory level, 54 percent scored at the incorrect or minimal level, and 20 percent of the students did not attempt to solve the problem. However, our analysis of a convenience sample of students' responses to this task revealed a slightly more optimistic picture of students' mathematical reasoning. We found the use of mathematically correct reasoning in a large number of students' responses in this sample, although not all of the explanations were complete. A careful look at the NAEP scoring guide indicates that there were no clear provisions for giving appropriate credit to those responses that showed some evidence of correct mathematical reasoning in the context of the problem if they did not present complete explanations or if they contained computational or other errors. For example, students who attempted to describe the pattern using an arithmetic method to find the number of tiles of the 20th figure and did not emphasize the shape of the figure were probably scored at the partial, or even minimal, level.

The analysis of students' responses to the Extend Pattern of Tiles task is a good example of the kinds of insights into students' mathematical thinking that ECR tasks can provide. The detailed analysis of the task allowed for the identification of problem-solving strategies and approaches as well as some of the difficulties students face when asked to solve ECR tasks. The analysis suggests that students can engage in rich mathematical problem solving, finding connections between algebraic and geometric patterns and generating and describing general patterns and functions. Nonetheless, our analysis of students' responses, when taken together with the low performance levels reported by NAEP, clearly indicates that many students were unable or unwilling to engage with the problem and that many more have difficulty making generalizations and communicating their reasoning, despite their demonstration of some appropriate reasoning on the problem—a finding which is in line with other indicators of the low performance of students on problem solving.

NAEP EXTENDED CONSTRUCTED-RESPONSE TASKS: CONCLUDING COMMENTS

The picture of students' mathematical problem solving, reasoning, and communication that is presented by the overall NAEP performance data and by our qualitative analysis of student responses to selected extended constructed-response tasks from the 1996 NAEP mathematics assessment shows that student performance is unsatisfactory. NAEP's quantitative analysis showed that students performed poorly on all ECR tasks at all grade levels. The qualitative analysis confirmed the inadequacy of most of the responses that students provided.

Although the 1996 NAEP ECR task performance data are discouraging, it is important to acknowledge that there are several reasons why these results might represent underestimates of students' capacity to respond successfully to such tasks. First, the high rate of omits may be due, at least in part, to a lack of time available to respond to the task. It is likely that many students work through the tasks in the item block in order. Thus, the placement of the ECR tasks at the ends of item blocks increases the likelihood that students found themselves with too little time to respond to a complex task at the end of an item block.

A second factor likely to contribute to the lack of high-quality responses, and especially to the high omit rate, is a lack of student motivation to engage in the necessary work. Because there is little or no incentive for students to perform well on the NAEP assessment, they may be less motivated to engage in the relatively greater amount of work involved in responding to ECR tasks than to respond to multiple-choice items, for example. It seems reasonable to view this lack of motivation as a likely contributing factor in the exceedingly high rate of omits by twelfth-grade students on the ECR tasks.

A third factor contributing to the general lack of student success relates more to the tasks and scoring guides than to the students. For at least some of the tasks, there were ways in which the assumptions or constraints reflected in the task wording and in the expectations evident in the scoring guide did not appear to be shared by students. In some cases, students responded in a manner that they may have deemed reasonable but which was not so judged by raters using the scoring guides. Moreover, some nuances related to the cognitive complexity or mathematical sophistication of students' responses were not always attended to in the assignment of level by the scoring guide.

Although these factors may account for some of the poor performance, it is clear that they do not account for most of it. Our look at student performance on these NAEP ECR tasks makes it abundantly clear that students at all the grade levels have genuine difficulty explaining and justifying their thinking in ways that are mathematically adequate. Although it is possible that some of the tasks may have been too "simple," in that the answer was easy to obtain and most of the credit was assigned to the quality of the accompanying written explanation, it is by no means clear that greater task complexity would have evoked higher-level reasoning and better communication by students. Moreover, the tendency of these tasks to ask for explanation as an "add on" rather than as a natural part of the problem solution probably also adds to the difficulty. As has been noted in more general considerations of explicit verbalization of problem-solving processes (for example, Ericsson and Simon 1984), having students explain their solution *post hoc*—after solving a problem in which their cognitive attention has been on obtaining the solution rather than on tracking the process—may not optimally elicit students' thinking. Nevertheless, it is more likely that a solution to the problem of better performance on extended constructed-response tasks lies not primarily in improved task and rubric design but rather in better and different school mathematics instruction.

Judging from the low success and high omit rates on these tasks and from the finding of differential performance by students from different race/ethnicity (or socioeconomic) groups, it appears that current calls for greater emphasis in school mathematics education on developing *all* students' capacities for problem solving, reasoning, and communication are aimed at the correct target. When many students have regular, sustained opportunities to solve complex problems and communicate their solution methods, their understandings of important concepts, and their arguments and rationales, there is a greater likelihood for significant improvement in performance on tasks like the 1996 NAEP ECR tasks.

The 1996 NAEP data on classroom instruction (see chapter 5 by Grouws and Smith) indicate that teachers at grades 4 and 8 are far less likely to address "developing reasoning and analytic ability to solve unique problems" and "learning how to communicate ideas in mathematics effectively" than they are to address "learning mathematics facts and concepts" and "learning skills and procedures needed to solve routine problems." As Grouws and Smith note, higher student proficiency scores, particularly at grade 8, are associated with increased teacher attention to reasoning and analytic skills in unique problem situations and also to communication of mathematical ideas.

If U.S. teachers of mathematics devote more of their instructional attention to the communication of mathematical thinking and reasoning, perhaps this will be an area in which U.S. students can become as good as students anywhere in the world. The data considered in this chapter suggest that we have a long way to go, but other data are more encouraging about the possibility. For example, the analysis of extended performance tasks in TIMSS revealed that students all over the world, even in the high-performing countries, had difficulty with tasks requiring explanation: "Regardless of context, items requiring explanations were consistently more difficult than other types of questions" (Harmon et al. 1997, p. 126). Similarly, Cai (1995) compared Chinese students' performance on three types of tasks, one of which was very similar to the ECR tasks used in NAEP, and found that Chinese students performed less well on these tasks that required explanation than they did on the other types of tasks. Interestingly, he also found that the performance difference between Chinese students and a comparable population of U.S. students was lower on these tasks than on the other types of tasks. In fact, U.S. students actually outperformed Chinese students on several of the tasks used in Cai's study.

We hope that the examination of students' performance on NAEP extended constructed-response tasks suggests something for all of our readers to do. Teachers and students need to work together to achieve improved performance on these types of tasks, which is likely to result only if mathematical thinking, reasoning, and communication are given regular, sustained attention in the mathematics classroom. Those charged with the initial preparation and ongoing professional development support of teachers are invited to address this kind of mathematics teaching and this mathematical goal for students explicitly. Test designers are asked to attend to the issues raised by our qualitative analysis of students' responses in order to design better tasks and more appropriate scoring rubrics. Finally, researchers and other scholars interested in mathematics learning are encouraged to consider students' responses to the NAEP ECR tasks as a rich source of information.

REFERENCES

Cai, Jinfa. *A Cognitive Analysis of U.S. and Chinese Students' Mathematical Performance on Tasks Involving Computation, Simple Problem Solving, and Complex Problem Solving.* Journal for Research in Mathematics Education, Monograph no. 7. Reston, Va.: National Council of Teachers of Mathematics, 1995.

Dossey, John A., Ina V. S. Mullis, and Chancey O. Jones. *Can Students Do Mathematical Problem Solving?: Results from Constructed-Response Questions in NAEP's 1992 Mathematics Assessment.* Washington, D.C.: National Center for Education Statistics, 1993.

Ericsson, K. Anders, and Herbert A. Simon. *Protocol Analysis: Verbal Reports As Data.* Cambridge, Mass.: MIT Press, 1984.

Harmon, Maryellen, Teresa A. Smith, Michael O. Martin, Dana L. Kelly, Albert E. Beaton, Ina V. S. Mullis, Eugenio J. Gonzalez, and Graham Orpwood. *Performance Assessment in IEA's Third International Mathematics and Science Study (TIMSS).* Chestnut Hill, Mass.: TIMSS International Study Center, Boston College, 1997.

Lehrer, Richard, Michale Jenkins, and Helen Osana. "Longitudinal Study of Children's Reasoning about Space and Geometry." In *Designing Learning Environments for Developing Understanding of Geometry and Space,* edited by Richard Lehrer and Daniel Chazan, pp. 137–67. Mahwah, N.J.: Lawrence Erlbaum Associates, 1998.

Magone, Maria E., Jinfa Cai, Edward A. Silver, and Ning Wang. "Validating the Cognitive Complexity and Content Quality of a Mathematics Performance Assessment." *International Journal of Educational Research* 21 (April 1994): 317–40.

Magone, Maria. "A Study of Gender Differences in Responses to a Mathematics Performance Assessment Instrument Consisting of Extended Constructed-response Tasks." Doctoral diss., University of Pittsburgh, 1996.

Marshall, Sandra P. "Sex Differences in Children's Mathematics Achievement: Solving Computations and Story Problems." *Journal of Educational Psychology* 76 (April 1984): 194–204.

———. "Sex Differences in Mathematical Errors: An Analysis of Distracter Choices." *Journal for Research in Mathematics Education* 14 (November 1983): 325–36.

Mitchell, Julia H., Evelyn F. Hawkins, Pamela M. Jakwerth, Frances B. Stancavage, and John A. Dossey. *Student Work and Teacher Practices in Mathematics.* Washington, D.C.: National Center for Education Statistics, 1999.

van Hiele, Pierre M. *Structure and Insight: A Theory of Mathematics Education.* Orlando, Fla.: Academic Press, 1986.

12

Students' Performance on Thematically Related NAEP Tasks

Patricia Ann Kenney and Mary M. Lindquist

AMONG the many recommendations for mathematics instruction in the *Professional Standards for Teaching Mathematics* (NCTM 1991, p. 24) is one that advocates the use of worthwhile mathematical tasks that "provide the stimulus for students to think about particular concepts and procedures, their connections with other mathematical ideas, and their applications to real-world contexts." In the mathematics classroom, teachers can select and organize sets of worthwhile tasks so that they are related to one another. For example, sets of tasks can assess students' understanding of a particular mathematics concept such as proportional reasoning across content areas (for example, measurement and geometry) or within a single area. Tasks can also be related by a common context. In large-scale assessments, however, such sets of related tasks are rare.

The 1996 NAEP mathematics assessment was an exception in that it included sets of items that were developed around a common context or theme. These thematically related sets were called *theme blocks,* and they formed the basis of the Theme Study, a special study in the 1996 assessment for grades 4, 8, and 12. In NAEP, special studies that focus on particular areas of interest to mathematics educators and researchers are periodically conducted in addition to the main assessment. The focus of the Theme Study was mathematics-in-context (Mitchell et al. 1999), that is, student performance on sets of items that "related to some aspect of a rich problem setting that served as a unifying theme for the [set]" (Reese et al. 1997, p. 79). There were five different theme blocks: two administered at each NAEP grade level, with one block common to grades 8 and 12. Within a theme block at a particular grade level, all items were based on a thematic context that was thought to be familiar and interesting to most students, such as studying butterflies for a science fair project (grade 4), building a doghouse (grade 8), and buying or leasing a car (grade 12).

In this chapter we first present a summary of particular features of the Theme Study, including the characteristics of the sample of students and how the theme blocks were administered, and then we focus on selected performance results on items in the three released theme blocks, one block from each of grades 4, 8, and 12. After presenting overall performance results for the released theme blocks, we conclude with a discussion about what value was added to NAEP by including thematically related sets of tasks in the 1996 assessment.

Highlights

- The theme blocks, an innovation in the 1996 NAEP mathematics assessment, differed from the item blocks in the main NAEP assessment in that they contained items structured around a common context or theme, most items were nonroutine problems, fewer items were in each block, and more time was allotted.

- The performance results show that there were lower omission and off-task rates for students in grades 4, 8, and 12 on the theme block items than on those in main NAEP. This suggests that the context of the items and extra time allotment may have had some effect on student motivation.

- Despite lower omission and off-task rates, performance on the theme block items was low, especially for many items in the blocks for grades 4 and 8 and for items at all three grade levels requiring students to explain their reasoning. Also, particular topics such as proportional reasoning at grade 4 and the relationship between a constant perimeter and a changing area at grade 8 were difficult for students.

- In general, results from the theme blocks provided few if any completely new insights into how students solve complex problems that were not already available from the main NAEP assessment. To improve the utility of the theme blocks, the tasks should be related in ways that go beyond context. For example, all tasks could be developed to assess a particular concept but at different levels (giving a definition, applying the concept in different settings, generalizing the concept) or to assess concept such as ratio and proportion across content topics (measurement, geometry, and algebra).

FEATURES OF THE THEME STUDY

The Theme Study had certain features that were similar to those of the main NAEP assessment (that is, the set of items given to the national sample of students) and other features that were different from main NAEP. The students in grades 4, 8, and 12 who participated in the Theme Study were selected according to standard NAEP sampling procedures, but no student participated in both main NAEP and the Theme Study. A comparison of students in the main NAEP sample and those in the Theme Study sample revealed that the samples were nearly identical on a variety of variables (for example, gender, race/ethnicity, participation in Title I programs) (Mitchell et al. 1999).

There were important differences between the Theme Study and main NAEP in item format and timing. With respect to item format, most of the theme block items were constructed-response items that required students to show their work and explain their answers. In fact, of the forty-two items in the five theme blocks, 25 percent were multiple-choice items and 75 percent were constructed-response items. This distribution of items is different from that in main NAEP, where 55 percent and 45 percent of the items were in multiple-choice and constructed-response formats, respectively.

Because constructed-response items are typically more complex than multiple-choice items, students either need more time to work on them or fewer items to answer within a given time period. In the Theme Study, students had both fewer items and more time allotted for working on them. Each student worked on one theme block during a 30-minute time period; the number of items to be completed in that time period ranged from six to eleven items, with an average of eight items across the five blocks. This stands in contrast to the main-NAEP item blocks that have an average of twelve items per block (across all blocks in main NAEP) and the requirement that all blocks be completed in 15 minutes. It was thought that having fewer items and giving students more time to work on them would provide NAEP with the opportunity to "explore the use of questions that were more detailed and [more] complex than questions in the main NAEP assessment" (Mitchell et al. 1999, p. 27).

PERFORMANCE ON THE RELEASED THEME BLOCKS

In the next three sections of this chapter, we focus on selected students' performance results on items in the three released theme blocks, one block for each grade level. The appendixes to this chapter contain copies of the items and ancillary materials for each released theme block. We chose to include the released theme blocks so that readers could refer to particular items and materials as they are discussed and so that readers could have access to them for possible use in instruction, assessment, and research.

The Butterfly Booth

Of the two theme blocks administered to fourth-grade students, one block (hereafter referred to as the "Butterfly Booth") was set in the context of a science fair project on butterflies. The items in the Butterfly Booth block are in appendix A, pp. 364–67. On the first page of the block was a banner that read "Coming Soon to Oakville School—Science Fair—May 6–10," followed by a short introductory paragraph introducing the context to the student: "Each class in Oakville School will have a booth at the Science Fair. Your class is planning to have a Butterfly Booth."

Following the introductory information were six items that assessed students' understanding of a variety of mathematical topics such as symmetry, measurement in the metric system, reasoning about division, and proportional reasoning. All items were in constructed-response format, and most were classified by NAEP test developers in the mathematical ability category of Problem Solving. The items were not routine problems but, instead, were problems in which students had to work with more than one piece of information and that they could solve in several ways. Students had the use of a four-function calculator, a 6-inch ruler that also included a 15-centimeter scale, a Butterfly Information Sheet containing life-size, colorful pictures of three kinds of butterflies (Monarch, Black Swallowtail, Common Blue), and a set of fifteen colored cutouts of Common Blue butterflies. (A copy of the Butterfly Information Sheet appears in appendix A, p. 367.)

Overall Performance on the Items in the "Butterfly Booth" Block

Table 12.1 contains a description of each item and selected performance results. Students' responses to the constructed-response tasks were scored on the basis of either five, four, or three score levels, depending on the complexity of the item, the combinations of possible responses, and the depth of student responses. Because of the variety of score levels, it was somewhat difficult to compare performance across the set of items. For the purposes of this section, we focus on the percent of students scoring at the highest score level (complete or satisfactory) for items with three or four score levels and at the top two score levels (satisfactory and extended) for extended constructed-response items.

Looking at these particular score levels, then, performance across the six items ranged from 40 percent correct on item 2, which asked students to obtain two wingspan measurements in centimeters, to 1 percent correct on item 6, a complex item involving concepts in both measurement and patterns. These results suggest that nearly all of the items in the Butterfly Booth

Table 12.1
Performance on Items in the "Butterfly Booth" Block: Grade 4

Item Description	Percent Responding at Highest Score Level(s)
1. Draw four missing markings on pictures for two butterflies to make each butterfly symmetrical.	28
2. Obtain correct measurements in centimeters for the wingspans of two butterflies.	40
3. Determine and show the greatest number of butterflies that can be stored in a case and the number of cases needed to hold 28 butterflies; explain how answer was obtained.	17[a]
4. Determine the maximum number of butterfly models that can be made from a given number of parts (wings, bodies, antennae); show or explain how answer was obtained.	3
5. Given that 2 caterpillars eat 5 leaves per day, determine the number of leaves needed each day to feed 12 caterpillars; show or explain how answer was obtained.	6
6. Find the number of each type of butterfly needed to create a repeating pattern on a banner that is 130 centimeters long; show how answer was obtained. [Note: measurements from item 2 were needed in this problem.]	1

[a]Percent of students scoring at either the satisfactory or extended level.
Note: Copies of the items described in the table appear in appendix A. All items were constructed-response items, but were scored on the basis of either three (item 5), four (items 1, 4, and 6), or five (items 2 and 3) score levels. For items 2 through 6, the omission rates were 4 percent or less.

block were difficult for students in grade 4, and especially the last four items, which asked students to explain how they got their answer. The last three items were particularly difficult, as about 60 percent of the students gave incorrect answers for item 4 and nearly 90 percent of the students gave incorrect answers for items 5 and 6.

Despite the fact that performance was low on the theme-block items, there is evidence that students tried to answer the items. In particular, for all but the first item, the average rate of omissions (that is, items that students left blank) was about 3 percent, a rate that was quite low for NAEP constructed-response questions. Moreover, the omission rate for the last item in the block (item 6) was only 1 percent, which was far less than the average omission rate of about 15 percent for extended constructed-response questions in the main NAEP assessment at grade 4, as reported by Silver, Alacaci, and Stylianou in chapter 11. Also, fewer than 10 percent of the responses were off-task; that is, responses that did not relate to the item (for example, "I don't like this test."). These results suggest that the thematic context about butterflies engaged the students in ways that encouraged them to attempt to answer even the most difficult questions and that students in grade 4 probably had sufficient time to work on all six items in the block.

It was somewhat surprising that the first item, which asked students to draw four missing markings on pictures of two butterflies to make the markings symmetrical, had an omission rate of 33 percent. The symmetry item is reproduced in appendix A, p. 364. The usual pattern for NAEP items is that the first item in a block has the lowest omission rate; that is, most students try to answer the first item. Although any explanations for the high omission rate would be speculative at best, it is not unreasonable to think that some fourth-grade students did not understanding the meaning of *symmetrical*. Performance on NAEP items from previous assessments shows that when the concept of symmetry is presented informally and using pictures, students try to answer the items and tend to answer them correctly. For example, on a secure multiple-choice item in which symmetry was presented informally and pictorially as folding a figure (for example, a stick figure of a person) along a vertical dashed line so that both parts match, 91 percent of students in grade 4 chose the correct response, and only 2 percent omitted it. Comparing performance on this secure item and the butterfly-symmetry item suggests that students were more likely to answer and performed at higher levels when an informal word such as *folding* was used rather than the mathematical term.

Although according to the NAEP scoring standards overall performance on the Butterfly Booth block was low, there is evidence that students had at least some knowledge of the mathematics assessed in the items. However, students' mathematical knowledge was at times incomplete, and students often had difficulty explaining or illustrating their reasoning. To gain some insight into how fourth-grade students attempted to solve the tasks in the Butterfly Booth block and the kinds of errors they made, we focus on two items: one that assessed students' ability to measure a real object and another that assessed proportional reasoning.

Measurement in the Metric System

Results from the 1992 NAEP mathematics assessment showed that most fourth-grade students were able to obtain correct linear measurements of geometric figures (for example, the longest side of a rectangle) using a centimeter ruler (Kenney and Kouba 1997). An item in the Butterfly Booth block, described as item 2 in Table 12.1 and reproduced in appendix A (p. 364), required students to measure the wingspans of two butterflies, using pictorial representations from the Butterfly Information Sheet, which also appears in appendix A, p. 367. To answer the question, students had to understand the definition of *wingspan,* illustrated for the Monarch Butterfly, and then measure the wingspans of the other two butterflies. Only 40 percent of the students obtained two correct wingspan measurements of 7 cm and 3 cm for the Black Swallowtail and Common Blue butterflies, respectively.

Performance on the Butterfly Booth measurement item was lower than that for other measurement items in the 1996 NAEP mathematics assessment. For example, performance was higher on items that required students to measure one line segment or a series of connected line segments in centimeters, with about 55 percent of students in grade 4 answering those items correctly. These comparisons suggest that students are more successful at measuring the length of given line segments than they are at measuring a dimension not explicitly shown on the object, such as wingspan.

Proportional Reasoning

As Wearne and Kouba note in chapter 7, mathematics educators view proportional reasoning as having a central role in the mathematics curriculum, particularly at the middle school level. However, the rudiments of proportional reasoning are also an important part of elementary school mathematics (Reys, Suydam, and Lindquist 1995). Although there were no ratio and proportion items given to fourth-grade students on the main NAEP assessment, the Butterfly Booth block includes an item that could be solved using proportional reasoning—item 5 described in table 12.1 and reproduced in appendix A (p. 366). This item asked students to find the number of leaves needed to feed 12 caterpillars if it takes 5 leaves to feed 2 caterpillars. Performance results show that 6 percent of the students have a correct answer with a correct explanation, and another 7 percent either gave a correct answer without an explanation or showed a correct method with a computational error. Most of the students, 86 percent, gave an incorrect response.

NAEP performance results provide little information on the strategies students used to solve this problem or the kinds of errors students made. To gain some insight into how students approached this problem, we examined a set of 175 responses from students in the NAEP Theme Study sample and chose some examples to illustrate correct and incorrect responses. The responses were from a convenience sample of responses to be used for illustrative purposes; they do not necessarily represent the full range of responses on any particular item in the theme blocks. The examples of student work appear in figure 12.1. Responses 1, 2, and 3 illustrate strategies such as drawing pictures or creating charts to show the relationship between leaves and caterpillars or explaining the relationship between leaves and caterpillars in words. It is interesting that the student who produced response 3 viewed the proportion in fractional terms and found the unit value; that is, this student noted that each caterpillar needed 2 1/2 leaves each day and used this value to solve the problem.

A fourth-grade class needs five leaves each day to feed its 2 caterpillars. How many leaves would they need each day for 12 caterpillars?

Answer: _____

Use drawings, words, or numbers to show how you got your answer.

Response 1	Response 2
Answer: 30 leaves	Answer: 30
	C 2 4 6 8 10 12
	L 5 10 15 20 25 30

Response 3	Response 4
Answer: 30	Answer: 15 catapillars
because for two catapillers each catapiller gets to eat 2 ½ leaves. So for 12 catapillers they would get 30 leaves.	They added 10 catapillers, so I added 10 leaves.

Response 5

Answer: 15

I looked at the numbers 5 and 2 and I added 3+2=5 so I went three more numbers that 12.

Fig. 12.1. Sample responses to the proportional reasoning item

With respect to incorrect answers and faulty strategies, there were a number of responses in the sample set with an answer of 60, accompanied by the multiplication sentence $5 \times 12 = 60$. Perhaps students who gave this answer understood that the problem involved a multiplicative situation, but they did not have a complete understanding of the situation in the problem: 5 leaves for 1 caterpillar instead of 5 leaves for every 2 caterpillars. We also saw responses in the sample set that gave 19 as the answer, accompanied by the addition sentence $5 + 2 + 12 = 19$ using the numbers in the item. In their work with NAEP results, Kouba, Zawojewski, and Strutchens (1997) refer to this as the "when in doubt, add" strategy.

Another kind of error, which is well documented in the literature on proportional reasoning, involves students' thinking about proportion in an additive way. This way of thinking, which results in an answer of 15 leaves, is illustrated by responses 4 and 5 in figure 12.1. In response 4, the student probably focused on the fact that there were 10 more caterpillars rather than 6 times as many caterpillars. Thus, he or she just added 10 more leaves to the original 5 leaves rather than multiplying the original number by 6. Similarly, in response 5, the student may have reasoned that since there were 3 more leaves than caterpillars in the original statement, there should be 3 more leaves for the 12 caterpillars, or 15 leaves.

Summary of the "Butterfly Booth" Block

As a whole, this theme block presented some interesting mathematics tasks for students in grade 4, and the low omission rate on all but one item and the low percentage of off-task responses suggest that the students were engaged in the problems and tried to answer nearly all of them. This is an unusual pattern for constructed-response items in NAEP, which often have double-digit omission rates. Also, the low omission and off-task rates suggest that 30 minutes was probably enough time for students to work on the problems in this block.

Performance across the items, however, was at very low levels. Although students in grade 4 attempted to answer them, nearly 90 percent of the answers, on some items, were incorrect. Some possible explanations for the low performance levels include the inexperience of fourth-grade students in providing written explanations to mathematics problems and the multistep nature of most of the items in the Butterfly Booth block. Despite the low performance levels, there were some encouraging signs of students trying to make sense of the mathematics in those items. For example, the students who used addition in the proportionally-based problem (see responses 3 and 4 in fig. 12.1) appeared to be looking for patterns and thinking about the situation. This stands in contrast to those students who merely performed an operation on the numbers in the problem, especially those who added the numbers in the problem.

Building a Doghouse

One of the two theme blocks administered to eighth-grade students was set in the context of building a doghouse (hereafter referred to as "Building a Doghouse"). The items in this block appear in appendix B, pp. 368–73. On the first page of the block was a two-dimensional representation of the three-dimensional doghouse and an introduction to the thematic context: "Julie wants to build a doghouse like the one shown in the picture above. She has asked you to help her to build the doghouse." Before beginning to work on the items in the block, students had to build a model doghouse using materials in a kit. In the kit was a sheet of sturdy paper containing the perforated pieces (called "pushouts") of the seven parts of the scale model doghouse, two pushouts that when folded together formed a model of the doghouse, and a pushout that represented the door opening of the doghouse. (A copy of the sheet of pushouts appears in appendix B, p. 373).

After the students punched out the pieces and assembled the model doghouse, they answered ten items, some of which were based on the model, the full-sized doghouse, a combination of both representations, and other aspects of building a doghouse. The items in this block employed a variety of formats, including one multiple-response item, three multiple-choice items, four short constructed-response items, and two extended constructed-response items. Most of the ten items assessed concepts in the NAEP content strands of Measurement and Geometry and Spatial Sense and in the ability categories of Conceptual Understanding and Problem Solving. Most of the items were nonroutine, multipart problems that required students to integrate information and explain their answers. While working on items in this block, students were provided with scientific calculators, combination ruler/protractors, and the kit of pushouts described above.

Overall Performance on the Items in the "Building a Doghouse" Block

Table 12.2 contains a description of each item and performance results representing the percent of correct responses for all items except the extended constructed-response task that used five score levels. For that task, the percent of students scoring at either the satisfactory or extended level is given. For all but the first item, fewer than half of the students answered correctly, with performance especially low for items that required students to work with a scale that related measurements on the model doghouse to those on the actual doghouse (14 percent and 19 percent correct for items 4 and 7, respectively) and on the two extended constructed-response questions (2 percent of responses at the satisfactory or at the satisfactory or extended levels for items 9 and 10).

The items in the Building a Doghouse block had an average omission rate (percent of items students left blank) of about 10 percent and an average

Table 12.2
Performance on Items in the "Building a Doghouse" Block: Grade 8

Item Description	Percent Correct
1. Determine which of five measurements are needed to decide whether the doghouse will be large enough for a dog.	55[a]
2. Choose the minimum number of pieces (four walls, two roof pieces, and floor) that need to be measured before being cut.	42
3. Obtain measurements in inches for three dimensions on the model of the doghouse (longer side of rectangular floor; shorter side of rectangular floor; height from floor to peak of roof).	46
4. Given a scale, explain what you would do to each measurement in item 3 to find the measurements in feet of the actual doghouse.	14
5. Given five scales, choose the scale that would produce the largest doghouse.	35
6. Given a scale, determine how many times as tall the actual house will be as compared to the height of the model.	35
7. Locate a door on the doghouse according to given conditions.	19
8. Using graph paper, show how the seven doghouse pieces can be cut from plywood sheets so that the *fewest* sheets are used.	48
9. Determine the decrease in the actual height of the doghouse when the pitch of the roof is decreased from 53° to 30°.	2
10. Determine the largest area that can be enclosed with 36 feet of fencing; support answer by showing work.	<1[b]

[a]Percent of students answering all five statements correctly.
[b]Percent of students scoring at either the satisfactory or extended level.
Note: Copies of the items described in the table appear in appendix B. Item 1 was a multiple-response item, item 2, 5, and 6 were multiple-choice items, items 9 and 10 were extended constructed-response items, and the rest were short constructed-response items.

off-task rate of about 2 percent for the constructed-response questions. The extended constructed-response items had the highest omission rates, and the multiple-choice items had the lowest rates. The average omission rate of 10 percent for this theme block was lower than the 18 percent omission rate for main NAEP, suggesting that the thematic link between items may have engaged students in ways that encouraged them to try to answer the questions. However, when compared to the 3 percent omission rate for the fourth-grade theme block, the rate for the Building a Doghouse block was nearly three times as high. This pattern of fourth-grade students being more willing than eighth-grade students to try to answer the items is the same as that in main NAEP. The high omission rates for the last two items (24 percent for item 9 and 27 percent for item 10) suggests that these items were particularly difficult for students or that the 30-minute time limit may not have been adequate for students, or both.

The next sections contain descriptions of performance on two items in the Building a Doghouse block. We focus on these items because performance on them could be linked to performance on other items from prior NAEP mathematics assessments or to discussions in other chapters in this volume. Additional information about performance on these items was obtained by examining student work from a convenience sample of 144 responses to items in this theme block.

Obtaining Measurements from a Three-Dimensional Model

Results from the 1992 NAEP mathematics assessment showed that nearly 80 percent of the eighth-grade students correctly obtained linear measurements using a ruler (Kenney and Kouba 1997). However, these measurements, such as measuring the diagonal or longest side of a given rectangle, were done on two-dimensional geometric figures. One item in the Building a Doghouse block, described as item 3 in table 12.2 and reproduced in appendix B (pp. 369), asked students to measure specific dimensions on the scale-model doghouse they had assembled from the two pushout pieces A and B. [The pushout pieces, A and B, appear in appendix B, pp. 373.] On the model, the measurements of the longer side of the rectangular floor and the height from the floor to the peak of the roof were each 2 inches; the measurement of the shorter side of the floor was 1 1/2 inches. Results showed that 46 percent of students in grade 8 correctly determined all three measurements, and about 70 percent got at least one measurement correct.

The way in which NAEP results are reported did not identify which of the three measurements gave students the most difficulty. To gain insight on which measurements were more difficult to obtain than others, we examined a sample of 144 responses to this item from students in the Theme Study sample. Again, these responses were from a convenience sample of responses to be used for illustrative purposes; they do not necessarily represent the full range of responses on any particular item in the theme blocks. We found that the measurements for the lengths of the longer and shorter sides of the floor were obtained by most students in that sample, but just over half correctly measured the height from the floor to the peak of the roof. This suggests that the correct measurement of the distance from floor to roof was somewhat more difficult to obtain than the other two measurements. But why was obtaining this measurement more difficult? One possible answer may be found in the handling of the model of the doghouse, which was not taped together. If students tried to obtain the measurements using the assembled model, they may have had difficulty simultaneously holding the model and using the ruler. Measuring the dimensions of the bottom of the model was probably not as difficult as measuring the front part of the model from the bottom to the peak of the roof. If students realized that the untaped model could be disassembled and the pieces flattened, then measuring the front part of the model, from the bottom to the peak of the roof, became much easier.

Maximum Area

Many middle school students have difficulty with the concepts of area and perimeter. In fact, results from the 1992 NAEP mathematics assessment show that many eighth-grade students have an incomplete conceptual understanding of area, sometimes confuse area and perimeter, and have difficulty applying area concepts to complex situations (Kenney and Kouba 1997). Recent curricular projects have included more experiences in which students contrast the two measures and investigate important mathematical ideas such as how figures with the same perimeter can have different areas.

The last question in the Building a Doghouse block, an extended constructed-response task described as item 10 in table 12.2 and reproduced in appendix B (pp. 372), involves the interplay between a constant perimeter and a changing area. In particular, this item asked students to determine the maximum area that can be enclosed with 36 feet of fencing. This problem has appeared in one form or another in a variety of curriculum materials and textbooks for the middle school. However, performance on this item suggests that students in the theme block sample were not at all familiar with this kind of problem, in that fewer than 1 percent of the students scored at the extended or satisfactory levels, nearly 40 percent gave an incorrect response, and nearly one-third omitted this item.

In at least one important way, however, the NAEP results for this item could represent an underestimate of student performance. In particular, the scoring guide was based on the requirement that in order to be scored as "extended" or "satisfactory," a response had to contain *all* possible integer length and width combinations or at least show substantial progress toward identifying all possible combinations. An example of such a response at the extended level appears as response 1 in figure 12.2.

But where does this requirement to produce all possible combinations leave the student who has done this problem before in mathematics class and who is convinced that the square will always produce the largest area when compared to all other four-sided figures? Such a student might produce an answer like response 2 in the figure. According to the NAEP scoring guide, this response would probably be scored as a "partial" response. Thus, despite the explicit directions about "showing work that will convince Julie that your area is the largest," students may not have completely understood the need to show all or most of the nine rectangular figures that have a perimeter of 36 feet, and the areas of those rectangles, or seen that just saying that "a 9×9 foot square with an area of 81 square feet" was not sufficient.

Julie wants to fence in an area in her yard for her dog. After paying for the materials to build her doghouse, she can afford to buy only 36 feet of fencing.

She is considering various shapes for the enclosed area. However, she wants all of her shapes to have 4 sides that are whole number lengths and contain 4 right angles. All 4 sides are to have fencing.

What is the largest area that Julie can enclose with 36 feet of fencing?

Support your answer by showing work that would convince Julie that your area is the largest.

Response 1

The largest area would be a square with sides of 9 ft. which would allow for 81 ft.² of area.

This is because of the following:

Width	Length	Area	Perimeter
9	9	81	36
10	8	80	36
11	7	77	36
12	6	72	36
13	5	65	36
14	4	56	36
15	3	45	36
16	2	32	36
19	1	17	36

Response 2

A Square 9 feet on each side is the largest because The area is 81 feet² and every other way tried was 80 feet and lower.

Fig. 12.2. Sample responses to the maximum area item

Thus, we are left to wonder what portion of the 29 percent of students who produced partial responses may have actually understood the problem.

Summary of the "Building a Doghouse" Block

NAEP results from the Building a Doghouse block provide evidence that the context of building a doghouse was somewhat engaging to students in grade 8. In particular, the low omission and off-task rates for the short constructed-response items suggest that most students at least tried to answer some of the questions. However, the response pattern for the last two items in this block was similar to that for extended constructed-response items in NAEP: that is, about 25 percent of the students who provided answers to the first eight items in the block did not try to answer the extended questions. This suggests that the 30-minute time period may not have been long enough for students to work on the ten items. However, despite the fact that eighth-grade students tried to answer most of the items in this theme block, performance on most items in the block was very low. This is particularly true for items that assessed students' understanding of ratio as applied to a scale, that involved concepts of area and perimeter, or that required students to provide a written explanation.

Buying a Car

One of the two theme blocks administered to twelfth-grade students was set in the context of buying and leasing a car (hereafter referred to as "Buying a Car"). The seven items in the Buying a Car block are shown in appendix C (pp. 374–76). The first page of the block sets the context for the items, all of which involved aspects of purchasing or leasing a car, such as the down payment amount, interest rate, and monthly payments on a loan. The items consisted of a variety of formats including two multiple-choice items, three short constructed-response items, and two extended constructed-response items, one of which was scored at five levels. All seven items assessed concepts in the NAEP content strands of Number Sense, Properties, and Operations (particularly, percents) or Algebra and Functions (particularly, evaluation of algebraic expressions and solving equations), and most items were classified in the ability category Procedural Knowledge. Twelfth-grade students were provided with scientific calculators, but no additional materials, such as manipulatives, models, or ruler/protractors were needed to solve the problems in this theme block.

Overall Performance on the Items in the "Buying a Car" Block

Table 12.3 contains a description of each item and performance results representing the percent-correct results for all items except the extended constructed-response task using five score levels. For that task, the percent of students scoring at either the satisfactory or extended level is given. Performance was high, about 80 percent correct, on the first three items. These items assessed procedures that should be familiar to students in grade 12, such as calculating a simple percent (item 1), finding the total amount paid for a car (item 2), and determining the difference between the amount paid for a car and the selling price (item 3). The other four items in the block were constructed-response questions involving multiple calculations, complex formulas, or algebraic expressions. Performance on these items was much lower than on the first three items, ranging from 34 percent correct on item 4 to 16 percent on item 7. On the extended constructed-response item (item 5), however, the 23 percent of students who scored at either the satisfactory or extended level was higher than performance on most extended questions given at grade 12 in main NAEP.

Table 12.3
Performance on Items in the "Buying a Car" Block: Grade 12

Item Description	Percent Correct
1. Choose the correct amount for a down payment of 20 percent based on a car's selling price of $16,500.	82
2. Choose the correct total amount paid for a car based on monthly payments and the down payment.	83
3. On the basis of the answer in item 2 above, find the difference between the total amount paid for the car and its original selling price.	80
4. Calculate the amount to be financed for a two-year lease on a car.	34
5. Given a formula for calculating a monthly car payment, determine the total cost of buying a particular car; show appropriate work.	23[a]
6. Given appropriate algebraic expressions, find the amount saved by leasing the car instead of buying it; show how answer was obtained.	27
7. Given appropriate algebraic expressions, find the selling price of a car for which the leasing cost and the buying cost are the same; show how answer was obtained.	16

[a]Percent of students scoring at either the satisfactory or extended level.
Note: Copies of the items described in the table appear in appendix C. Items 1 and 2 were multiple-choice items, items 3 and 4 were short constructed-response items, and the rest were extended constructed-response items.

With respect to the percent of omitted responses, the Buying a Car block items had an average omission rate of about 5 percent, with even the omission rates for both the short and extended constructed-response items in single digits. The average omission rate was much lower than the 24 percent rate for main NAEP at grade 12, and the off-task rate was less than 5 percent, suggesting again that the thematic link between items may have engaged students in ways that encouraged them to try to answer the questions.

There were no items in the Buying a Car block that lend themselves to further discussion in this chapter. One reason for this has to do with the rather high performance on the first three items in this block. Students did well on these simple, rather straightforward items, with percent-correct performance at about 80 percent. The kinds of errors that were revealed through distracter analysis for the multiple-choice questions, or analyses of incorrect responses, to the short constructed-response questions were not surprising. For the most part, students who gave incorrect answers were making computational errors or were completing only one part of a two-part problem.

Another reason for not analyzing these items further has to do with the procedural nature of the other four constructed-response items. None of the items asked students to provide an explanation, but instead they all required students to use a definition ("residual value" in item 4), a given formula (item 5), or given algebraic expressions (items 6 and 7) to solve particular problems. The procedural nature of the constructed-response items and the scoring standards based on the answers and work shown likely contributed to error patterns that were not unexpected. Some of the error patterns included answers without accompanying work, answers that included computational errors, and attempts to use the numbers given in the item but in a disorganized way.

Summary of the "Buying a Car" Block

A consistent pattern that has emerged from the omission and off-task rates across the fourth-grade, eighth-grade, and now the twelfth-grade theme blocks is that there is something about the theme blocks that encouraged students to try the items. Perhaps it is the context of buying a car that is appealing to students in grade 12, the 30-minute time period, or a combination of both. Performance results in this theme block suggest that students in grade 12 can execute simple procedures and solve single-step problems. However, as formulas become more complex (as in item 5) or as algebraic notation is introduced (as in items 6 and 7), performance levels decreased.

THE THEME STUDY: WHAT VALUE WAS ADDED TO THE 1996 NAEP MATHEMATICS ASSESSMENT?

The Theme Study was an innovation in the 1996 NAEP mathematics assessment in that it was the first attempt in recent NAEP assessments to organize an entire block of items around a unifying context, or theme, and to allow students more time (30 minutes) to work on the items. It was thought that relating the items thematically and in a real-world context might be more interesting and motivating to students. Also, by giving students twice as much time to work on the theme block items than the 15-minute time limit in the main assessment, NAEP had the opportunity to include items that were thought to be more detailed and complex.

There is evidence from the theme block results that the thematic link between items and the allotment of more time had some effect on student motivation to work on the items. In particular, when the theme blocks and main NAEP were compared on the basis of omission rates, the theme block rates were lower for all three grade levels. This suggests that students in the theme block samples tried to answer the items, although their performance on some items, especially the extended constructed-response items, was at very low levels.

In addition to learning that context and time may have increased students' willingness to try problems, it is important to consider other findings that are suggested by students' performance on the theme blocks. What value did this special study add to the 1996 NAEP mathematics assessment? After studying the report on the Theme Study (Mitchell et al. 1999) and after looking at the items and performance results ourselves, our conclusion is that the additional benefits of doing the special study of mathematics-in-context were limited to the confirmation of results that were already available from performance on main NAEP. For example, it is well documented that students at all three grades have difficulty explaining their answers in writing (for example, Dossey, Mullis and Jones 1993), and the results from the Theme Study confirmed this. Also, the theme block results provided further evidence that students' performance on single-step problems (for example, calculating a simple percent) is higher than that on multistep problems, a finding that pervades analyses of NAEP results (for example, Kouba, Zawojewski, and Strutchens 1997).

Thus, the Theme Study provided few, if any, completely new insights into how students solve complex problems. Perhaps we would have learned more about complex problem solving if the theme block items had been developed as true "item families." The 1996 NAEP mathematics framework document recommends that the assessment include families of items, to "measure the breadth and depth of student knowledge in mathematics" (National Assessment Governing Board [NAGB] 1994, p. 5), and two kinds of families are proposed: vertical and horizontal. We conclude this chapter by discussing the potential benefits of reconceptualizing the theme blocks as either vertical or horizontal families of items for future NAEP mathematics assessments.

According to NAEP, vertical families include items or tasks that measure students' understanding of a single important mathematics concept in a content strand but at different levels, such as giving a definition, applying the concept in both familiar and novel settings, and generalizing knowledge about the concept to represent a new level of understanding. This description of a vertical family is reminiscent of the description of the "superitems" developed according to the Structure of the Learned Outcomes (SOLO) taxonomy (Collis, Romberg and Jurdak 1986; Romberg, Zarinnia, and Collis 1990; Wilson and Chavarria 1993). In the SOLO taxonomy, superitems involve the assessment of an important mathematics concept (for example, symmetry, area and perimeter, functions) in a hierarchical fashion; that is, the set of questions progresses from lower-level concepts such as using definitions or doing simple calculations to higher-order thinking such as generalizing the results from the previous questions. Analysis of students' responses to such superitems can provide insights into the level of students' problem-solving ability. In the mathematics education literature (for example, Wilson 1990), there are a number of examples of superitems developed according to the SOLO taxonomy and they can provide NAEP test developers with ideas for creating vertical item families.

The other kind of item family described in the NAEP mathematics framework—horizontal family—involves understanding of a concept or principle across the various content strands in NAEP within a grade level or across grade levels. For example, the concept of ratio and proportion can be assessed in a variety of contexts such as number, measurement, geometry, probability, and algebra. The Building a Doghouse block at grade 8 contains some items that represent a first step in developing a horizontal family. In particular, about half of the items in that block involved the applied use of ratio as a scale between the model doghouse and the actual doghouse across content topics in the NAEP strands of Measurement and Geometry and Spatial Sense. It was unfortunate that the results from these items were not analyzed in any connected way. For example, NAEP results did not link performance on the Doghouse block items that had students measure three dimensions of the model doghouse (see item 3 in appendix B, p. 369) and then explain how they would transform those measurements into measurements on the actual doghouse (see item 4, appendix B, p. 369).

As was the case for vertical families, it would be wise for NAEP test developers to consider examples from the mathematics education literature that have some of the same features described in the NAEP mathematics framework for horizontal item families. In their framework for assessing conceptual understanding, Zawojewski and Silver (1998) propose that one way to ascertain the robustness of students' understanding of important mathematical concepts is to vary the context (for example, content topics such as measurement, geometry, and number) in which the concept is presented. Their examples of "constellations," or collections of tasks that assess related aspects of a target concept might be useful to future development of related items in NAEP.

From the results reported in this chapter, it appears that the theme-block format based on a thematically-related set of items is a promising direction for the NAEP mathematics assessment. However, there is evidence that the theme blocks need refining before they are repeated in future NAEP assessment. In particular, although the notion of building the items around a common theme and within a real-world context could be retained, the view of these related sets of items should be expanded so that other connections exist between items (Kenney 2000). The scoring of the items also needs to be structured so that results from one item can be related to results from other items. By linking items in ways that go beyond context, the blocks can evolve into true item families that have the potential to contribute in important ways to identifying what American students know and can do in mathematics, particularly in the area of complex problem solving.

REFERENCES

Collis, Kevin F., Thomas A. Romberg, and Murad E. Jerdak. "A Technique for Assessing Mathematical Problem-Solving Ability." *Journal for Research in Mathematics Education* 17 (May 1986): 206–21.

Dossey, John A., Ina V. S. Mullis, and Chancey O. Jones. *Can Students Do Mathematical Problem Solving? Results from Constructed-Response Questions in NAEP's 1992 Mathematics Assessment.* Washington, D.C.: National Center for Education Statistics, 1993.

Kenney, Patricia Ann. "Families of Items in the NAEP Mathematics Assessment." In *Grading the Nation's Report Card: Research from the Evaluation of NAEP,* edited by James Pellegrino, Lee R. Jones, and Karen J. Mitchell, pp. 5–43. Committee on the Evaluation of the National and State Assessments of Educational Progress, Board on Testing and Assessment. Washington, D.C.: National Academy Press, 2000.

Kenney, Patricia Ann, and Vicky L. Kouba. "What Do Students Know about Measurement?" In *Results from the Sixth Mathematics Assessment of the National Assessment of Educational Progress,* edited by Patricia Ann Kenney and Edward A. Silver, pp. 141–63. Reston, Va.: National Council of Teachers of Mathematics, 1997.

Kouba, Vicky L., Judith S. Zawojewski, and Marilyn E. Strutchens. "What Do Students Know about Numbers and Operations?" In *Results from the Sixth Mathematics Assessment of the National Assessment of Educational Progress,* edited by Patricia Ann Kenney and Edward A. Silver, pp. 87–140. Reston, Va.: National Council of Teachers of Mathematics, 1997.

Mitchell, Julia H., Evelyn F. Hawkins, Frances B. Stancavage, and John A. Dossey. *Estimation Skills, Mathematics-in-Context, and Advanced Skills in Mathematics: Results from Three Studies of the National Assessment of Educational Progress 1996 Mathematics Assessment.* Washington, D.C.: National Center for Education Statistics, 1999.

National Assessment Governing Board. *Mathematics Framework for the 1996 National Assessment of Educational Progress.* Washington, D.C.: National Assessment Governing Board, 1994.

National Council of Teachers of Mathematics. *Professional Standards for Teaching Mathematics.* Reston, Va.: National Council of Teachers of Mathematics, 1991.

Reese, Clyde M., Karen E. Miller, John Mazzeo, and John A. Dossey. *NAEP 1996 Mathematics Report Card for the Nation and the States: Findings from the National Assessment of Educational Progress.* Washington, D.C.: National Center for Education Statistics, 1997.

Reys, Robert E., Marilyn N. Suydam, and Mary Montgomery Lindquist. *Helping Children Learn Mathematics.* 4th ed. Boston: Allyn & Bacon, 1995.

Romberg, Thomas A., E. Anna Zarinnia, and Kevin F. Collis. "A New World of Assessment in Mathematics." In *Assessing Higher Order Thinking in Mathematics,* edited by Gerald Kulm, pp. 21–38. Washington, D.C.: American Association for the Advancement of Science, 1990.

Wilson, Linda Dager, and Silvia Chavarria. "Superitem Tests as a Classroom Assessment Tool." In *Assessment in the Mathematics Classroom,* 1993 Yearbook of the National Council of Teachers of Mathematics, edited by Norman L. Webb, pp. 135–42. Reston, Va.: National Council of Teachers of Mathematics, 1993.

Wilson, Mark. "Investigation of Structured Problem-Solving Items." In *Assessing Higher Order Thinking in Mathematics,* edited by Gerald Kulm, pp. 187–203. Washington, D.C.: American Association for the Advancement of Science, 1990.

Zawojewski, Judith A., and Edward A. Silver. "Assessing Conceptual Understanding." In *Classroom Assessment in Mathematics: Views from a National Science Foundation Working Conference,* edited by George W. Bright and Jeane M. Joyner, pp. 287–95. Lanham, Md.: University Press of America, 1998.

Appendix A: The Butterfly Booth (Grade 4)

All pictures and other materials have been reduced from their original sizes.

Use the packet you have been given to help you answer the questions in this section. (Authors' note: Contents of packet appears on p. 347)

Each class in Oakville School will have a booth at the Science Fair. Your class is planning to have a Butterfly Booth.

Your class has a lot to do to get ready for the Science Fair. You need to make decorations for the booth, plan activities, and order materials.

1. The butterfly booth will be decorated with butterfly drawings. Draw only the missing markings on each picture to make each butterfly symmetrical.

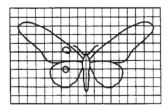 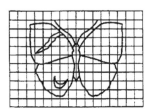

2. **Take the Butterfly Information Sheet from your packet.** (Authors' note: This sheet appears on p. 367.)

   On the Butterfly Information Sheet the wingspan of the Monarch butterfly is shown.

   Use your ruler to measure the <u>wingspans</u> of the other two butterflies on the sheet, the Black Swallowtail butterfly and the Common Blue butterfly, to the nearest centimeter.

   Black Swallowtail   Wingspan: _____ centimeters
   Common Blue   Wingspan: _____ centimeters

3. **Take the butterfly cutouts from your packet.** (Authors' note: Pictures of these cutouts appear on p. 367.)

   What is the greatest number of Common Blue butterflies that can be stored in the case below? (When you put butterflies in the case, you can't stack them. The butterflies can touch, but they can't overlap at all.)

   Answer: _____

   Show how the butterflies fit in the case.

   Storage Case

   12 centimeters

   6 centimeters

   How many storage cases would you need to store 28 Common Blue butterflies?

   Answer: _____

   Use drawings, words, or numbers to explain how you got your answer.

4. The children who visit your booth are going to build models of butterflies. For each model, they will need the following:

   4 wing pieces   1 body   2 antennae

   When the model is put together, it looks like this:

   If the class has a supply of 29 wings, 8 bodies, and 13 antennae, how many complete butterfly models can be made?

   Answer: _____

   Use drawings, words, or numbers to explain how you got your answer.

5. A fourth-grade class needs 5 leaves each day to feed its 2 caterpillars. How many leaves would they need each day for 12 caterpillars?
Answer: _____

Use drawings, words, or numbers to show how you got your answer.

6. **Use the Butterfly Information Sheet and your answer from question 2 to solve this question.**

   Your class has decided to have a banner that will be 130 centimeters long. This banner will have a repeating pattern of one Monarch butterfly followed by two Black Swallowtail butterflies, as shown here.

   This part keeps repeating across the banner.

   The butterflies will just touch but will not overlap.

   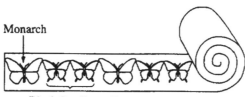

   How many of each type of butterfly are needed for the banner?

   Monarch _____

   Black Swallowtail _____

   Show how you got your answers.

Packet Contents for the Butterfly Booth Theme Block

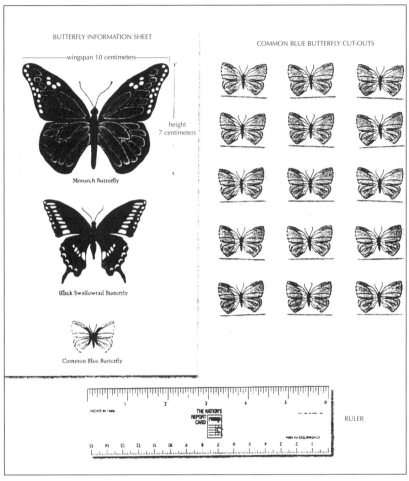

Source: Mitchell et al. (1999, p. 32).

Appendix B: Building a Doghouse (Grade 8)

All pictures and other materials have been reduced from their original sizes.

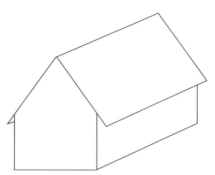

Julie wants to build a doghouse like the one shown in the picture above. She has asked you to help her build the doghouse.

The kit you have been given contains a model of a doghouse like the one Julie wants to build. (Authors' note: Kit of pushouts appears on p. 373) Please put the model together now by following these instructions.

1. Separate pieces A and B from the paper. **Do not separate any other pieces from this paper until you are told to do so.**
2. Fold up the four walls on piece A so that they form right angles with the rectangular floor.
3. Fold the roof (piece B) in half, and set it on top of the house. The edges of the roof will extend slightly beyond the walls.

Note: When Julie builds the house, the roof will be made up of two identical pieces of wood, since wood cannot be folded the way you folded the piece of paper just now to make the roof.

You may also use your calculator and ruler/protractor to help answer the questions in this part.

1. Consider each of the following measurements. Will knowing the measurement help Julie to determine whether the doghouse she plans to build will be large enough for her dog to sleep in and to go in and out comfortably? (Answer "Yes" or "No" for each part.)

a. The length of the floor	○ Yes	○ No
b. The height of the house	○ Yes	○ No
c. The weight of the house	○ Yes	○ No
d. The width of the rectangular floor	○ Yes	○ No
e. The width and height of the door's opening	○Yes	○ No

2. Seven pieces—four walls, two roof pieces, and the floor—make up the doghouse. Since some of the pieces are exactly the same in size and shape, Julie does not need to measure every piece. She can measure and cut a piece and then make identical pieces <u>without measuring</u> by tracing an outline of the cut piece onto the wood and then cutting out the traced shape.

    How many of the seven pieces does Julie need to <u>measure</u> before she cuts?

    A. Two
    B. Three
    C. Four
    D. Five
    E. Seven

3. The model of the doghouse that you put together is a smaller version of the actual house. Measure the following lengths, in inches, **on your model** and record your results in the spaces below.

Longer side of rectangular floor	_____ inches
Shorter side of rectangular floor	_____ inches
Height from floor to highest point of roof	_____ inches

    > 1 inch represents 1 1/2 feet (18 inches).

4. Explain what you would need to do to each of your measurements in question 3 to find the measurements <u>in feet</u> of the actual house.

    (You do not need to find each of the actual measurements.)

5. Of the following scales, which one would produce the largest doghouse?

    A. 2 inches on model represents 5 feet on actual house.
    B. 1 inch on model represents 3 feet on actual house.
    C. 1 inch on model represents 1 1/2 feet on actual house.
    D. 1/2 inch on model represents 1 foot on actual house.
    E. 1/2 inch on model represents 3/4 foot on actual house.

| 1 inch represents 1 1/2 feet (18 inches). |

6. The height of the actual house will be how many times as tall as the height of the model?

    A.   1 1/2
    B.   9
    C.   18
    D.   24
    E.   27

7. **You will now need piece C to answer this question. Separate piece C from the paper.** (Authors' note: Piece C appears on p. 373.)

| 2 inches represents 1 foot (12 inches). |

Piece C represents a scale model of the door for the doghouse. The front wall of the doghouse, shown below, as well as piece C has been drawn to a different scale than the one used in the previous question. The scale is 2 inches represents 1 foot.

On the drawing below, use piece C to locate the door on the wall so that it will be 1/2 foot above the floor level of the doghouse (to keep the water out) and centered exactly between the vertical edges of the wall. When you have correctly positioned the door, trace its location on the drawing. (Disregard the thickness of the wood that will be used to build the doghouse.)

8.

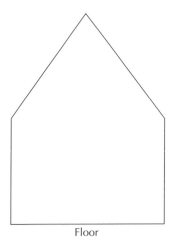

Floor

Separate the <u>remaining</u> seven pieces from the paper, and use <u>only those seven pieces</u> to help you answer the following question. (Authors' note: Pieces appear on p. 373).

> 1 inch represents 1 1/2 feet (18 inches).

Julie plans to use plywood to build her doghouse, using the scale above. The plywood is sold in rectangular sheets that are each 4 feet wide and 8 feet long. She wants to determine the fewest number of sheets that she will need.

On the grids below, the plywood sheets have been drawn to the same scale as the seven pieces. Show how the seven pieces (four walls, two roof pieces, and the floor) could be cut from the plywood sheets so that the *fewest* number of sheets are used. This should be done by tracing the pieces on the sheets.

(Note: There may be more sheets shown than you will need to use.)

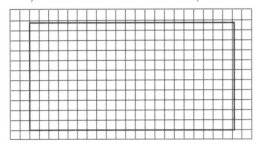

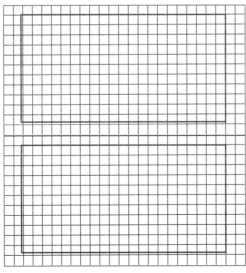

This question requires you to show your work and explain your reasoning. You may use drawings, words, and numbers in your explanation. Your answer should be clear enough so that another person could read it and understand your thinking. It is important that you show all of your work.

> 1 inch represents 1 1/2 feet (18 inches).

9. The drawing below shows a wall of Julie's doghouse. The **pitch** is defined as the slope of the roof; it can also be described as the angle formed between the roof and a horizontal line, as shown in the drawing.

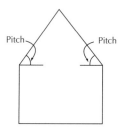

The pitch of the roof in the drawing is slightly more than 53 degrees, which is the same as the roof pitch on your model.

Julie read in a book that the best air flow inside a doghouse occurs when the roof pitch is 30°. If the height of the doghouse is measured from the floor to the highest point on the roof, by about how many feet is the actual height of the doghouse decreased when the pitch is decreased from 53° to 30°?

Show how you got your answer. (You may find it helpful to mark on the drawing.)

This question requires you to show your work and explain your reasoning. You may use drawings, words, and numbers in your explanation. Your answer should be clear enough so that another person could read it and understand your thinking. It is important that you show all of your work.

10. Julie wants to fence in an area in her yard for her dog. After paying for the materials to build her doghouse, she can afford to buy only 36 feet of fencing.

She is considering various different shapes for the enclosed area. However, she wants all of her shapes to have 4 sides that are whole number lengths and contain 4 right angles. All 4 sides are to have fencing.

What is the largest area that Julie can enclose with 36 feet of fencing?

Support your answer by showing work that would convince Julie that your area is the largest.

Sheet of Pushouts for the Building a Doghouse Theme Block

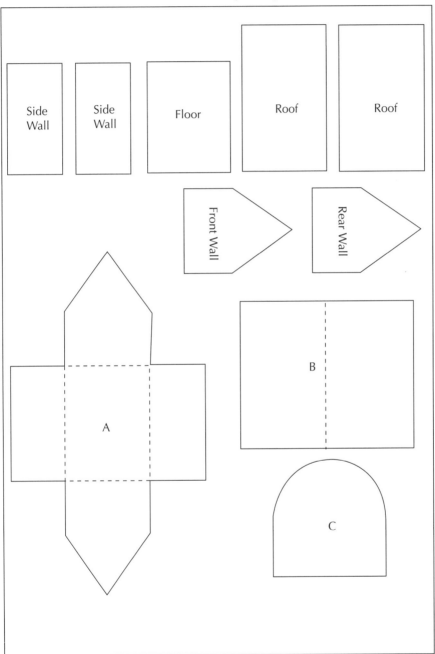

Source: Mitchell et al. (1999, p. 68).

Appendix C: Buying a Car (Grade 12)

A family is trying to decide how to pay for the new car they want. Should they buy the car with a loan or should they lease it? If they lease the car, then they must return the car to the dealer at the end of the lease period. If they buy the car, then at the end of the loan period the car belongs to them. The monthly payments for the car, whether they buy it or lease it, will depend upon the size of their down payment. If they make a big down-payment now, they will have smaller monthly payments, but they might like to use the down-payment money to buy something else. How will the interest rate and the period of time for the loan or lease affect the total amount they will pay?

All the problems in this test are related to financial questions similar to those asked above that involve buying or leasing a car.

1. Donna decides to buy a new car that is selling for $16,500. If she is required to pay 20 percent of the selling price as a down payment, what is the number of dollars required for the down payment?

    A. $330
    B. $1,650
    C. $3,300
    D. $4,125
    E. $13,200

Questions 2–3 refer to the following information.
Bill purchased a new car that was selling for a price of $19,200. He paid $4,800 as a down payment and obtained a 36-month car loan to finance the remainder of the selling price.

2. If his monthly payments were $451, what was the total amount, including the down payment, that Bill paid for the car?

    A. $11,436
    B. $16,236
    C. $19,200
    D. $21,036
    E. $24,000

Questions 2-3 refer to the following information.

Bill purchased a new car that was selling for a price of $19,200. He paid $4,800 as a down payment and obtained a 36-month car loan to finance the remainder of the selling price.

3. By how much did the total amount Bill paid for the car exceed the selling price of the car?

4. When leasing a new car, the amount to be financed is the amount left when the down payment and the residual value are subtracted from the selling price. (The residual value is the value of the car at the end of the lease period.) The residual value for a two-year-old car, whose original selling price was $18,700, is estimated to be 53 percent of the selling price. If the amount of the down payment is $1,500, what is the number of dollars to be financed for a two-year lease? Show the work that led to your answer.

This question requires you to show your work and explain your reasoning. You may use drawings, words, and numbers in your explanation. Your answer should be clear enough so that another person could read it and understand your thinking. It is important that you show all of your work.

5. A formula used to calculate a monthly car payment is

$$MP = A \left[\frac{\left(\frac{r}{12}\right)\left(1 + \frac{r}{12}\right)^n}{\left(1 + \frac{r}{12}\right)^n - 1} \right]$$

where *MP* is the monthly payment,
        A is the number of dollars to be financed,
        r is the annual percent rate (e.g., if the annual percent rate is 7%, then *r* is 0.07), and
        n is the length of the loan in months.

Use the formula shown above to help calculate the <u>total cost</u> of purchasing an $18,000 car if there is a down payment of $4,000 and the remaining $14,000 is to be financed at an annual percent rate of 8 percent for 48 months.

Show the work that led to your answer.

Note: in your calculations, use 1.3757 as the approximate value of $\left(1 + \frac{0.08}{12}\right)^{48}$.

Questions 6–7 refer to the following information.

Mary is interested in leasing or buying a particular car model. She has determined that the cost of leasing this car for 4 years is $0.78p + 450$, where p is the selling price. Mary has also determined that the cost of buying this car, including finance charges, is $1.2p + 80$. At the end of 4 years, when the car is paid off, it can be sold for $0.4p$.

6. Mary decided to buy the car at a selling price of $20,000. She kept the car for 4 years and then sold the car to recover some of the costs. If Mary had decided to lease the same car for 4 years, what amount of money would she have saved? Show the work that led to your answer.

7. For the particular car model that Mary is interested in, determine the selling price for which the car costs the same to buy and resell at the end of 4 years as its costs to lease the car for 4 years. Show the work that led to your answer.

13

The Performance of Students Taking Advanced Mathematics Courses

Jeremy Kilpatrick and Judith Lynn Gieger

AN ADEQUATE assessment of what American students know and can do in mathematics is complicated by the differentiation of the curriculum that occurs around grade 8, if not before. After grade 7, some students proceed on to challenging courses whose content spans many areas of mathematics, including algebra, geometry, statistics, trigonometry, and elementary functions. Others follow a narrower track, usually repeating work in the arithmetic of whole numbers, fractions, and decimals or taking courses in consumer mathematics, technical mathematics, and the like. By grade 12, some students are studying advanced calculus, college-level statistics, or computer science; others are taking less advanced courses; and many are enrolled in no mathematics course at all.

This differentiation of the mathematics curriculum makes it difficult to assess the performance of secondary school students in the United States and to compare it with that of students in other countries—in many of which all students study the same mathematics in secondary school. Because U.S. students follow so many different tracks through and out of mathematics, designers of tests intended for representative state or national samples of secondary students are compelled to exclude topics being studied by students who are advanced in mathematics because questions on those topics would

---

We are grateful to Jeff Haberstroh (Educational Testing Service) and John Dossey (Illinois State University) for their special assistance, and to the American Institutes for Research for providing us with a draft copy of their report on the Advanced Study. We would also like to thank Vilma Mesa of the University of Georgia, who helped with the comparison of TIMSS and NAEP items.

be unsuited to the majority of the other students. So as not to be unfair, even tests designed for students presumably prepared for college, such as the Scholastic Assessment Test I and the American College Tests, seldom include content beyond the rudiments of algebra and geometry.

The National Assessment Governing Board (NAGB) and the National Center for Education Statistics (NCES) decided to undertake a special study as part of the 1996 NAEP mathematics assessment "to extend and enhance the information from the national main mathematics and science assessments" (Ballator 1996, p. 39). For the first time, a collection of items would be used that were especially suited to those students in grades 8 and 12 who were advanced in mathematics. In the technical proposal for the 1996 NAEP assessment, the Educational Testing Service (ETS) spelled out the argument against attempting to assess the performance of such students within the main assessment itself:

> An important goal for an assessment is to learn what the best prepared students know and can do within a subject area. Unfortunately, many of the students in our country have not been exposed to the curriculum and instruction required to be able to answer questions based on the knowledge acquired by taking advanced mathematics courses. Thus, to include many items measuring topics in elementary algebra (at grade 8) and in statistics, advanced algebra, and elementary functions including trigonometry (at grade 12), in the main assessment would be inefficient, since most of the assessed students would not be capable of answering such items. Further, the motivation of these less well-prepared students might be negatively affected. (NAEP 1996, p. 104)

In 1993, as preparation for the special study (hereafter referred to as the Advanced Study), ETS developed, field tested, and refined two assessments, one at grade 8 and the other at grade 12. No special frameworks or detailed specifications were developed for these assessments. They were designed simply to provide information on performance in algebra at grade 8 and in precalculus mathematics at grade 12.

This chapter provides information on how the students were chosen for the 1996 NAEP Advanced Study, on some characteristics that distinguished them from students who did not meet the criteria for selection, and on the assessment procedures. It then discusses the performance of the Advanced Study students at each grade, relating it to the performance of students in the Third International Mathematics and Science Study (TIMSS).

STUDENT CHARACTERISTICS AND ASSESSMENT PROCEDURES

The criteria for the Advanced Study at grade 8 were that students had to be enrolled in an elementary algebra course (for example, first-year algebra,

Highlights

- At grade 8, students taking algebra or a course beyond algebra were more likely to be White, and less likely to be Hispanic, than were students not taking such courses. At grade 12, the pattern of racial/ethnic differences had shifted so that students taking a precalculus or calculus course were more likely to be Asian or Pacific Islander, and less likely to be Black, than were students not taking such courses.

- At grade 8 but not at grade 12, students taking advanced mathematics courses were more likely to be female than were students not taking such courses. At both grades, taking advanced mathematics courses was associated with various forms of socioeconomic advantage.

- Students advanced in mathematics at grade 8 performed at much higher levels than eighth-grade students in the main NAEP assessment on a set of items dealing with concepts of elementary algebra. On items not taken by students in the main assessment, the performance of the advanced students was mediocre to low. Items asking for the interpretation and use of linear relationships presented in context were especially difficult, as were items asking students to provide explanations or justifications for their reasoning. High percentages of omissions and irrelevant comments on the items suggested a lack of motivation.

- Students advanced in mathematics at grade 12 performed at much higher levels than twelfth-grade students in the main NAEP assessment on a set of items dealing with algebra, functions, geometry, and statistics. Performance on the items in the twelfth-grade Advanced Study tended to be low, especially on the items dealing with algebra and functions and on other items requiring reasoning and justification. Lack of motivation may have been a more serious problem than at grade 8.

- At grade 8 the performance of advanced students in the United States appeared to be at roughly the same level as that of average students in high-achieving countries in the Third International Mathematics and Science Study (TIMSS). At grade 12, comparison with advanced students in other countries was complicated by differences in the assessments used, but U.S. performance was low on both the TIMSS and the NAEP Advanced Study assessments.

algebra 1) or a more advanced course (for example, geometry). In the 1990 NAEP mathematics assessment, 16 percent of the students in grade 8 had reported taking an algebra course, and that number had grown to 20 percent by 1992 (Dossey et al. 1994, p. 125). At the time of the 1996 assessment, the number was about 25 percent (Mitchell, Hawkins, Jakwerth et al. 1999, p. 214). The most common placement of algebra, however, is still at grade 9. Consequently, students in the Advanced Study sample at grade 8 can be considered to have been advanced beyond the usual eighth-grade mathematics course.

The criteria were rather different for the Advanced Study at grade 12. Students had to be taking one of the following courses: precalculus, calculus and analytic geometry, calculus, or Advanced Placement (AP) calculus. The first course is a college preparatory course taken by twelfth-grade students but offered at lower grades as well. It is not an advanced course for students in grade 12. Only about 15 percent of U.S. students, however, complete grade 12 having taken a course in precalculus or calculus (John Dossey, personal communication, 31 July 1998; in contrast, see Mitchell, Hawkins, Jakwerth et al. 1999, p. 223, for data suggesting that 15 percent may be an underestimate). Other advanced courses, such as AP computer science and AP statistics, were not included on the list. The twelfth-grade students in the Advanced Study sample may be described as students who were enrolled in one of several college-preparatory mathematics courses offered in the last year of high school.

When the samples of students were selected for the 1996 NAEP mathematics assessment, a sample meeting the above criteria was drawn at each of grades 8 and 12. These were the samples used in the Advanced Study. Each sample for the main NAEP assessment at grades 8 and 12 included a nationally representative group of students who were eligible for the Advanced Study but were not selected as well as a nationally representative group of students who were not eligible.

Some comparisons in this chapter are between students in the Advanced Study sample and that group of students in the main NAEP sample who were not eligible for the Advanced Study. Such comparisons are useful in examining factors associated with taking advanced courses. Other comparisons—those dealing with performance on items included in both the Advanced Study and the main NAEP assessments—are between students in the Advanced Study sample and all of the students in the main NAEP sample. These comparisons allow the patterns of performance observed in the Advanced Study sample to be examined in light of patterns identified in other chapters of this book. In this chapter, data are not reported separately on the students in the main sample who were eligible for the Advanced Study but were not selected. Because the Advanced Study sample and the subsample of eligible students in the main sample were selected on the same criteria, differences in their demographic characteristics ought not to have been greater than chance. Additional data on demographic characteristics can be found in chapter 4 of Mitchell, Hawkins, Stancavage, and Dossey (1999).

Student Characteristics at Grade 8

The eighth-grade students in the Advanced Study differed from those students not eligible for the study in some characteristics and not in others. Table 13.1 contains demographic data at grade 8 with respect to self-identified race/ethnicity and gender. Students who were White comprised a significantly greater percent of students in the Advanced Study than of those who were not eligible, and students who were Hispanic comprised a significantly smaller percent of students in the Advanced Study than of those not eligible. The percents of Blacks and Asian/Pacific Islanders were not significantly different between the two groups. Thus, at grade 8 some racial/ethnic groups showed differential participation in, which likely resulted from differential access to, advanced mathematics courses such as algebra. There were also group differences by gender. The percent of students in the Advanced Study who were female was slightly but significantly greater than the percent of those not eligible who were female. This result suggests a gender difference in participation in advanced mathematics courses at grade 8 that may be related to the often reported tendency for female students to have received somewhat higher mathematics grades than their male counterparts in the elementary and middle school years (Kimball 1989, p. 199; Sadker and Sadker 1994, p. 156).

Table 13.1
Percent of Eighth-Grade Students by Demographic Category in the Advanced Study versus Those Not Eligible

Demographic Category	Advanced Study	Not Eligible for Advanced Study
Race/Ethnicity		
White	71	62
Black	14	18
Hispanic	6	13
Asian/Pacific Islander	6	6
Gender		
Male	48	53
Female	52	47

There were also some striking and significant differences in a variety of socioeconomic characteristics between the eighth-grade students in the Advanced Study and those not eligible for the study. In brief, when students in the Advanced Study were compared with those ineligible for it, the advanced students were less likely to participate in the Federal Free/Reduced-Price Lunch program, less likely to be in Title I programs, more likely to have parents who graduated from college, and more likely to report the presence in their home of large numbers of educational items such as newspapers, books, and encyclopedias than the students who were not eligible (Mitchell, Hawkins, Stancavage, and Dossey 1999; tables 4.1 and 4.2, pp. 141–42). These differences suggest that participation in advanced mathematics courses at grade 8 is associated with various forms of socioeconomic advantage that may well be related to a school's resources as well as to its racial/ethnic composition.

Student Characteristics at Grade 12

As at grade 8, the students in the Advanced Study at grade 12 differed in some respects and not in others from their counterparts who were not eligible. Table 13.2 contains demographic data for twelfth-grade students. At grade 12, the pattern of difference for the race/ethnicity categories was different from that at grade 8. The percent of twelfth-grade students in the Advanced Study who were White was greater than the percent not eligible for the study who were White. The difference was smaller than the one at grade 8 and was not statistically significant. Further, the corresponding difference in the opposite direction for Hispanic twelfth-grade students was not significant. The significant differences at grade 12 were for the Black and the Asian/Pacific Islander students. Black students were less than half as likely, and Asian/Pacific Islanders more than twice as likely, to have been in the Advanced Study than to have been ineligible for it. Thus, from grade 8 to grade 12, participation in advanced mathematics courses appears increasingly to advantage Asian/Pacific Islander students and to disadvantage Black students. Specifically, the percent of Asian/Pacific Islander students in the Advanced Study across the two grades rose from 6 to 10 percent, and the percent of Black students in the Advanced Study across the two grades fell from 14 to 7 percent. Also, the percent of White students in the Advanced Study rose slightly from 71 to 74 percent between grades 8 and 12, and the percent of Hispanic students rose slightly from 6 to 8 percent. As these differences grow, the advantage for White students and the disadvantage for Hispanic students decrease. With respect to gender differences, the greater percent of female eighth-grade students in the Advanced Study compared with the percent who were ineligible disappeared at grade 12. Instead, the data show a slight, but nonsignificant, tendency for the percent of twelfth-grade students taking advanced courses who are male to be greater than that of those who were not.

Much the same pattern of difference in socioeconomic characteristics between students in the Advanced Study and those who were not eligible appeared at grade 12 as at grade 8 (Mitchell, Hawkins, Stancavage, and Dossey 1999; tables 4.13 and 4.14, pp. 165–66). Except for the data on Title I participation, which was very low at grade 12, the results show the same

Table 13.2
Percent of Twelfth-Grade Students by Demographic Category in the Advanced Study versus Those Not Eligible

Demographic Category	Advanced Study	Not Eligible for Advanced Study
Race/Ethnicity		
White	74	68
Black	7	15
Hispanic	8	11
Asian/Pacific Islander	10	4
Gender		
Male	51	48
Female	49	52

sort of striking socioeconomic advantage associated with advancement beyond the usual mathematics course at grade 12 that was seen at grade 8.

Assessment Procedures

Each student taking the Advanced Study assessment responded to three blocks of items. Two of the blocks, those constituting the Advanced Study proper, were composed entirely of specially developed items on advanced topics. The majority of the items in each of these two blocks required a constructed response, in contrast to the extensive use of multiple-choice items in the main NAEP assessment. Each of the blocks was 20 minutes in length at grade 8 and 30 minutes at grade 12. The third block at each grade was a so-called linking block and contained some of the same items that were given to students in the main NAEP sample. The 15-minute linking blocks enabled the performance of the advanced students to be compared with that of a representative sample of all students from the main NAEP at that grade.

Whereas in the main NAEP assessment students were given calculators to use on certain blocks of items, students in the Advanced Study could use scientific or graphing calculators in answering all items. They were told that they could bring their own calculators, and if they did not do so, they were provided with a scientific calculator. On each item they were asked to indicate whether or not they had used a calculator. Unfortunately, data are not available from NAEP on the reported calculator use.

PERFORMANCE AT GRADE 8

In this section, the performance of the eighth-grade students in the Advanced Study is considered from several angles. First, their performance on the items in the linking block is compared with that of all students in the main sample. Second, their performance on the Advanced Study items is examined by content topic. Finally, some possible connections with the performance of eighth-grade students in the TIMSS are explored.

Linkage to the Main Sample

Data from the linking block permit a comparison of the performance of students taking the Advanced Study assessment with that of all students in the main NAEP sample (which, as noted above, included students eligible, but not chosen, for the Advanced Study as well as students not eligible for the study). Table 13.3 summarizes the eighth-grade students' performance on the thirteen linking-block items, all of which dealt with algebra and functions.

The students in the Advanced Study performed better on every linking item than did the students in the main NAEP assessment, which is not surprising in view of the advanced students' better preparation in algebra than most of the main sample. The difference ranged from 8 to 34 percentage points, averaging around 21 percentage points. As table 13.3 shows, the greatest differences were on items involving equations, with the next greatest differences on items involving inequalities. Both topics are given considerable attention in the usual first-year algebra course and would not have been studied extensively by most of the students in the main assessment. In particular, two of the equation items involved linear equations in two variables, which are almost always reserved for first-year algebra. Item 7, described in table 13.3, was something of an exception to this pattern of difference, with a less than average difference (18 percentage points) between the two groups. It is not clear why the Advanced Study students were not more successful on the item. Students were given a simple linear equation in two variables to be solved for one of the variables. All they had to do was to subtract the same term from each side of the equation. The pattern of incorrect responses suggests that quite a few eighth-grade students who have taken, or are taking, first-year algebra still have trouble managing addition and subtraction of terms when solving linear equations.

The two linking items on which there was the least difference in performance between the two groups both involved algebraic reasoning. Item 12, described in table 13.3, was answered correctly by 83 percent of the students in the main sample. It required reasoning with terms such as *more* and *fewer* to construct an ordering of three quantities. Since this sort of reasoning is not dealt with explicitly in first-year algebra as usually taught, it is perhaps not surprising that no more than 91 percent of the Advanced Study students responded correctly. There was not much room for improvement anyway. The other reasoning item, item 13, was much more difficult—the most

Table 13.3
Performance on Linking-Block Items at Grade 8 by Sample

Item Description	Percent Correct	
	Advanced Study	Main NAEP
Patterns		
1. Find the next several numbers in a given pattern and write the rule used.	68	51
2. Extend a pattern of sums and and choose the correct answer.	56	32
Informal Algebra		
3. Select the solution to an equation presented pictorially.	86	72
4. Choose the coordinates of a vertex of a polygon when the coordinates of the polygon's other vertices are given.	70	43
Equations		
5. Solve a simple radical equation.	80	50
6. Determine the solution to a linear equation in two variables.	75	41
7. Solve a linear equation in two variables for one variable in terms of the other.	68	50
8. Determine which equation in two variables fits the value given in a table.	65	37
Inequalities		
9. Choose the value that solves a simple linear inequality.	79	53
10. Choose the graph that represents the solution to a simple linear inequality.	59	33
Functions		
11. From a verbal description, select an expression for a linear function described in a real-world context.	79	58
Algebraic Reasoning		
12. Given information about pairwise relationships between quantities, reason about their order.	91	83
13. Given a linear function, reason about the effects of changes in either x or $f(x)$ on the other.	41	33

Note: Item 1 was a regular constructed-response item, and the rest were multiple-choice items.

difficult linking item for the Advanced Study students and one of the most difficult for the students taking the main assessment. It involved reasoning about the effects of increasing one variable on another when they are linked in a linear function. Most of the Advanced Study students who answered incorrectly seemed to have had difficulty distinguishing between additive and multiplicative change. That distinction should have been clarified as a consequence of studying various linear relations in the algebra course, but it was obviously still a stumbling block for these students.

In general, the results on the linking-block items show that the Advanced Study students appeared to have benefited from the study of elementary algebra primarily by gaining facility in solving linear equations and inequalities as well as in judging solutions to them. They may also have improved in dealing with patterns and relationships, informal algebra, and functions. There were very few items in each of these areas, however. Although there were only two items on algebraic reasoning, it does appear that these advanced students found proportional reasoning a serious challenge.

Advanced Study Items

The twenty-two items in the Advanced Study at grade 8 concentrated on topics from the NAEP content strand Algebra and Functions, with particular attention to what the students knew about graphical, numerical, symbolic, and verbal representations for quantities and relationships. Performance on items in the algebra and functions category is discussed first, followed by the other categories.

Algebra and Functions Items

There were eighteen items in the Advanced Study that dealt with the topic of algebra and functions. They can be classified according to the type of representation given and the type of representation expected for the solution. The representations were symbolic, graphical, numerical, or verbal. Nine pairings of representation given and representation expected could be found in the items, as shown in table 13.4. There were no obvious patterns of item difficulty associated with the use of different representations. The number of items in each category was too small to support any firm conclusions, but some conjectures are suggested by the data. The Advanced Study students apparently found the interpretation and use of linear graphs to be rather difficult, as shown by the percent-correct values for items 12, 13, and 14 in the table. Items asking for a choice of symbolic expressions, conversely, tended to be rather easy, as shown by the percent-correct values for items 1 and 15.

Performance on the constructed-response items on algebra and functions tended to be poorer than on the multiple-choice items. The percent correct on the constructed-response items described in table 13.4 ranged from 8 percent (on item 18, an extended constructed-response item that involved graphing and predicting) to 57 percent (on item 16, which asked for a number satisfying a verbal description). On six of the nine constructed-response items in the table (items 4, 12, 13, 14, 16, and 17), at least 10 percent of the students either omitted the item or wrote irrelevant comments (designated as "off-task" responses by NAEP) such as "I don't like this test." Some of these students may not have been taking the assessment seriously. Others

Table 13.4
Performance on Algebra and Functions Items by Eighth-Grade Students in the Advanced Study

Item Description	Percent Correct
Symbolic to Symbolic	
1. Choose an algebraic expression equivalent to a given expression.	83
Symbolic to Numerical	
2. Choose the correct value for one unknown using corresponding entries of equal matrices to derive two equations.	69
3. Solve a radical equation.	60
4. Estimate the solution of an exponential equation.	52
5. Choose one coordinate of a point on the graph of a line, given the other coordinate and the equation of the line.	36
6. Describe the solution of a linear equation in one variable.	20
Graphical to Symbolic	
7. Choose the equation that represents the translation of the graph of a given function.	11
Graphical to Numerical	
8. Find the polar coordinates of a point, given its relation to given points.	36
9. Find the length of one side of a right triangle in the coordinate plane.	24
Graphical to Verbal	
10. Choose the correct description of the point at which two graphs intersect.	57
Symbolic and Graphical to Numerical	
11. Find all whole-number solutions for the unknown part of a line segment, given relations among the parts of the segment and restrictions on the length.	22
12. Use the y-intercept of the graph of an equation to solve a problem.	13
13. Use the x-intercept of the graph of an equation to solve a problem.	21
14. Use an equation and a boundary condition to find a maximum value.	27
Verbal to Symbolic	
15. Choose which of several algebraic expressions represents a given situation.	81
Verbal to Numerical	
16. Find the number satisfying a given description.	57
17. Find the time at which two objects meet, given the distance between them and the rates at which they move toward each other.	19
Verbal and Numerical to Graphical and Numerical	
18. Use a relationship expressed in words, along with a table of values, to draw a graph and predict a value.	8

Note: Items 1–3, 5, 7, 9, 10, and 15 were multiple-choice items, item 18 was an extended constructed-response item, and the rest were short constructed-response items. The reported percents for the constructed-response items are for the highest score levels.
Items 12, 13, and 14, all released items, are shown in figure 13.1 and table 13.5. Item 17, also a released item, is shown in table 13.6.

may have found the challenge of providing an answer, as opposed to choosing one, an unfamiliar and difficult task. Still others may have found the content of an item unfamiliar or difficult. We cannot know how serious the problem of lack of motivation was, but to the extent that these students did not engage in the tasks posed by the items, their abilities were probably underestimated by the Advanced Study assessment.

Four items from the Advanced Study at grade 8 have been released. All were short constructed-response items having responses scored as either correct, partially correct, or incorrect. The first three released items (described as items 12, 13, and 14 in table 13.4) were a set of related items which dealt with the same situation. A graph was presented of the linear relation between the number of cars washed by an eighth-grade class and the profit they earned in dollars. Figure 13.1 shows the situation as it was explained and illustrated for the students. The three items, hereafter called the Car Wash items, followed, and in each the students had to use the given symbolic and graphical information to obtain a numerical result. They could either interpret the graph or use the given equation.

The next three questions refer to the following information and graph.

The eighth-grade class at Carter School is going to hold a car wash to raise money for a class trip. The class determined that it had purchased enough supplies to wash at most 50 cars. The graph below shows the relationship between the number of cars washed and the profit earned in dollars. The line that can be drawn through these points is represented by the equation $y = 2x - 8$.

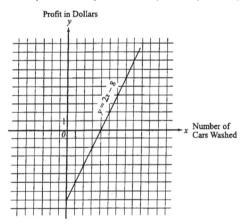

<u>Profit</u> is defined as the amount of money collected from the people whose cars are washed <u>minus</u> the amount of money that was spent on car-wash supplies.

Fig. 13.1. General directions and graph for the Car Wash items

The three Car Wash items appear in table 13.5 and are labeled as items 1–3. The solution to the first item (the fixed costs, or the amount spent before the first car was washed) can be found either by reading the graph to find the y-intercept at $y = -8$ or by substituting $x = 0$ into the equation of the line, $y = 2x - 8$, and solving to find $y = -8$. To be scored as correct, a student's response had to indicate that the amount was 8 and to contain an explanation that showed a correct understanding of the y-intercept in the graph or the equation. A response was scored as partial if the student either (*a*) responded with the value 8 but gave no explanation or an incorrect explanation or (*b*) responded with no value or an incorrect value but provided an explanation showing that he or she understood that the y-intercept answers the question posed. For example, one student whose response received a partial score answered "8 dollars" but then wrote as an explanation, "I looked at the y is the amount of money made and the $2x$ is how much of the money in all. I saw the 8 which is how much spent on supplies." Another student with a partial score answered "$9.00" but then explained, "Because I looked on the y-axis, which represent how much they were in debt." Responses of –8 or some other value were classified as incorrect. The poor performance on item 12 is difficult to understand, given that the students presumably either were studying or had already studied linear equations and their graphs in the first-year algebra course. Because about one-quarter of the students either did not respond or wrote irrelevant comments, even though these three items came early in the assessment block (on which the students had 20 minutes to respond to thirteen items), it appears that many students may not have been motivated to make a serious attempt at answering the question.

The second Car Wash item, item 2 in table 13.5, required the student to find the x-intercept of the graph or the value of x for which $y = 0$. The value could be found either by reading $x = 4$ from the graph or by solving the equation $y = 2x - 8$ for $y = 0$. Some students responded with a value of 5 cars, since that is the first value for which the profit is positive. Values of either 4 or 5 were accepted, but again, to be counted as correct, the response had to contain an explanation. Performance on item 2, although low, was not as low as it was on item 1, probably because item 2 did not require as much interpretation of what was needed and because students could solve the equation at $y = 0$ by simply asking what number multiplied by 2 gives 8.

The final Car Wash item, item 3 in table 13.5, had the highest percentage of correct responses of the three, and few students' responses received a score of partial. A correct answer required the students to use the information that at most 50 cars could be washed and to find the profit associated with $x = 50$ by substituting that value into $y = 2x - 8$ to get 92. Some students whose answers were scored incorrect apparently forgot about the $8 amount spent on supplies and simply multiplied 50 × 2 to arrive at an answer of 100; others drew a graph to make an estimate.

Table 13.5
Car Wash Items

Item	Percent Responding Grade 8
1. According to the graph, how much money did the class spend on car wash supplies? Explain how you found your answer.	
Correct response of 8 with correct explanation	13
Partial response (for example, 8 with no explanation; incorrect value but explanation indicates answer involves y-intercept)	26
Any incorrect response	36
Omitted	21
2. How many cars do the students have to wash before the class would start to earn a profit? Explain your answer.	
Correct response of 4 or 5 with correct explanation	21
Partial response (for example, 4 or 5 with no explanation; incorrect value but explanation indicates answer involves x-intercept)	19
Any incorrect response	39
Omitted	18
3. What is the greatest profit (in dollars) that the class can expect to earn? Show your work.	
Correct response of 92	27
Partial response (for example, incorrect value with correct process)	4
Any incorrect response	40
Omitted	26

Note: Percents may not add to 100 because of rounding or off-task responses.

Across the three Car Wash items, the percent of students giving incorrect responses was high, the percent who offered correct explanations to accompany their answers was moderate to low, and the percent failing to make some sort of relevant response was at least as great as the percent giving fully correct responses. Interpreting given data and relationships in solving a realistic if simplified problem and then explaining the solution apparently presented a greater challenge to these students than dealing with a strictly mathematical problem that could be solved by routine procedures and that did not require them to show or explain their work.

The fourth released item from the Advanced Study blocks at grade 8 was the Hot Air Balloon item. It is described in table 13.4 as item 17 and shown in table 13.6. It dealt with a situation in which two hot-air balloons are moving in opposite directions at constant rates. Given the rates of change in elevation and the distance between starting points, the students were to find the time in seconds for the two balloons to reach the same altitude. The easiest solution is to use the relationship *distance equals rate times time (d = rt)* and to solve either the single equation $5t = 1000 - 3t$ or the system of equations, $y = 5t$ and $y = 1000 - 3t$. However, a student could also obtain a correct solution by such means as constructing a table of values, estimating, or using trial and error.

Table 13.6
Hot Air Balloon

Item	Percent Responding Grade 8
A hot-air balloon begins rising at the rate of 3 feet per second. At the same time, a second hot-air balloon that is 1,000 feet above the first balloon begins to descend at the rate of 5 feet per second. In how many seconds will the balloons reach the same altitude?	
Correct response with or without work or explanation	19
Partial response showing some work toward an answer	7
Any incorrect response	58
Omitted	13

Note: Percents may not add to 100 because of rounding or off-task responses.

Problems in which two objects are moving toward or away from each other are common in first-year algebra and are generally perceived by students as difficult. Many traditional algebra books have entire sections of exercises devoted to "distance-rate-time" problems. The eighth-grade Advanced Study students appeared to have had difficulty with the Hot Air Balloon item. A major hurdle appears to have been formulating the problem algebraically. Only 19 percent of the students found the correct value for the time in seconds, even though they did not have to show their work. An example of a solution in which the work was shown is the correct response shown in figure 13.2. The student not only drew a figure but also used a table to set up and solve a correct equation. The use of such tables is often recommended in first-year algebra courses to help students organize the given and unknown data and thereby to derive an equation. The numbers and relationships in this item were sufficiently complicated that trial and error or estimation methods did not appear to work well. Only 7 percent of the students' responses received a partial score. Some of these responses consisted of a correct equation with no solution. Others, like the partial response in figure 13.2, consisted of lists of heights of the balloons for each few seconds, but the student whose list is shown obviously had difficulty figuring out how to coordinate the information. Incorrect responses ranged from unexplained wrong answers ("2 minutes, 6 seconds") to incorrect equations with no solutions. The incorrect response in figure 13.2 is of a third type: a drawing with quantities labeled but no solution indicated. The algebra courses these students were taking or had taken did not appear to have equipped most of them with the ability to handle this problem successfully, even though they had probably encountered other problems of this type.

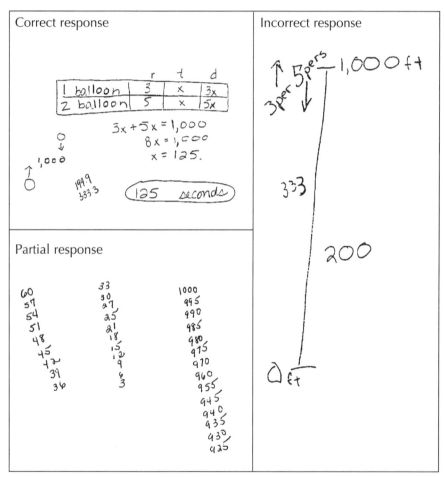

Fig. 13.2. Examples of correct, partial, and incorrect responses to Hot Air Balloon

Unfortunately for the purpose of generalizing about these advanced students' ability to handle algebraic word problems, there was only one other word problem of the conventional sort in the Advanced Study assessment at grade 8, the secure item 16 described in table 13.4. The secure item asked students to find a number, given that two different operations on it yield the same number. For example, if $n \div 2 = n - 2$, then $n = 4$. Such problems are also common in first-year algebra courses. They typically appear very early in the course, when students are first learning to solve linear equations. They ordinarily involve small whole numbers, which makes trial and error a feasible strategy, and they yield equations that are much simpler to set up and solve than do distance-rate-time problems. Given the simplicity of the problem, the level of performance was not high: 57 percent of the students answered correctly. On the evidence available, one can only conclude that these eighth-grade students had apparently not developed high levels of facility in using the algebra they had studied to solve word problems involving linear equations.

Other Items

Descriptions of the four items on the Advanced Study assessment at grade 8 that were not classified as algebra and functions items are given in table 13.7. The items ranged over three other NAEP content strands: Geometry and Spatial Sense; Number Sense, Properties, and Operations; and Data Analysis, Statistics, and Probability. There was one item on geometry (item 1 in table 13.7), an extended constructed-response item giving the diameter and cost of various circular items and one person's incorrect reasoning as to how to get the most for a specific number of dollars. The students were asked to explain what was incorrect about the reasoning. Their level of performance on this item was the lowest in the entire Advanced Study assessment. Fewer than 1 student in 100 got the answer completely correct, and about one-third either omitted the item or wrote irrelevant comments. In contrast, the one item on number (item 4 in the table) was answered correctly by 64 percent of the students. The item gave the ratio between two subsets of a specified set and the total number of elements in the set and asked for the number of elements in one subset. Finally, there were two items on data analysis (items 2 and 3 in the table). Both referred to a frequency table giving the results of a survey of students' preferences. One, answered correctly by 45 percent of the students, was a multiple-choice item asking students to convert an entry in the table to a percentage of its row total. The second item, answered correctly by only 17 percent of the students, gave the percent that two entries were of their row total and asked for the numbers and calculations used to obtain that percent. The performance on these items confirms the general picture obtained from the algebra and functions items: moderate to high levels of performance on items involving the straightforward application of routine procedures, such as solving a given equation, and a low level of performance on items that required more complex processes, such as writing an equation to model a situation, especially when an explanation of the reasoning was requested.

Table 13.7
Performance by Eighth-Grade Students on Advanced Study Items Other than Algebra and Functions

Item Description	Percent Correct
Geometry and Spatial Sense	
1. Explain incorrect reasoning in a problem involving diameters and areas of circles.	<1[a]
Data Analysis, Statistics, and Probability	
2. Find a percentage from tabular data.	45
3. Explain a student's correct reasoning in obtaining a percentage from tabular data.	17
Number Sense, Properties, and Operations	
4. Solve a problem involving ratios.	64

[a]Percent of students responding at the *extended* level.
Note: Item 1 was an extended constructed-response item, item 2 was a multiple-choice item, and the remaining two were short constructed-response items.

Connections to TIMSS

TIMSS included an assessment of the mathematics achievement of students in grades 7 and 8 in forty-one countries in 1994–1995 (Beaton et al. 1996). On that assessment, U.S. eighth-grade students performed slightly, but not significantly, below the average for the other countries. The results from the NAEP linking block in table 13.3 show that the main NAEP sample averaged 49 percent correct, which was just under the U.S. sample's average of 51 percent correct on the TIMSS algebra items. The Advanced Study sample's average score on the linking block was 71 percent correct, a figure that corresponds roughly to the performance of high-ranking countries such as Japan, Korea, and Hong Kong on the TIMSS algebra items (Beaton et al. 1996, p. 41). It is tempting to conclude from these data that if all U.S. eighth-grade students were to take algebra in grade 8, they might reach levels of performance well above those observed in countries in which algebra is a substantial part of the curriculum for all students at or before grade 8.

Comparisons between the TIMSS and the NAEP Advanced Study assessments, however, are highly problematic. The assessment instruments had no items in common, and their emphases were different. Unlike the Advanced Study assessment at grade 8, for example, fewer than one-fifth of the items on the TIMSS test were classified as algebra items. Among the TIMSS items, however, some were similar to NAEP items, including two items from the linking block. For example, in a TIMSS item and a NAEP item, shown in figure 13.3, students had to choose a correct function rule, given a relationship in verbal form between two variables. The international and U.S. averages on the TIMSS item were 47 and 45 percent correct, respectively, compared with averages on the NAEP linking-block item of 58 percent correct for the U.S. main sample and 79 percent correct for the advanced students. Another TIMSS item, shown in figure 13.4(a), presented a sequence of three similar geometric figures with increasing numbers of components and asked how many components the eighth figure in the sequence would have (the students first had to figure out the number of components in the second and third figures). One secure NAEP item in the linking block (item 2 in table 13.3) used the same sequence but presented it as a pattern of three sums rather than a pattern of four figures, and the question concerned the number of components in the twelfth sum in the sequence. The international and U.S. averages on the TIMSS item were 26 and 22 percent correct, respectively, compared with averages on the linking-block item of 32 percent correct for the U.S. main sample and 56 percent correct for the advanced students. Again, these two items were not identical. The TIMSS items may well have posed extra challenges that the NAEP items did not. Nonetheless, the pattern of performance on these two items fits the pattern observed on the TIMSS algebra items and the NAEP Advanced Study linking-block items as a whole.

TIMSS Item

Juan has 5 fewer hats than Maria, and Clarissa has 3 times as many hats as Juan. If Maria has n hats, which of these represents the number of hats that Clarissa has?

    A. $5 - 3n$
    B. $3n$
    C. $n - 5$
    D. $3n - 5$
    E. $3(n - 5)$

NAEP Item

A plumber charges customers $48 for each hour worked plus an additional $9 for travel. If h represents the number of hours worked, which of the following expressions could be used to calculate the plumber's total charge in dollars?

    A. $48 + 9 + h$
    B. $48 \times 9 \times h$
    C. $48 + (9 \times h)$
    D. $(48 \times 9) + h$
    E. $(48 \times h) + 9$

Fig. 13.3. TIMSS and NAEP algebra items, Grade 8

TIMSS Pattern Item
Here is a sequence of three similar triangles. All of the small triangles are congruent.

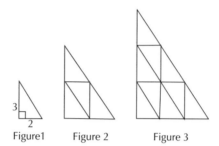

Figure 1   Figure 2   Figure 3

a. Complete the chart by finding how many small triangles make up each figure.

Figure	Number of small triangles
1	1
2	
3	

b. The sequence of similar triangles is extended to the 8th Figure. How many small triangles would be needed for Figure 8?

(a)

TIMSS Ratio Item
A class has 28 students. The ratio of girls to boys is 4:3. How many girls are in the class?
Answer: _____

(b)

Fig. 13.4. TIMSS items, Grade 8

Only one item in the Advanced Study assessment itself was similar to a TIMSS item, but the similarity was quite strong. The TIMSS item is shown in figure 13.4(b); the NAEP item (item 4 in table 13.7) is secure. In both items, the number of students in a class and the ratio of girls to boys were given, and the number of girls was to be calculated. The numbers and ratios were different as were the ways in which the ratios were reported. Nevertheless, the items appeared to require much the same reasoning with numbers that were about the same magnitude. The international and U.S. averages on the TIMSS item were 37 and 30 percent correct, respectively, compared with 64 percent correct on the NAEP item for the advanced students. Although weak, the evidence from TIMSS is consistent in suggesting that had the eighth-grade students in the Advanced Study been participating as a national group in TIMSS, their performance would likely have been at the highest levels.

PERFORMANCE AT GRADE 12

In this section, the performance of the twelfth-grade students in the Advanced Study is examined in a similar fashion to that of the performance of eighth-grade students. First is a comparison of their performance on the items in the linking block with the performance of the main sample, and then their performance on the Advanced Study items is examined by content topic. Some context is provided by examining data from TIMSS.

Linkage to the Main Sample

Data from the linking block permit a comparison of the performance of students taking the Advanced Study with that of the students in the main NAEP sample. Table 13.8 summarizes the twelfth-grade students' performance on the fifteen linking-block items, most of which dealt with algebra and functions.

The students in the Advanced Study performed better than the twelfth-grade students in the main NAEP sample on all items but one, item 9 in table 13.8, an algebra and functions item that asked them to apply trigonometric concepts to a geometric figure. In both samples, the point biserial coefficient was low (.31 for the Advanced Study sample, .28 for the main sample), suggesting that the item was not strongly related to others in the linking block. Many high-scoring students, whether taking advanced courses or not, were apparently lured by the figure into applying a sine ratio to the wrong triangle. Performance on the linking items was a little lower on the average for the twelfth-grade students in both samples than for the eighth-grade students in the corresponding samples. The differences in percent correct for the advanced students and for the main NAEP sample ranged from 0 to 33 percentage points, averaging around 17 percentage points, less than the 21 percentage point difference at grade 8. These results suggest that although, as might be expected, the advanced students performed at a higher level than the students in the main NAEP sample, the items in the linking block at grade 12 may not have been as closely tied to the advanced courses as they were at grade 8. At grade 12, the content of the advanced courses was likely to have been much more varied than at grade 8.

The items on which students in the Advanced Study at grade 12 showed much greater proficiency than students in the main NAEP assessment were

Table 13.8
Performance on Linking-Block Items at Grade 12 by Sample

Item Description	Percent Correct	
	Advanced Study	Main NAEP
Algebra and Functions		
1. Use the graph of a function to choose the value of x, given $f(x)$.	89	68
2. Identify a graph to represent data in a table.	85	76
3. Given the graphs of $f(x)$ and $g(x)$, choose the value for which $f(x) = g(x)$.	84	63
4. Use substitution and select the values for several simple quadratic equations.	84	58
5. Determine the set of ordered pairs that is a solution to a nonlinear equation.	68	53
6. Use the graph of a function to choose the value of $f(x)$.	67	35
7. Choose the symbolic expression equivalent to a symbolic exponential expression.	66	46
8. Given an equation related to a real-world setting, substitute values to solve the equation.	63	30
9. Apply trigonometric concepts to a geometry problem.	43	43
10. Apply rules of logarithms to simplify an expression.	38	24
Geometry and Spatial Sense		
11. Given the coordinates of three vertices of a parallelogram, find the coordinates of the fourth.	80	63
12. Use similar triangles to solve for an unknown side.	51	37
13. Identify a special right triangle.	37	24
Data Analysis, Statistics, and Probability		
14. Identify the least standard deviation across data sets.	62	45
15. Read a box plot and select the correct value.	46	38

Note: All items were multiple-choice items.

in the algebra and functions category. Differences in proficiency were especially great for those items that required solving equations, evaluating functions, or substituting values into equations. As at grade 8, the advanced students' greater proficiency seemed to lie primarily in performing routine algebraic manipulations. Unlike at grade 8, however, none of the linking items at grade 12 called for students to set up an equation for a given situation or otherwise engage in reasoning about a complex problem. The topics represented by the items on which the advanced twelfth-grade students showed little or no greater proficiency than the main NAEP sample dealt with elementary concepts and procedures from trigonometry, exploratory data analysis, and graphical representation of data. These topics may not have been emphasized in the courses taken by the advanced students.

Advanced Study Items

The items on the Advanced Study assessment at grade 12 spanned the five NAEP content strands, but concentrated on topics in the Algebra and Functions and Geometry and Spatial Sense strands. Table 13.9 gives the performance data for all twenty-two items. Performance tended to be low, with an average of 28 percent correct, and omissions were high. On thirteen items, 10 percent or more of the students omitted the item, and on five of those items (items 6, 7, 12, 14, and 21 in table 13.9), the omissions were above 20 percent. Either these items were too difficult for the students, the students were not sufficiently motivated to attempt them, or both.

Algebra and Functions

On the seven items on algebra and functions, the percent correct ranged from 2 percent (item 7 in table 13.9) to 39 percent (item 1), with an average across these items of 14 percent correct. One of these items (item 1) has been released.

This item, called the Ferris Wheel item, dealt with trigonometric functions. The item is shown in table 13.10 along with the distribution of students' responses. To receive the highest (extended) of the five score levels for this item, students had to sketch a curve that took into account the three conditions that (*a*) at time $t = 0$, $h = 5$, (*b*) h could vary only from 5 to 35 feet, and (*c*) the *period* (time for a complete rotation) was 15 seconds. The curve had to extend over the 45-second interval requested. The remaining score levels, which were not counted as correct, were given to responses in which one or more of these conditions were missing. Several points should be noted about the item. First, although it asked for both a solution and an explanation of the reasoning, the scoring was based only on the sketch that the student drew. Second, the term *graph* was used to refer to both the curve to be drawn and the grid on which it was to be drawn. Some students may have been confused by this ambiguity. The study of trigonometric, or circular, functions is usually given substantial attention in precalculus and calculus courses, yet fewer than half of these twelfth-grade students showed that they could successfully coordinate the multiple conditions on the function to be sketched. Some of the relatively high performance on this item compared with the other algebra and functions items may be attributed to the response requested. Students had only to graph a function rather than solve an equation or write an explanation of their work. If an algebra and functions item requested an explanation or a nonstandard procedure, performance was low and the percent of omissions was high. Performance rose when the item requested a graph or the solution of an equation, but even then it was not high.

Table 13.9
Performance by Twelfth-Grade Students in the Advanced Study

Item Description	Percent Correct
Algebra and Functions	
1. Sketch a trigonometric graph.	39
2. Solve an inequality involving absolute value.	21
3. Use trigonometric ratios to find an unknown angle in a triangle.	16
4. Find the terms in a exponential function given the x- and y-intercepts.	12
5. Find the value that minimizes a composite function and explain why it is a minimum.	6
6. Explain why a given expression is always divisible by a given number.	3
7. Find coefficient values in a quadratic equation that yield no real roots.	2
Geometry and Spatial Sense	
8. Select the resultant vector on a graph.	63
9. Select the slope of a line parallel to a line whose equation is given.	56
10. Find a point on a graph resulting from reflections in two given lines.	33
11. Find the distance between points on a polar coordinate plane.	27
12. Given two ordered pairs satisfying a linear function, find the function value for a given domain value.	20
13. Determine the measure of an angle formed by the intersection of two lines whose equations are given.	15
14. Compare the volumes of square pyramids and give linear dimensions.	5
Data Analysis, Statistics, and Probability	
15. Given tabular data, select the probability of an event	81
16. Given tabular data, select the probability of an event.	68
17. Select a weighted mean for two data sets with given averages and frequencies.	30
18. Select the probability of an event occurring a number of consecutive times under the condition of selection without replacement.	26
19. Given tabular data, find an expected value.	24
Measurement	
20. Find the circumference of a circle that circumscribes a square.	36
21. Find the area of a region bounded by lines of given lengths.	15
Number Sense, Properties, and Operations	
22. Extend a pattern of three-dimensional figures and use it to solve a problem.	11

Note: Items 8, 9, 15–18, and 20 were multiple-choice items; items 1, 3, 6, 14, and 22 were extended constructed-response items; and the rest were short constructed-response items. The reported percents for the constructed-response items are for the highest score levels.

Items 1, 8, and 14 are released items and are shown in tables 13.10, 13.11, and 13.12, respectively.

Table 13.10
Ferris Wheel

Item	Percent Responding Grade 12

[General directions]
This question requires you to show your work and explain your reasoning. You may use drawings, words, and numbers in your explanation. Your answer should be clear enough so that another person could read it and understand your thinking. It is important that you show all of your work.

The Ferris wheel above is 30 feet in diameter and 5 feet above the ground. It turns at a steady rate of one revolution each 30 seconds. The graph below shows a person's distance from the ground as a function of time if the person is at the top of the Ferris wheel at time 0.

On the same graph draw a second curve that shows a person's distance from the ground, as a function of time, if that person is at the bottom of the Ferris wheel at time 0 and if the Ferris wheel turns at a steady rate of one revolution each 15 seconds. Sketch the graph from time equals 0 to time equals 45 seconds.

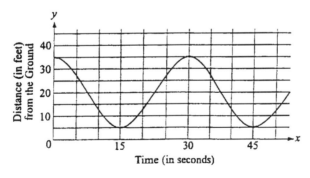

Extended response (curve satisfies the three conditions and extends over 45 seconds)	39
Satisfactory response (curve satisfies conditions but does not extend over 45 seconds)	<1
Partial response (curve satisfies two conditions)	18
Minimal response (curve satisfies one condition)	13
Incorrect response	19
Omitted	10

Geometry and Spatial Sense

Performance on the seven items on geometry and spatial sense ranged from 5 percent correct (item 14 in table 13.9) to 63 percent correct (item 8 in table 13.9). The average percent correct across these items was 31 percent, which meant that these items were, in general, easier than the algebra and function items. Of the three released items on geometry and spatial sense at grade 12, two were among the easiest items (items 8 and 9) and one (item 14) the most difficult in this category. The item referred to as Find Resultant Vector (item 8 in table 13.9) was the easiest item. The item and the advanced students' performance on it are shown in table 13.11. Students had to pay attention to the directions and magnitudes of the vectors being added, but essentially only one distracter (choice E) proved very attractive. Students choosing that distracter presumably took "southwest" to mean "west." The addition of vectors is not always given much attention in precalculus and calculus courses, but the item does not demand much technical knowledge of vectors.

Table 13.11
Find the Resultant Vector

Item	Percent Responding Grade 12
A ship travels due south for 40 miles and then southwest for 30 miles. Which of the vectors in the figure above best represents the result of the ship's movement from its starting point?	
A. a	2
B. b	4
C. c	4
D. d*	63
E. e	27

*Indicates correct response.

The most difficult released item, called Square Pyramid, is shown in table 13.12 along with the students' performance. The item asked students to order three square pyramids by volume, given that the altitude of one is equal to the lateral edge of the second and to the slant height of the third. Because all three pyramids have the same base, the order of their volumes is the same as the order of their heights, h. The solution, therefore, depended on determining the height of each pyramid. Pyramid R has height 10. Because the distance from the center of the base to one vertex of the base is $5\sqrt{2}$, the height of Pyramid Q is, by the Pythagorean theorem, also $5\sqrt{2}$, or about 7.07 units. Because the distance from the center of the base to the midpoint of one side of the base is 5, the height of Pyramid R is, again by the Pythagorean theorem, $5\sqrt{3}$, or about 8.66 units. Therefore the correct order, from smallest to largest volume, is Q, R, P. Students might well have used visual reasoning to determine the order (the smaller the angle the 10-unit piece makes with the base, the smaller the altitude and, hence, the smaller the volume), but to be considered correct a response required mathematical evidence.

The highest score level (extended) required that the student give the correct ordering of volumes along with supporting evidence. The three volumes had to be given correctly if the student did not indicate that the order depends only on the height. If the student either found two of the three volumes correctly, found the heights of Q and R correctly, or correctly compared the three heights, a score level of satisfactory was given. A score level of partial was given to students who either found the heights of Q or R correctly or found the volume of Q or R correctly. A score level of minimal was given to students who showed or stated that the pyramids had the same base area, who showed or stated that one-third of the base area was the same for all three pyramids, or who found the volume of P correctly.

The Square Pyramid item was the only item in the Advanced Study that required reasoning about three-dimensional figures. Although the students were given the formula they needed, few appeared able to apply the Pythagorean theorem appropriately to cross sections of the pyramids to order their volumes. The high percentages of omissions (20 percent) and irrelevant, off-task comments (3 percent) suggest that many of these advanced students either did not know how to attack the problem or did not care to make an attempt. Perhaps they lacked experience in solving problems involving geometric solids.

Data Analysis, Statistics, and Probability

Performance on two probability items, items 15 and 16 in table 13.9, was the highest of the entire Advanced Study assessment at grade 12. The items asked students to determine the probability of an event, given a table of data. The calculations of probability were straightforward; all the students

Table 13.12
Square Pyramids

Item	Percent Responding Grade 12

[General directions]
This question requires you to show your work and explain your reasoning. You may use drawings, words, and numbers in your explanation. Your answer should be clear enough so that another person could read it and understand your thinking. It is important that you show all of your work.

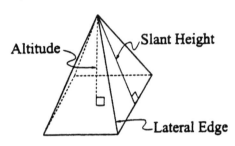

The figure above shows the altitude, lateral edge, and slant height of a square pyramid. Each of three square pyramids P, Q, and R (not shown) has a base with side 10. In each pyramid the lateral edges are of equal lengths. Pyramids P, Q, and R have the following characteristics.

    a. The altitude of pyramid P is 10.
    b. The lateral edge of pyramid Q is 10.
    c. The slant height of pyramid R is 10.

List the pyramids by order of volume from smallest volume to largest volume. Support your conclusion with mathematical evidence.

[The formula for the volume of a pyramid with base B and height (altitude) h is $V = 1/3\ Bh$.]

Extended response (correct ordering and supporting evidence)	5
Satisfactory response (for example, incorrect ordering but correct volumes for two of three pyramids)	5
Partial response (for example, correct solution for either height or volume of pyramid Q or R)	5
Minimal response (for example, statement or work showing that the pyramids have the same base area)	11
Incorrect response	52
Omitted	20

Note: Percents may not add to 100 because of rounding or off-task responses.

needed to do was to select the appropriate values from the table and find their quotient. The students might well have encountered problems of this sort in middle school. Item 17 also involved a straightforward calculation, but weighted means were probably less likely to have been included in these students' high school mathematics courses. Only 3 out of 10 students performed the calculation correctly, with 6 out of 10 students choosing the unweighted mean instead. When students were given the same table as used for items 15 and 16 and asked instead to find an expected value (item 19), only about one in four students answered correctly, and 15 percent either omitted the item or wrote irrelevant comments. They also had difficulty with the calculation of an event involving selection without replacement (item 18); again, only about one in four answered correctly. These advanced students, therefore, could manage simple probability concepts relatively well but apparently had more difficulty when the problem required complex computations of probability or expected value.

Measurement

The advanced students were not very successful with either of the two measurement items (items 20 and 21 in table 13.9). They had more success in selecting a circumference than in calculating the area of a triangular region even though they were given a figure for the triangle item and not for the circle item. Because the circle item was a multiple-choice item and the triangle item a regular constructed-response item, much or all of the difference may be attributable to guessing. The equations of two of the lines bounding the triangular region, however, contained fractional coefficients and required a constructed response; that may have been enough to convince almost a third of the students not to attempt the item (that is, 28 percent omitted it, and 4 percent wrote irrelevant comments).

Number Sense, Properties, and Operations

On item 22 in table 13.9, the requirement both to extend a pattern in a realistic situation and then to solve a problem that used the pattern, giving "mathematical evidence" to support the solution, proved very difficult for the advanced students. Only 11 percent received full credit for their response. Interestingly, even though this item came at the end of an item block and required an extended response, more than 90 percent of the students attempted it. The distribution of scores suggests that most students either gave a completely incorrect response or were not able both to extend the pattern and to figure out how far it should go. Once again, the request for some justification or explanation may have resulted in low performance.

The results for the Advanced Study items at grade 12 suggest that these advanced students were not, for the most part, able to solve problems requiring that they reason their way through several steps to a solution. And as at grade 8, their performance was invariably much lower when an explanation was requested. The items on which the advanced students at grade 12 had the most success were multiple-choice items on probability that required the reading of a table and multiple-choice items on geometry that dealt with elementary concepts of lines and vectors. The students' generally low level of performance indicates a lack of proficiency in dealing with the challenging mathematics in this assessment, but their numerous omissions and irrelevant comments on some items also suggest a lack of engagement with the assessment.

Connections to TIMSS

In principle, comparisons between the Advanced Study and TIMSS should be easier at grade 12 than at grade 8. TIMSS included an assessment of the mathematics achievement of students taking advanced mathematics courses in the last year of secondary school, and therefore the U.S. TIMSS and the Advanced Study samples ought to have been roughly comparable in academic preparation. The TIMSS Advanced Mathematics Assessment yielded data on students from sixteen countries in 1994–1995 (Mullis et al. 1998). On that assessment, U.S. twelfth-grade students taking precalculus or calculus performed significantly below the average for the other countries. On the NAEP linking-block items in table 13.8, the Advanced Study sample averaged 64 percent correct, which was well above the U.S. sample's average of 38 percent correct on the TIMSS items (NCTM 1998). The items in the linking block, however, tended to be more homogeneous in content and somewhat more concerned with elementary concepts and procedures than the TIMSS items were.

One item, shown in figure 13.5(a), on the TIMSS assessment was slightly similar to an item in the NAEP linking block at grade 12. The international and U.S. averages on this item were 50 and 39 percent correct, respectively. The similar item in the NAEP assessment, item 11 in table 13.8, included a figure of a parallelogram and asked for the coordinate of the fourth vertex. Performance was higher on this item in both the Advanced Study sample and the main NAEP sample (80 and 63 percent correct, respectively). Not only did the linking-block item use a simpler geometric shape and provide a figure, but all the points were in the first quadrant and the item avoided vector notation. The performance difference between the Advanced Study sample and the TIMSS sample seems due to the additional complications in the TIMSS item.

Performance on the Advanced Study items shown in table 13.9 averaged 28 percent correct, as noted above—a level of difficulty that might suggest that they were more like the TIMSS items than the linking items were. Again, however, an inspection of the items revealed that individually and

> **TIMSS Vector Item**
>   The rectangular coordinates of three points in a plane are $Q(-3, -1)$, $R(-2, 3)$, and $S(1, -3)$. A fourth point T is chosen so that $ST = 2QR$. The y-coordinate of T is
>   A. -11
>   B. -7
>   C. -1
>   D. 1
>   E. 5
>
> (a)
>
> ---
>
> **TIMSS Probability item**
>   One thousand people selected at random were questioned about smoking and drinking. The results of this survey are summarized in the table below. Calculate the probability that a randomly selected respondent drinks and smokes.
>
> | | Smokers | Non-smokers |
> |---|---|---|
> | Drinkers | 320 | 530 |
> | Non-drinkers | 20 | 130 |
>
> (b)

Fig. 13.5. TIMSS items, Grade 12

collectively they, too, were rather different from the TIMSS items. More than a quarter of the TIMSS assessment, for example, consisted of items on calculus or on "validation and structure." The TIMSS assessment included several pages of relevant mathematical formulas for use with the items, whereas formulas were not always provided for the Advanced Study items. And almost three-quarters of the TIMSS items were multiple-choice items, compared with fewer than one-third of the Advanced Study items.

Only one TIMSS item, shown in figure 13.5(b), resembled a NAEP Advanced Study item. The international and U.S. averages on the TIMSS item were 51 and 52 percent correct, respectively. On the Advanced Study item, item 15 in table 13.9, a 3-by-2 table of frequencies with row, column, and grand totals was given, and students were asked to select a probability requiring division of a row total by the grand total. The high level of performance on this item by the Advanced Study sample, 81 percent correct, is puzzling when compared with the U.S. TIMSS sample's average of 52 percent correct on the TIMSS item. Presumably these two samples were very similar: both were selected according to comparable criteria, and both represented about 14 or 15 percent of all U.S. twelfth-grade students (John Dossey, personal communication, 31 July 1998; NCES 1998, p. 38). Apparently, some features of the Advanced Study item, including the numbers used (which were smaller than the numbers used in the TIMSS item) and the fact that it was a multiple-choice item, made it an easier item for the students in the Advanced Study sample. All in all, we can draw no definite conclusions from this comparison with TIMSS, but the data do suggest that the items in the two assessments were not measuring the same accomplishments in mathematics. The Advanced Study assessment appears to have been much more challenging at grade 12 than the TIMSS assessment was, and the performance of advanced U.S. students was low on both assessments.

CONCLUSION

The 1996 Advanced Study at grades 8 and 12 provided information that was not available in the past about the performance of students taking advanced mathematics courses. A simple summary of that performance is that advanced students do much better than average on the usual NAEP items, as one might have expected, but that their performance is still quite modest on challenging items dealing with more-advanced content, especially at grade 12. Advanced students at both grades seem to have difficulty explaining or justifying their responses. At both grades, but especially at grade 12, there were high percentages of omissions on the constructed-response items even though ample time was allowed, indicating that many students did not spend that time engaged with the items. Comparisons with data from TIMSS suggested that the performance of Advanced Study students at grade 8 but not at grade 12 was like that of students in high-performing TIMSS countries.

The Advanced Study assessment appeared to have been better suited to the courses the students were taking at grade 8 than it was at grade 12. These advanced students, especially at grade 12, had presumably learned a variety of mathematical concepts and procedures that were not assessed. Nonetheless, the results of the study show convincingly that acceleration into so-called advanced courses not only is unequally distributed across demographic groups but also apparently fails to yield high levels of accomplishment in communicating and reasoning about mathematics and in using mathematics to solve complex problems.

REFERENCES

Ballator, Nada. *The NAEP Guide: A Description of the Content and Methods of the 1994 and 1996 Assessments*, rev. ed. Washington, D.C.: National Center for Education Statistics, 1996.

Beaton, Albert E., Ina V. S. Mullis, Michael O. Martin, Eugenio J. Gonzalez, Dana L. Kelley, and Teresa A. Smith. *Mathematics Achievement in the Middle School Years: IEA's Third International Mathematics and Science Study (TIMSS)*. Boston: Center for the Study of Testing, Evaluation, and Educational Policy, Boston College, 1996.

Dossey, John A., Ina V. S. Mullis, Steven Gorman, and Andrew S. Latham. *How School Mathematics Functions.* (Report No. 23-FR-02). Washington, D.C.: National Center for Education Statistics, 1994.

Kimball, Meredith M. "A New Perspective on Women's Math Achievement." *Psychological Bulletin* 105 (March 1989): 198–214.

Mitchell, Julia H., Evelyn F. Hawkins, Pamela M. Jakwerth, Frances B. Stancavage, and John A. Dossey. *Student Work and Teacher Practices in Mathematics.* Washington, D.C.: National Center for Education Statistics, 1999.

Mitchell, Julia H., Evelyn F. Hawkins, Frances B. Stancavage, and John A. Dossey. *Estimation Skills, Mathematics-in-Context, and Advanced Skills in Mathematics.* Washington, D.C.: National Center for Education Statistics, 1999.

Mullis, Ina V. S., Michael O. Martin, Albert E. Beaton, Eugenio J. Gonzalez, Dana L. Kelley, and Teresa A. Smith. *Mathematics Achievement in the Final Year of Secondary School: IEA's Third International Mathematics and Science Study (TIMSS)*. Boston: Center for the Study of Testing, Evaluation, and Educational Policy, Boston College, 1998.

National Assessment of Educational Progress. *1996 Assessment Development: Technical Proposal.* Vol. 1. Princeton, N.J.: Educational Testing Service, 1996.

National Council of Teachers of Mathematics. *U.S. Mathematics Teachers Respond to the Third International Mathematics and Science Study: Grade 12 Results.* 1998. Available at www.nctm.org/publications/releases/1998/02/timss.12.reaction.

National Center for Education Statistics. *Pursuing Excellence: A Study of U.S. Twelfth-Grade Mathematics and Science Achievement in International Context* (NCES 98-049). Washington, D.C.: U.S. Government Printing Office, 1998.

Sadker, Myra, and David Sadker. *Failing at Fairness: How America's Schools Cheat Girls.* New York: Macmillan, 1994.

List of NAEP-Related Publications: Fifth, Sixth, and Seventh Mathematics Assessments

SEVENTH NAEP MATHEMATICS ASSESSMENT, 1996

NCES/NAEP Publications

Allen, Nancy L., Frank Jenkins, Edward Kulick, and Christine A. Zelenak. *Technical Report of the NAEP 1996 State Assessment Program in Mathematics.* NCES 97-951. Washington, D.C.: National Center for Education Statistics, 1997.

Ballatore, Nada. *The NAEP Guide: A Description of the Content and Methods of the 1994 and 1996 Assessments.* Rev. ed. NCES 97-586. Washington, D.C.: National Center for Education Statistics, 1996.

Campbell, Jay R., Kristin E. Voelkl, and Patricia L. Donahue. *NAEP 1996 Trends in Academic Progress.* NCES 97-985. Washington, D.C.: National Center for Education Statistics, 1997.

Hawkins, Evelyn F., Frances B. Stancavage, and John A. Dossey. *School Policies and Practices Affecting Instruction in Mathematics: Findings from the National Assessment of Educational Progress.* NCES 98-494. Washington, D.C.: National Center for Education Statistics, 1999.

Mitchell, Julia H., Evelyn F. Hawkins, Pamela M. Jakwerth, Frances B. Stancavage, and John A. Dossey. *Student Work and Teacher Practices in Mathematics.* NCES 99-453. Washington, D.C.: National Center for Education Statistics, March 1999.

Mitchell, Julia H., Evelyn F. Hawkins, Frances B. Stancavage, and John A. Dossey. *Estimation Skills, Mathematics-in-Context, and Advanced Skills in Mathematics: Results from Three Studies of the National Assessment of Educational Progress 1996 Mathematics Assessment.* NCES 99-503. Washington, D.C.: National Center for Education Statistics, 1999.

National Assessment Governing Board. *Mathematics Framework for the 1996 National Assessment of Educational Progress.* Washington, D.C.: National Assessment Governing Board, 1994.

Reese, Clyde, Karen E. Miller, John Mazzeo, and John A. Dossey. *NAEP 1996 Mathematics Report Card for the Nation and the States.* NCES 97-488. Washington, D.C.: National Center for Education Statistics, 1997.

Shaughnessey, Catherine A., Jennifer E. Nelson, and Norma A. Norris. *NAEP 1996 Mathematics Cross-State Data Compendium for the Grade 4 and Grade 8 Assessment: Findings from the State Assessment in Mathematics of the National Assessment of Educational Progress.* NCES 98-481. Washington, D.C.: National Center for Education Statistics, 1998.

NCTM Publications

Silver, Edward A., and Patricia Ann Kenney, eds. *Results from the Seventh Mathematics Assessment of the National Assessment of Educational Progress.* Reston, Va.: National Council of Teachers of Mathematics, 2000.

Strutchens, Marilyn E. "Data Collection: Getting to Know Your Students' Attitudes." *Mathematics Teaching in the Middle School* 4 (March 1999): 382–84.

Zawojewski, Judith S., and J. Michael Shaughnessy. "Mean and Median: Are They Really So Easy?" *Mathematics Teaching in the Middle School* 5 (March 2000): 436–40.

———. "Secondary Students' Performance on Data and Chance in the 1996 NAEP." *Mathematics Teacher* 92 (November 1999): 713–18.

Other Publications

Kenney, Patricia Ann. "Families of Items in the NAEP Mathematics Assessment." In *Grading the Nation's Report Card: Research from the Evaluation of NAEP,* edited by James W. Pellegrino, Lee R. Jones, and Karen J. Mitchell Washington, D.C.: National Academy Press, 2000.

Wenglinsky, Harold. *Does It Compute?: The Relationship between Educational Technology and Student Achievement in Mathematics.* Princeton, N.J.: Policy Information Center, Educational Testing Service, 1998.

SIXTH NAEP MATHEMATICS ASSESSMENT, 1992

NCES/NAEP Publications

Dossey, John A., Ina V. S. Mullis, Steven Gorman, and Andrew S. Latham. *How School Mathematics Functions: Perspectives from the NAEP 1990 and 1992 Assessments.* 23-FR-02. Washington, D.C.: National Center for Education Statistics, 1994.

Dossey, John A., Ina V. S. Mullis, and Chancey O. Jones. *Can Students Do Mathematical Problem Solving?: Results from Constructed-Response Questions in NAEP's 1992 Mathematics Assessment.* 23-FR-01. Washington, D.C.: National Center for Education Statistics, 1993.

Johnson, Eugene, and James E. Carlson. *The NAEP 1992 Technical Report.* 23-TR-20. Washington, D.C.: National Center for Education Statistics, 1994.

Johnson, Eugene, John Mazzeo, and Deborah L. Kline. *Technical Report of the NAEP 1992 Trial State Assessment Program in Mathematics.* Princeton, N.J.: Educational Testing Service, National Assessment of Educational Progress, 1993.

Mullis, Ina V. S. *The NAEP Guide: A Description of the Content and Methods of the 1990 and 1992 Assessments.* 21-TR-01. Washington, D.C.: National Center for Education Statistics, 1991.

Mullis, Ina V. S., ed. *America's Mathematics Problem: Raising Student Achievement.* 23-FR-03. Washington, D.C.: National Center for Education Statistics, 1994.

Mullis, Ina V. S., Frank Jenkins, and Eugene G. Johnson. *Effective Schools in Mathematics: Perspectives from the NAEP 1992 Assessment.* 23-RR-01. Washington, D.C.: National Center for Education Statistics, 1994.

Mullis, Ina V. S., John A. Dossey, Jay R. Campbell, Claudia A. Gentile, Christine O'Sullivan, and Andrew S. Latham. *NAEP 1992 Trends in Academic Progress.* 23-TR-01. Washington, D.C.: National Center for Education Statistics, 1994.

Mullis, Ina V. S., John A. Dossey, Eugene H. Owen, and Gary W. Phillips. *NAEP 1992 Mathematics Report Card for the Nation and the States: Data from the National and Trial State Assessments.* 23-ST-02. Washington, D.C.: National Center for Education Statistics, 1993.

National Center for Education Statistics. *Data Compendium for the NAEP 1992 Mathematics Assessment of the Nation and the States.* 23-ST-04. Washington, D.C.: National Center for Education Statistics, 1993.

NCTM Publications

Blume, Glendon W., Judith S. Zawojewski, Edward A. Silver, and Patricia Ann Kenney. "Focusing on Worthwhile Mathematical Tasks in Professional Development: Using a Task from the National Assessment of Educational Progress." *Mathematics Teacher* 91 (February 1998): 156–61.

Kenney, Patricia Ann, and Edward A. Silver. "Probing the Foundations of Algebra: Grade 4 Pattern Items in NAEP." *Teaching Children Mathematics* 3 (February 1997): 268–74.

Kenney, Patricia Ann, and Edward A. Silver, eds. *Results from the Sixth Mathematics Assessment of the National Assessment of Educational Progress.* Reston, Va.: National Council of Teachers of Mathematics, 1997.

Kenney, Patricia Ann, Judith S. Zawojewski, and Edward A. Silver. "Marcy's Dot Pattern." *Mathematics Teaching in the Middle School* 3 (May 1998): 474–77.

Stylianou, Despina A., Patricia Ann Kenney, Edward A. Silver, and Cengiz Alacaci. "Examining Written Communication in Assessment Tasks to Gain Insight into Students' Thinking." *Mathematics Teaching in the Middle School* 6 (October 2000): 136–44.

Other Publications

Kenney, Patricia Ann. "A Framework for the Qualitative Analysis of Student Responses to the Extended Constructed-Response Questions from the 1992 NAEP in Mathematics." In *Proceedings of the Seventeenth Annual Meeting of the North American Chapter of the International Group for the Psychology of Mathematics Education,* Vol. 1, edited by Douglas T. Owens, M. K. Reed, and G. M. Millsaps, pp. 175–80. Columbus, Ohio: ERIC Clearinghouse for Science, Mathematics, and Environmental Education, 1995.

Lindquist, Mary Montgomery, John A. Dossey, and Ina V. S. Mullis. *Reaching Standards: A Progress Report on Mathematics.* Princeton, N.J.: Policy Information Center, Educational Testing Service, 1995.

Silver, Edward A., and Patricia Ann Kenney. "The Content and Curricular Validity of the 1992 NAEP TSA in Mathematics." In *The Trial State Assessment: Prospects and Realities: Background Studies,* pp. 231–84. Stanford, Calif.: National Academy of Education, 1994.

———. "Expert Panel Review of the 1992 NAEP Mathematics Achievement Levels." In *Setting Performance Standards for Student Achievement: Background Studies,* pp. 215–81. Stanford, Calif.: National Academy of Education, 1993.

———. *Understanding Students' Mathematical Problem Solving: A Commentary on the 1992 NAEP Findings.* Pittsburgh, Pa.: Learning Research and Development Center, University of Pittsburgh, 1994.

FIFTH NAEP MATHEMATICS ASSESSMENT, 1990

NCES/NAEP Publications

Johnson, Eugene G., and Nancy L. Allen. *The NAEP 1990 Technical Report.* 21-TR-20. Princeton, N.J.: Educational Testing Service, National Assessment of Educational Progress, 1992.

Mullis, Ina V. S., Eugene H. Owen, and Gary W. Phillips. *America's Challenge: Accelerating Academic Achievement: A Summary of Findings from 20 Years of NAEP.* 19-OV-01. Washington, D.C.: National Center for Education Statistics, 1990.

Mullis, Ina V. S., John A. Dossey, Eugene H. Owen, and Gary W. Phillips. *The STATE of Mathematics Achievement: NAEP's 1990 Assessment of the Nation and the Trial Assessment of the States.* 21-ST-04. Washington, D.C.: National Center for Education Statistics, 1991.

National Assessment of Educational Progress. *Mathematics Objectives: 1990 Assessment.* 21-M-10. Princeton, N.J.: Educational Testing Service, National Assessment of Educational Progress, 1988.

NCTM Publications

Silver, Edward A., and Patricia Ann Kenney. "An Examination of Relationships between the 1990 NAEP Mathematics Items for Grade 8 and Selected Themes from the NCTM Standards." *Journal for Research in Mathematics Education* 24 (March 1993): 159–67.

Other Publications

Silver, Edward A., Patricia Ann Kenney, and Leslie Salmon-Cox. "The Content and Curricular Validity of the 1990 NAEP Mathematics Items: A Retrospective Analysis." In *Assessing Student Achievement in the States: Background Studies,* pp. 157–218. Stanford, Calif.: National Academy of Education, 1992.

Results from the Seventh
Mathematics Assessment of the
National Assessment of Educational
Progress

QA
13
.R47
2000

AIMS EDUCATION FOUNDATION
Research Library
Wiebe Education Center
P.O. Box 8120
Fresno, California 93747-8120
(209) 255-4094   DEMCO

DATE DUE

PRINTED IN U.S.A.